D0445464

Plant Engineers and Managers Guide to Energy Conservation

Fifth Edition

Plant Engineers and Managers Guide to Energy Conservation

Fifth Edition

Albert Thumann, P.E., C.E.M.

Published by
THE FAIRMONT PRESS, INC.
700 Indian Trail
Lilburn, GA 30247

Library of Congress Cataloging-in-Publication Data

Thumann, Albert.
Plant engineers and managers guide to energy conservation / by Albert Thumann. -- 5th ed.
p. cm.
Includes Index.
ISBN 0-88173-131-5
1. Factories--Energy conservation--Handbooks, manuals, etc.. I. Title.
TJ163.5.F3T48 1991 621.042--dc20 91-10717
CIP

Plant Engineers And Managers Guide To Energy Conservation / By Albert Thumann. -- 5th Edition.

Published by The Fairmont Press, Inc.
700 Indian Trail
Lilburn, GA 30247

Printed in Mexico

10 9 8 7 6 5 4 3 2 1

ISBN 0-88173-131-5 FP

ISBN 0-13-689639-1 PH

Distributed by Prentice-Hall, Inc.
A Simon & Schuster Company
Englewood Cliffs, NJ 07632

Prentice-Hall International (UK) Limited, London
Prentice-Hall of Australia Pty. Limited, Sydney
Prentice-Hall Canada Inc., Toronto
Prentice-Hall Hispanoamericana, S.A., Mexico
Prentice-Hall of India Private Limited, New Delhi
Prentice-Hall of Japan, Inc., Tokyo
Simon & Schuster Asia Pte. Ltd., Singapore
Editora Prentice-Hall do Brasil, Ltda., Rio de Janeiro

THIS BOOK IS DEDICATED TO THE ENGINEERS, ARCHITECTS, AND DESIGNERS WHO ARE IMPROVING ENERGY EFFICIENCY OF OPERATIONS IN A COST EFFECTIVE MANNER.

Preface

Plant engineers and managers of the 1990's are expected to apply the latest technologies and products to reduce utility costs. A recent survey conducted by the Association of Energy Engineers found that of those surveyed 93% indicated that energy management should play a key role in their organization's strategic plan; and 98% feel there is still opportunity to reduce energy consumption in their facilities further.

Since the first edition of this text was published in 1977, the emphasis on energy management has changed. Energy conservation programs have moved from applying general housekeeping measures to purchasing new equipment to reduce energy consumption. As a result, the plant engineer's role in energy management has increased. The plant engineer now has a multitude of proven new products and services at his disposal to reduce utility costs.

In the 1990's the plant engineer will be faced with new challenges. Almost half of the individuals surveyed by AEE believe that there is not sufficient electric generating capacity planned. As a result there will be increased emphasis on demand control, thermal storage, cogeneration and energy management systems.

This book is a comprehensive, easy to use guide which emphasizes the various options available on how to reduce energy consumption in a cost effective manner. The methodologies presented can be applied to both existing and new facilities.

Industry as a whole has done a good job in reducing energy consumption because doing so has made economic sense. As fuel prices continue to change and new products enter the marketplace, the plant engineer will have new opportunities to reduce energy costs further. Hopefully, this book will serve as a tool to guide the energy management program to its potential.

Albert Thumann

Contents

Introduction

The winds of change have once again brought energy issues into the forefront. There is now a new sense of urgency to reduce the United States' dependence on foreign oil which increased to 51.3% of total consumption. The crisis in the Gulf marks the beginning of a new era in which inexpensive and convenient sources of energy will disappear. The key element in any long-term corporate strategy will be to reduce energy consumption.

Even though the United States uses 22% fewer Btus (1978-1988) to generate each dollar of real gross domestic product, it still lags behind Japan and France. Companies such as 3M have made significant achievements in reducing energy use. For example, over the last 17 years, 3M has saved $732 million in energy costs and uses less than half as much energy to make a real dollar's worth of manufactured products as in 1973.

Decreasing reliance on oil and development of new power generating facilities offer a bridge to our energy future.

1

The Role of the Plant Engineer In Energy Management

Energy management is now considered part of every plant engineer's job. Today the plant engineer needs to keep abreast of changing energy factors which must be incorporated into the overall energy management program. The accomplishments of energy management have indeed been outstanding. In a 1988 opinion survey conducted by the Association of Energy Engineers, 14 percent of those responding indicated that they have saved their companies at least five million dollars in accumulated energy costs since being employed. Twenty-two percent had slashed energy costs 26 percent or more since the program was started.

Safety, maintenance and now energy management are some of the areas a plant engineer is expected to be knowledgeable in. The cook book and low cost-no cost energy conservation measures which were emphasized in the 1970s have been replaced with a more sophisticated approach.

The plant engineer of the 1990s must have a keen understanding of both the technical and managerial aspects of energy management in order to insure its success. When oil prices dropped in 1986 it was an opportunity in many plants to switch back to oil. As electric prices escalated it was an opportunity for many plants to install cogeneration facilities. Thus the energy management area is ever changing.

Energy management or energy utilization has replaced the simplistic house keeping measures approach.

The intent of this book is not to make you an expert in each subject, but to illustrate how the overall pieces fit together, Each chapter illustrates the various pieces that comprise an industrial energy utilization program. The energy manager is analogous to a system engineer. Only when the total picture is viewed will the solution become obvious. Of course, it should be noted that the energy manager must seek the advice of experts or specialists when required and use their expertise accordingly.

SURVEY OF WHAT INDUSTRY IS DOING

In a survey[1] of 318 manufacturing companies on the Fortune 1000 list, the following results were indicated.

- 52.1% of companies surveyed had a formal written energy plan; only 5.7% had formal written plans prior to the 1973 embargo by OPEC.
- In most companies, the senior energy executive ranks high in the corporate hierarchy; 21.7% indicate corporate energy responsibility with a vice president, 20.4% to a manager, and 16.4% to a director.
- Only 8.5% of all senior energy executives devoted all their efforts to energy matters.
- The budget and number of employees assigned to energy within a company is still modest: 35.2% indicated that fewer than 20 employees were directly involved in energy activities. Only one third of the respondents allocated over $100,000 to energy-related activities, primarily conservation. Unfortunately, 53.5% did not know the amount. This is, in part, because of the following.
    1. Energy activities are usually added on existing responsibilities.
    2. Energy activities are difficult to measure.
    3. Most companies are in their formative stages.
- 93% indicated that guidelines for surveying current energy consumption were included in their plan, 88.5% included specific recommendations as part of their plan, and 85% included a system for measuring savings.
- 77% established energy conservation committees or policy teams as part of their plan, and 71.6% added energy-related responsibilities to existing job descriptions.

Some conclusions that can be drawn from the study are:

1. Energy programs are relatively new for most companies and are still developing.
2. Budgets assigned to energy conservation are modest.
3. Most energy conservation programs are accomplished as an "add-on responsibility."
4. The survey, the analysis, and the economics are common to most companies' energy programs.

RESULTS OF INDUSTRIAL ENERGY UTILIZATION PROGRAMS

In the United States from 1971 to 1986 the energy intensity of industrial processes has declined approximately 2% per year. In a 1990 survey conducted by the Association of Energy Engineers, 1100 members responded to the accomplishments of energy cost reduction programs which are summarized below:

Estimated accumulated reduction in energy consumption since your program started? 16%: 5% or less 18%: 6-10% 15%: 11-15% 14%: 16-20% 12%:21-25% 9%: 26-30% 5%: 31-35% 4%: 36-40% 1%: 41-45% 4%: 46-50% 2%: 51% on up.

Estimated accumulated energy costs you have saved your company since your employment? 20%: Less than $100,000 29%: $100,001 to $500,000 14%: $500,001 to $1,000,000 10%: $1,000,001 to $2,000,000 $9%: $2,000,001 to $5,000,000 18%: Over $5,000,001.

ORGANIZATION FOR ENERGY UTILIZATION

A multi-divisional corporation usually organizes energy activities on a corporate and plant basis. On the plant basis, energy activities are in many instances added on to the duties of the plant manager.

An energy utilization program does not just happen. It needs a guiding force to "get the ball rolling." Production, energy costs, and raw material supplies are of great concern to plant managers; thus, they are usually the ones to initiate the program.

For a continual, ongoing program to develop, energy managers need to establish "the industrial audit program" for their facilities. The term "industrial audit" was introduced in most energy utilization programs in the late 1970s, yet it was rarely defined.

WHAT IS AN INDUSTRIAL ENERGY AUDIT?

The simplest definition for an energy audit is as follows: An energy audit serves the purpose of identifying where a building or plant facility uses energy and identifies energy conservation opportunities.

There is a direct relationship to the cost of the audit (amount of data collected and analyzed) and the number of energy conservation opportunities to be found. Thus, a first decision is made on the cost of the audit, which determines the type of audit to be performed.

The second decision is made on the type of facility. For example, a building audit may emphasize the building envelope, lighting, heating, and ventilation requirements. On the other hand, an audit of an industrial plant emphasizes the process requirements.

Most energy audits fall into three categories or types; namely, walk-through, *mini-audit*, or *maxi-audit*.

Walk-through. This type of audit is the least costly and identifies preliminary energy savings. A visual inspection of the facility is made to determine maintenance and operation energy saving opportunities plus collection of information to determine the need for a more detailed analysis.

Mini-audit. This type of audit requires tests and measurements to quantify energy uses and losses and determine the economics for changes.

Maxi-audit. This type of audit goes one step further than the mini-audit. It contains an evaluation of how much energy is used for each function, such as lighting or process. It also requires a model analysis, such as a computer simulation, to determine energy use patterns and predictions on a year-round basis, taking into account such variables as weather data.

The chief distinction between the mini-audit and the walk-through audit is that the mini-audit requires a quantification of energy uses and losses and determining the economics for change.

The chief distinction between the maxi-audit and the mini-audit is that the maxi-audit requires an accounting system for energy to be established and a computer simulation.

THE ENERGY UTILIZATION PROGRAM

The energy utilization program usually contains the following steps.

1. Determine energy uses and losses; refer to checklist, Table 1-1.
2. Implement actions for energy conservation, refer to checklist, Table 1-2.

TABLE 1-1 Checklist to Determine Energy Uses and Losses.

SURVEY ENERGY USES AND LOSSES

A. Conduct first survey aimed at identifying energy wastes that can be corrected by maintenance or operations actions, for example:
   1. Leaks of steam and other utilities
   2. Furnace burners out of adjustment
   3. Repair or addition of insulation required
   4. Equipment running when not needed

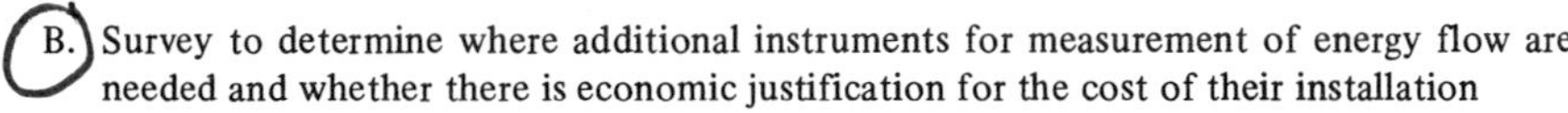

B. Survey to determine where additional instruments for measurement of energy flow are needed and whether there is economic justification for the cost of their installation

C. Develop an energy balance on each process to define in detail:
   1. Energy input as raw materials and utilities
   2. Energy consumed in waste disposal
   3. Energy credit for by-products
   4. Net energy charged to the main product
   5. Energy dissipated or wasted

   Note: Energy equivalents will need to be developed for all raw materials, fuels, and utilities, such as electric power, steam, etc., in order that all energy can be expressed on the common basis of Btu units.

D. Analyze all process energy balances in depth:
   1. Can waste heat be recovered to generate steam or to heat water or a raw material?
   2. Can a process step be eliminated or modified in some way to reduce energy use?
   3. Can an alternate raw material with lower energy content be used?
   4. Is there a way to improve yield?
   5. Is there justification for:
      a. Replacing old equipment with new equipment requiring less energy?
      b. Replacing an obsolete, inefficient process plant with a whole new and different process using less energy?

E. Conduct weekend and night surveys periodically

F. Plan surveys on specific systems and equipment, such as:
   1. Steam system
   2. Compressed air system
   3. Electric motors
   4. Natural gas lines
   5. Heating and air conditioning system

Source: NBS Handbook 115.

TABLE 1-2 Checklist for Energy Conservation Implementation.

IMPLEMENT ENERGY CONSERVATION ACTIONS

A. Correct energy wastes identified in the first survey by taking the necessary maintenance or operation actions

B. List all energy conservation projects evolving from energy balance analyses, surveys, etc. Evaluate and select projects for implementation:
1. Calculate annual energy savings for each project
2. Project future energy costs and calculate annual dollar savings
3. Estimate project capital or expense cost
4. Evaluate investment merit of projects using measures, such as return on investment, etc.
5. Assign priorities to projects based on investment merit
6. Select conservation projects for implementation and request capital authorization
7. Implement authorized projects

C. Review design of all capital projects, such as new plants, expansions, buildings, etc., to assure that efficient utilization of energy is incorporated in the design.
Note: Include consideration of energy availability in new equipment and plant decisions.

Source: NBS Handbook 115.

3. Continue to monitor energy conservation efforts; refer to checklist, Table 1-3.

Determine Energy Uses and Losses

Probably the most important aspect of an ongoing energy utilization program is to make individuals "accountable" for energy use.

Unfortunately, many energy managers find it difficult to economically justify "root metering." The savings as a result of increased accountability are difficult to measure.

Table 1-1 (B) indicates, as part of the initial survey, that a determination should be made as to who is responsible for which area or process and where "root metering" would have the biggest impact.

Implement Actions for Energy Conservation

Once energy usage is known potential energy conservation projects can be identified. Each project will be recommended on the basis of the annual energy savings projected and the initial investment required.

TABLE 1-3 Checklist to Develop Continuous Energy Conservation Efforts.

DEVELOP CONTINUING ENERGY CONSERVATION EFFORTS

A. Measure results:
   1. Chart energy use per unit of production by department
   2. Chart energy use per unit of production for the whole plant
   3. Monitor and analyze charts of Btu per unit of product, taking into consideration effects of complicating variables, such as outdoor ambient air temperature, level of production rate, product mix, etc.
      a. Compare Btu/product unit with past performance and theoretical Btu/product unit
      b. Observe the impact of energy saving actions and project implementation on decreasing the Btu/unit of product
      c. Investigate, identify, and correct the cause for increases that may occur in Btu unit of product, if feasible

B. Continue energy conservation committee activities
   1. Hold periodic meetings
   2. Each committee member is the communication link between the committee and the department supervisors represented
   3. Periodically update energy saving project lists
   4. Plan and participate in energy saving surveys
   5. Communicate energy conservation techniques
   6. Plan and conduct a continuing program of activities and communication to keep up interest in energy conservation
   7. Develop cooperation with community organizations in promoting energy conservation

C. Involve employees
   1. Service on energy conservation committee
   2. Energy conservation training course
   3. Handbook on energy conservation
   4. Suggestion awards plan
   5. Recognition for energy saving achievements
   6. Technical talks on lighting, insulation, steam traps, and other subjects
   7. "savEnergy" posters, decals, stickers
   8. Publicity in plant news, bulletins
   9. Publicity in public news media
   10. Letters on conservation to homes
   11. Talks to local organizations

D. Evaluate program
   1. Review progress in energy saving
   2. Evaluate original goals
   3. Consider program modifications
   4. Revise goals, as necessary

Source: NBS Handbook 115.

Continue to Monitor Energy Conservation Efforts

Energy usage needs to be tracked by using a common energy consumption base per unit of production. This tracking will allow quick identification of changes in energy consumption.

The remaining portion of this chapter will illustrate the language of energy conservation and its applications.

ENERGY ACCOUNTING

An important part of the overall energy auditing program is to be able to measure where you are, and determine where you are going. It is vital to establish an energy accounting system at the beginning of the program. Figures 1-1 through 1-3 illustrate how energy is used for a typical industrial plant. It is important to account for total consumption, cost, and how energy is used for each commodity such as steam, water, air, and natural gas. This procedure is required to develop the appropriate energy conservation strategy.

The top portion of Fig. 1-1 illustrates how much energy is used by fuel type and its relative percentage. The pie chart below shows how much is spent for each fuel type. Using a pie chart representation or nodal flow diagram can be very helpful in visualizing how energy is being used.

Figure 1-2, on the other hand, shows how much of the energy is used for each function such as lighting, process, and building heating and ventilation. Pie charts similar to the right-hand side of the figure should be made for each category such as air, steam, electricity, water, and natural gas.

Figure 1-3 illustrates an alternate representation for the steam distribution profile.

One of the more important aspects of energy management and conservation is measuring and accounting for energy consumption. At Carborundum an energy accounting and analysis system has been developed which is unique in industry, a simple but powerful analytical, management decision-making tool. The Office of Energy Programs of the U.S. Department of Commerce asked Carborundum to work with them in developing this system into a national system, hopefully to be used in the voluntary industrial conservation program. A number of major U.S. corporations are either using or are

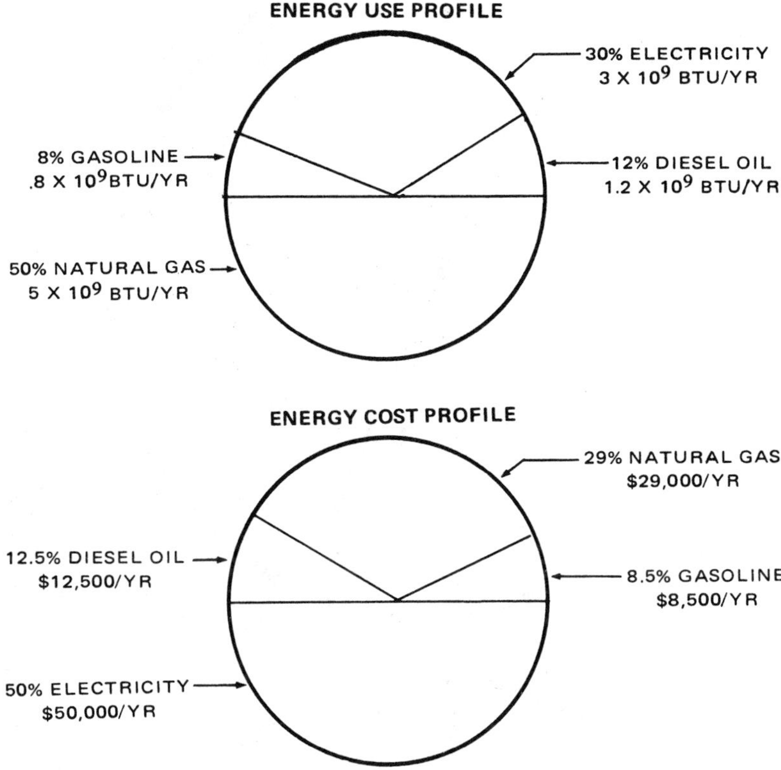

Fig. 1-1 Energy use and cost profile.

considering using the system proposed. The system is offered to those who want to use it.

Most energy accounting systems have been devised and are administered by engineers for engineers. The engineers' principal interest in developing these systems has been the display of energy consumed per unit of production. That ratio has been called "energy efficiency," and changes in energy efficiency are clearly energy conserved or wasted. The engineer focuses all of his attention on reducing energy consumed per unit of production.

An energy efficiency ratio alone, however, cannot answer the kinds of questions asked by business managers and/or government authorities:

- If we are conserving energy, why is our total energy consumption increasing?

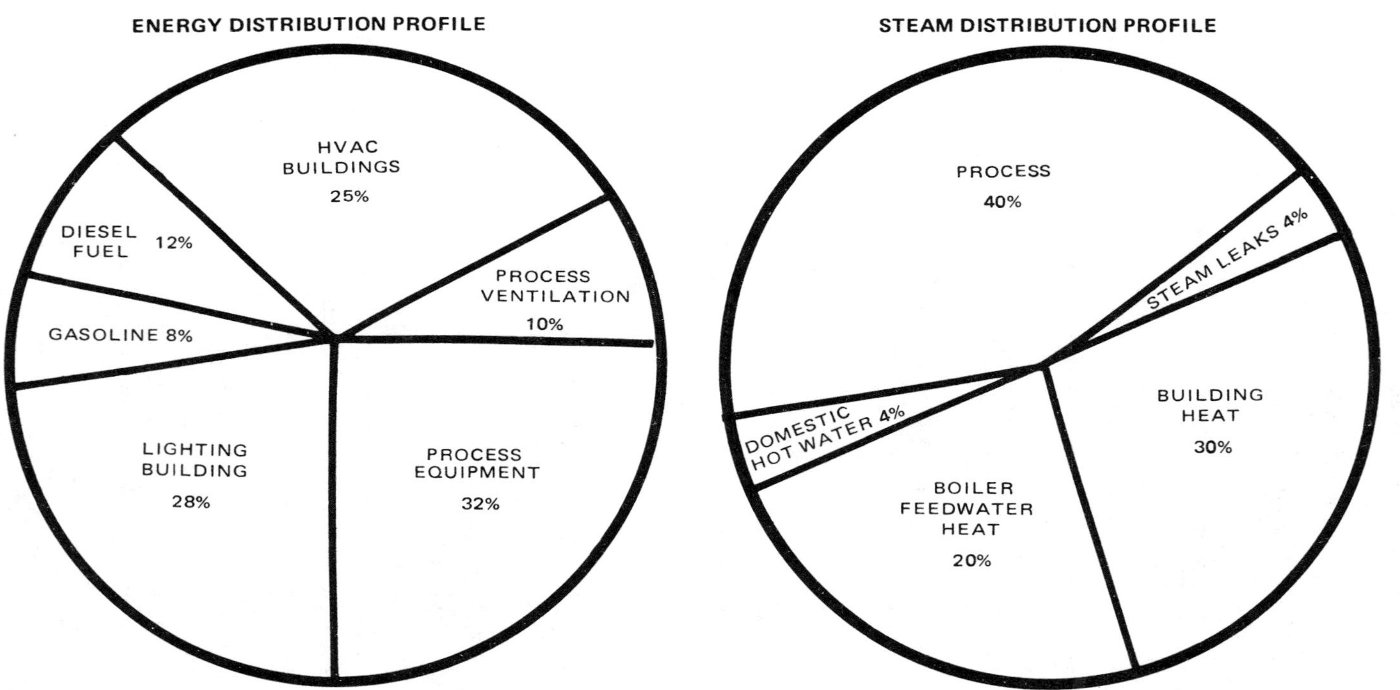

Fig. 1.2 Energy profile by function.

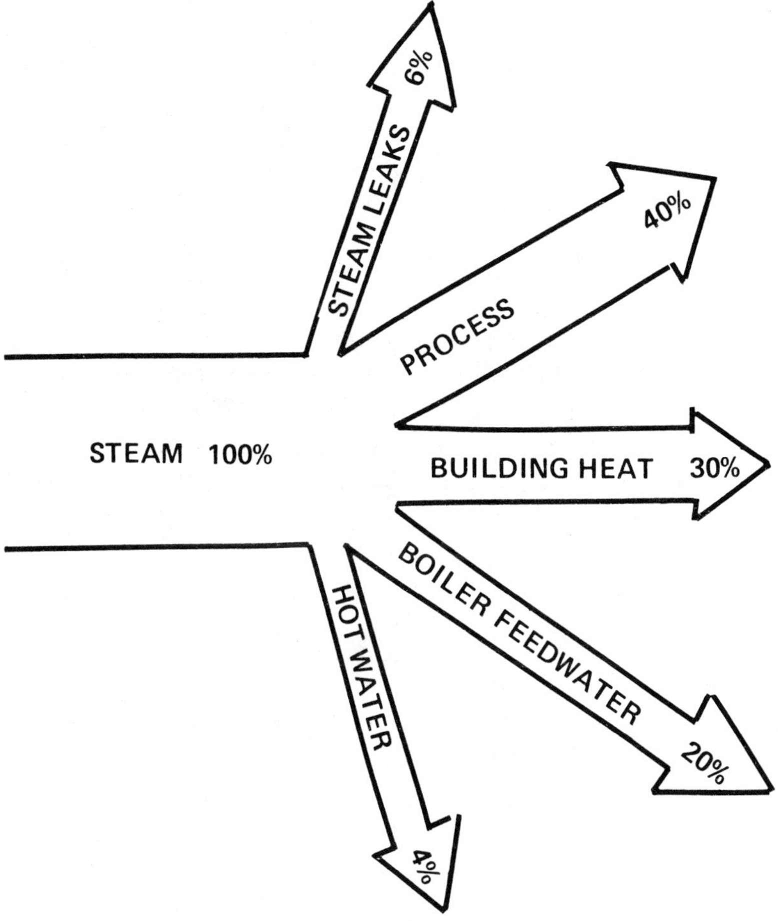

Fig. 1-3 Steam distribution nodal diagram.

- If we are wasting energy, why is our total energy consumption decreasing?
- If we have made no change in energy efficiency, why is our energy consumption changing?

Thus there is a need to evaluate several impacts, such as weather, volume mix, and pollution control, which affect energy use.

Weather Impact

The effect of weather changes (colder winter or hotter summer) on energy consumption is defined as the change in degree-days in the periods under discussion times the heating or cooling efficiency in the period used as the basis for analysis. In the Carborundum system, this translates into the difference in degree-days this year-to-date and last year-to-date times the energy used per degree-day last year-to-date. The monetary impact of weather is the impact calculated as above times the cost per unit of energy last year-to-date. That is, the impact of weather changes on energy use or cost is the difference between this period's weather and last, times the heating/cooling energy efficiency in the last or base period. The result ignores improvements in efficiency (identified later as energy conservation effects) and inflation (identified later as price effects), and isolates the effect of weather.

Volume/Mix Impact

The impact of volume and/or product mix changes is the amount of more (or less) energy that is used currently, as opposed to previously, solely as the result of producing more (or less) product or proportionately more (or less) energy-intense products.

Pollution Control Impact

The impact of the energy increase or decrease to control pollution in the current period versus any other time period is simply the difference in the energy used in the two periods. The financial impact is the impact calculated above multiplied by the cost per unit of energy in the last period. The result ignores conservation and price effects as before, and isolates the effect of pollution control.

"Other" Impacts

The impact of other energy uses, previously defined as experimental, start-up of product lines without history, of base loads, etc., is simply the difference in energy used in the two periods being compared. The economic impact is the impact calculated above multiplied by the cost per unit of energy in the prior period. Again, the result ignores conservation and price effects and isolates the effect of these "other" uses of energy.

Figure 1-4 illustrates the data input form used in the Carborundum system.

Carborundum Energy Accounting and Analysis System Data Input Form

Energy Management and Conservation Program

Plant ______

Division ______

Group ______

Plant Input Data

Today's Date ______

Period Covered ______

Description	Elec. kwh (000)*	Gas mcf	Oil gal. (000)*	Coal lbs. (000)*	Propane gal. (000)*	Other (000)*
Total Fuel Used						
Quantity						
Cost ($)						
** Conversion Factor	X					
Production						
Product 1 NAME						
Production Unit						
Quant. Prod. (000)						
Fuel Used						
Product 2 NAME						
Production Unit						
Quant. Prod. (000)						
Fuel Used						
Product 3 NAME						
Production Unit						
Quant. Prod. (000)						
Fuel Used						
Product 4 NAME						
Production Unit						
Quant. Prod. (000)						
Fuel Used						
Product 5 NAME						
Production Unit						
Quant. Prod. (000)						
Fuel Used						
Heating						
Degree Days						
Fuel Used						
Cooling						
Degree Days						
Fuel Used						
Pollution Control						
Fuel Used						
Other						
Fuel Used						
** Alternate Fuel						

*All fuels reported in thousands to two decimal places.

Fig. 1-4 Carborundum energy accounting and analysis system data input form.

TABLE 1-4 Heating Values for Various Fuels.

Fuel	Average Heating Value
Fuel Oil	
Kerosene	134,000 Btu/gal.
No. 2 Burner Fuel Oil	140,000 Btu/gal.
No. 4 Heavy Fuel Oil	144,000 Btu/gal.
No. 5 Heavy Fuel Oil	150,000 Btu/gal.
No. 6 Heavy Fuel Oil 2.7% sulfur	152,000 Btu/gal.
No. 6 Heavy Fuel Oil 0.3% sulfur	143,800 Btu/gal.
Coal	
Anthracite	13,900 Btu/lb.
Bituminous	14,000 Btu/lb.
Sub-bituminous	12,600 Btu/lb.
Lignite	11,000 Btu/lb.
Gas	
Natural	1,000 Btu/cu. ft
Liquefied butane	103,300 Btu/gal.
Liquefied propane	91,600 Btu/gal.

Source: Brick & Clay Record, October 1972.

TABLE 1-5 List of Conversion Factors.

1 U.S. barrel	= 42 U.S. gallons
1 atmosphere	= 14.7 pounds per square inch absolute (psia)
1 atmosphere	= 760 mm (29.92 in) mercury with density of 13.6 grams per cubic centimeter
1 pound per square inch	= 2.04 inches head of mercury
	= 2.31 feet head of water
1 inch head of water	= 5.20 pounds per square foot
1 foot head of water	= 0.433 pound per square inch
1 British thermal unit (Btu)	= heat required to raise the temperature of 1 pound of water by 1°F
1 therm	= 100,000 Btu
1 kilowatt (Kw)	= 1.341 horsepower (hp)
1 kilowatt-hour (Kwh)	= 1.34 horsepower-hour
1 horsepower (hp)	= 0.746 kilowatt (Kw)
1 horsepower-hour	= 0.746 kilowatt hour (Kwh)
1 horsepower-hour	= 2545 Btu
1 kilowatt-hour (Kwh)	= 3412 Btu
To generate 1 kilowatt-hour (Kwh) requires 10,000 Btu of fuel burned by average utility	
1 ton of refrigeration	= 12,000 Btu per hr
1 ton of refrigeration requires about 1 Kw (or 1.341 hp) in commercial air conditioning	
1 standard cubic foot is at standard conditions of 60°F and 14.7 psia.	
1 degree day	= 65°F minus mean temperature of the day, °F
1 year	= 8760 hours
1 year	= 365 days
1 MBtu	= 1 million Btu
1 Kw	= 1000 watts
1 trillion barrels	= 1×10^{12} barrels
1 KSCF	= 1000 standard cubic feet

Note: In these conversions, inches and feet of water are measured at 62°F (16.7°C), and inches and millimeters of mercury at 32°F (0°C).

THE LANGUAGE OF THE ENERGY MANAGER

In order to communicate energy conservation goals and to analyze the literature in the field, it is important to understand the language of the energy manager and how it is applied.

Each fuel has a heating value, expressed in terms of the British Thermal unit, Btu. The Btu is the heat required to raise the temperature of one pound of water 1°F. Table 1-4 illustrates the heating values of various fuels. To compare efficiencies of various fuels, it is best to convert fuel usage in terms of Btu's. Table 1-5 illustrates conversions used in energy conservation calculations.

TABLE 1-6 Heat of Combustion for Raw Materials.

	Formula	Gross Heat of Combustion Btu/lb
Raw Material		
Carbon	C	14,093
Hydrogen	H_2	61,095
Carbon monoxide	CO	4,347
Paraffin Series		
Methane	CH_4	23,875
Ethane	C_2H_4	22,323
Propane	C_3H_8	21,669
n-Butane	C_4H_{10}	21,321
Isobutane	C_4H_{10}	21,271
n-Pentane	C_5H_{12}	21,095
Isopentane	C_5H_{12}	21,047
Neopentane	C_5H_{12}	20,978
n-Hexane	C_6H_{14}	20,966
Olefin Series		
Ethylene	C_2H_4	21,636
Propylene	C_3H_6	21,048
n-Butene	C_4H_8	20,854
Isobutene	C_4H_8	20,737
n-Pentene	C_5H_{10}	20,720
Aromatic Series		
Benzene	C_6H_6	18,184
Toluene	C_7H_8	18,501
Xylene	C_8H_{10}	18,651
Miscellaneous Gases		
Acetylene	C_2H_2	21,502
Naphthalene	$C_{10}H_8$	17,303
Methyl alcohol	CH_3OH	10,258
Ethyl alcohol	C_2H_5OH	13,161
Ammonia	NH_3	9,667

Source: NBS Handbook 115.

Figure 1-5. Summary of Energy Consumption Measures for the Manufacturing Sector, 1985 *(Quadrillion Btu)*

Type of Consumption and Selected Industries	Total		Electricity	Fuel Oil	Natural Gas	Coal	Other[a]
	Quads	Percent					
Primary Energy Consumption[b]							
Paper and Allied Products	2.21	12.6	0.18	0.17	0.41	0.31	1.15
Chemicals and Allied Products	3.57	20.4	0.41	0.13	1.68	0.33	1.02
Petroleum and Coal Products	5.12	29.2	0.11	0.14	0.72	0.01	[c] 4.16
Primary Metals	2.63	15.0	0.48	0.05	0.69	1.13	0.27
All Other Manufacturing Industries	3.99	22.8	0.99	0.27	1.67	0.60	0.44
Total	17.52	100.0	2.17	0.76	5.17	2.38	7.05
Fuel Consumption to Produce Heat, Power, and Electricity[d]							
Paper and Allied Products	2.20	16.2	0.17	W	0.40	0.34	W
Chemicals and Allied Products	2.41	17.7	0.40	0.09	1.19	0.35	0.38
Petroleum and Coal Products	2.63	19.3	0.11	0.12	0.72	0.01	1.67
Primary Metals	2.39	17.5	0.47	0.05	0.69	0.09	1.09
All Other Manufacturing Industries	3.99	29.3	0.99	W	1.66	0.64	W
Total	13.62	100.0	2.11	0.69	4.66	1.43	4.73
Purchased Fuels and Electricity to Produce Heat, Power, and Electricity							
Paper and Allied Products	1.34	13.8	0.19	0.17	0.40	0.34	0.28
Chemicals and Allied Products	2.17	22.4	0.43	0.09	1.15	0.33	0.20
Petroleum and Coal Products	0.92	9.5	0.12	0.02	0.70	0.01	0.06
Primary Metals	1.54	15.9	0.48	0.05	0.69	0.09	0.22
All Other Manufacturing Industries	3.72	38.4	1.01	0.26	1.66	0.66	0.09
Total	9.69	100.0	2.23	0.59	4.60	1.42	0.85

[a] "Other" includes all other types of energy that respondents indicated were consumed.
[b] Includes feedstocks; does not include byproduct fuels.
[c] Includes feedstocks and raw materials for the production of nonenergy products, such as asphalt, regardless of the type of energy.
[d] Includes byproduct energy.

W = Withheld to avoid disclosing data for individual establishments. Data are included in higher level totals.

Note: Totals may not equal sum of components because of independent rounding.

Source: Energy Information Administration, Office of Energy Markets and End Use, Energy End Use Division, Form EIA-846(F), "1985 Manufacturing Energy Consumption Survey."

Figure 1-6. Energy Consumption in the Manufacturing Sector for 1985

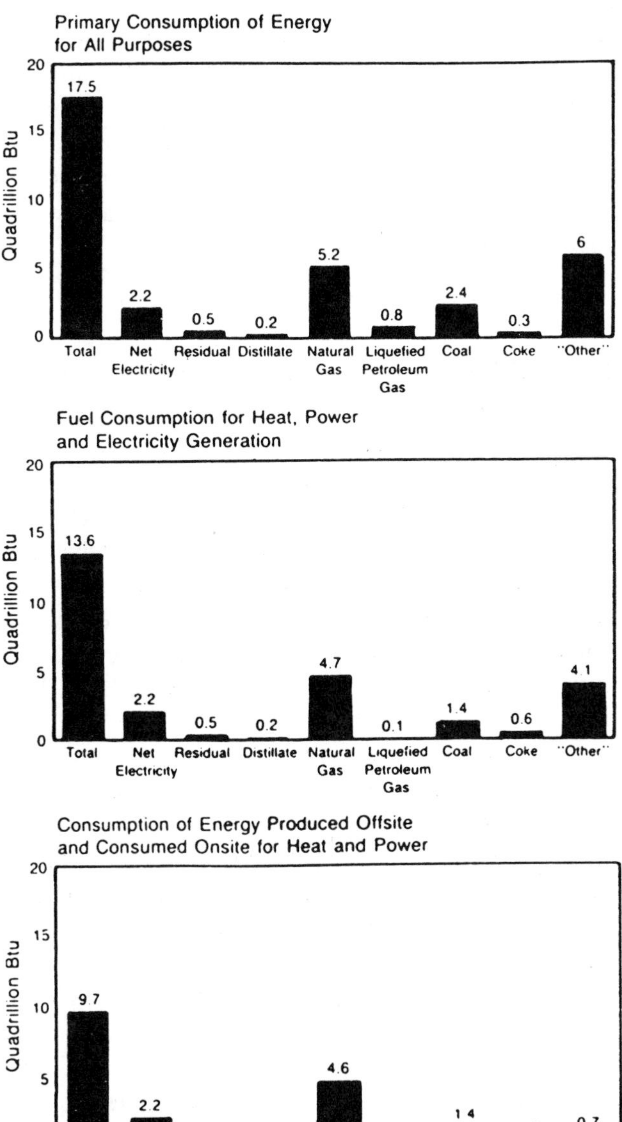

Note: The sum of "Fuel Consumption" and "Primary Energy Consumption for Nonfuel Purposes" does not equal "Total Primary Energy Consumption for All Purposes." Some nonfuel consumption results in byproduct fuels which are included in "Fuel Consumption."

When comparing the cost of fuels, the term "cents per therm" (100,000 Btu) is commonly used.

Knowing the energy content of the Plant's process is an important step in understanding how to reduce its cost. Using energy more efficiently reduces the product's cost, thus increasing profits. In order to account for the process energy content, all energy that enters and leaves a plant during a given period must be measured.

The energy content of various raw materials can be estimated by using the heating values indicated in Table 1-6.

CURRENT AND PROJECTED INDUSTRIAL ENERGY USE

Industry consumes energy for space conditioning, lighting, operation of machinery and processes, and as product feedstocks. Simply stated, the amount and type of energy consumed by industry depends upon the level of industrial output, the extent that less energy-intensive products can displace more energy-intensive products, the relative proportion energy contributes to production costs and the extent that other production inputs can displace energy, the cost of energy resources, and the energy efficiency of energy-using operations.

An estimated 20.4 quads of energy (one quad is 10^{15} Btus) were consumed by all U.S. industry in 1985; the estimated combined energy consumption of all sectors of the economy in 1985 is 55.4 quads. While the industrial share of U.S. energy use has declined over the last decade (from 42 percent in 1972 to about 37 percent in 1985), the industrial sector still consumes more energy than any other. Further, the share of the total energy consumed by industry is projected to increase somewhat in the coming decades, reversing the trend of the recent past.

During 1985, the total *primary energy consumption* by the manufacturing industries was 17.5 quads. This energy was used to produce heat and power and to generate electricity, and used as raw material input for manufacturing processes. **Primary energy consumption**, is the *total energy requirements* (including raw material inputs) of manufacturing industries necessary to *produce nonenergy goods.*

Refer to Figures 1-5 and 1-6 for summary of energy consumption data.

2

Energy Economic Decision Making

LIFE CYCLE COSTING

When a plant manager is assigned the role of energy manager, the first question to be asked is: "What is the economic basis for equipment purchases?"

Some companies use a simple payback method of two years or less to justify equipment purchases. Others require a life cycle cost analysis with no fuel price inflation considered. Still other companies allow for a complete life cycle cost analysis, including the impact for the fuel price inflation and the energy tax credit.

The energy manager's success is directly related to how he or she must justify energy utilization methods.

USING THE PAYBACK PERIOD METHOD

The payback period is the time require to recover the capital investment out of the earnings or savings. This method ignores all savings beyond the payback years, thus penalizing projects that have long life potentials for those that offer high savings for a relatively short period.

The payback period criterion is used when funds are limited and it is important to know how fast dollars will come back. The payback period is simply computed as:

$$\text{Payback period} = \frac{\text{initial investment}}{\text{after tax savings}} \qquad (2\text{-}1)$$

The energy manager who must justify energy equipment expenditures based on a payback period of one year or less has little chance for long-range success. Some companies have set higher payback periods for energy utilization methods. These longer payback periods are justified on the basis that:

- Fuel pricing will increase at a higher rate than the general inflation rate.
- The "risk analysis" for not implementing energy utilization measures may mean loss of production and losing a competitive edge.

USING LIFE CYCLE COSTING

Life cycle costing is an analysis of the total cost of a system, device, building, machine, etc., over its anticipated useful life. The name is new but the subject has, in the past, gone by such names as "engineering economic analysis" or "total owning and operating cost summaries."

Life cycle costing has brought about a new emphasis on the comprehensive identification of all costs associated with a system. The most commonly included costs are initial in place cost, operating costs, maintenance costs, and interest on the investment. Two factors enter into appraising the life of the system; namely, the expected physical life and the period of obsolescence. The lesser factor is governing time period. The effect of interest can then be calculated by using one of several formulas which take into account the time value of money.

When comparing alternative solutions to a particular problem, the system showing the lowest life cycle cost will usually be the first choice (performance requirements are assessed as equal in value).

Life cycle costing is a tool in value engineering. Other items, such as installation time, pollution effects, aesthetic considerations, delivery time, and owner preferences will temper the rule of always choosing the system with the lowst life cycle cost. Good overall judgement is still required.

The life cycle cost analysis still contains judgement factors pertaining to interest rates, useful life, and inflation rates. Even with the judgement element, life cycle costing is the most important tool in value engineering, since the results are quantified in terms of dollars.

As the price for energy changes, and as governmental incentives are initiated, processes or alternatives which were not economically feasible will be considered. This chapter will concentrate on the principles of the life cycle cost analysis as they apply to energy conservation decision making.

THE TIME VALUE OF MONEY

Most energy saving proposals require the investment of capital to accomplish them. By investing today in energy conservation, yearly operating dollars over the life of the investment will be saved. A dollar in hand today is more valuable than one to be received at some time in the future. For this reason, a *time value* must be placed on all cash flows into and out of the company.

Money transactions are thought of as a cash flow to or from a company. Investment decisions also take into account alternate investment opportunities and the minimum return on the investment. In order to compute the rate of return on an investment, it is necessary to find the interest rate which equates payments outcoming and incoming, present and future. The method used to find the rate of return is referred to as *discounted cash flow*.

INVESTMENT DECISION-MAKING

To make investment decisions, the energy manager must follow one simple principle: Relate annual cash flows and lump sum deposits to the same time base. The six factors used for investment decision making simply convert cash from one time base to another; since each company has various financial objectives, these factors can be used to solve *any* investment problem.

Single Payment Compound Amount—SPCA

The SPCA factor is used to determine the future amount S that a present sum P will accumulate at i percent interest, in n years. If P

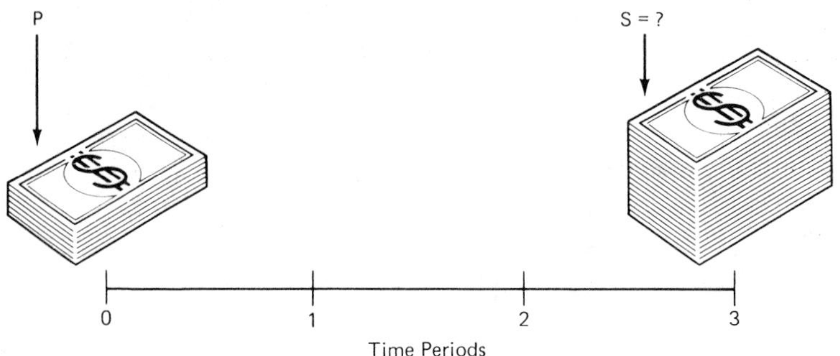

Fig. 2-1 Single payment compound amount (SPCA).

(present worth) is known, and S (future worth) is to be determined, then Equation 2-2 is used.

$$S = P \times (\text{SPCA})n_i \tag{2-2}$$

$$\text{SPCA} = (1 + i)^n \tag{2-3}$$

The SPCA can be computed by an interest formula, but usually its value is found by using the interest tables. Interest tables for interest rates of 10 to 50 percent are found at the conclusion of this chapter (Tables 2-1 through 2-8). In predicting future costs, there are many unknowns. For the accuracy of most calculations, interest rates are assumed to be compounded annually unless otherwise specified. Linear interpolation is commonly used to find values not listed in the interest tables.

Tables 2-9 through 2-12 can be used to determine the effect of fuel escalation on the life cycle cost analysis.

Single Payment Present Worth—SPPW

The SPPW factor is used to determine the present worth, P, that a future amount, S, will be at interest of i-percent, in n years. If S is known, and P is to be determined, then Equation 2-4 is used.

$$P = S \times (\text{SPPW})i_n \tag{2-4}$$

$$\text{SPPW} = \frac{1}{(1 + i)^n} \tag{2-5}$$

TABLE 2-1 10% Interest Factors.

Period n	Single-payment compound-amount (SPCA)	Single-payment present-worth (SPPW)	Uniform-series compound-amount (USCA)	Sinking-fund payment (SFP)	Capital recovery (CR)	Uniform-series present-worth (USPW)
	Future value of \$1 $(1+i)^n$	Present value of \$1 $\frac{1}{(1+i)^n}$	Future value of uniform series of \$1 $\frac{(1+i)^n-1}{i}$	Uniform series whose future value is \$1 $\frac{i}{(1+i)^n-1}$	Uniform series with present value of \$1 $\frac{i(1+i)^n}{(1+i)^n-1}$	Present value of uniform series of \$1 $\frac{(1+i)^n-1}{i(1+i)^n}$
1	1.100	0.9091	1.000	1.00000	1.10000	0.909
2	1.210	0.8264	2.100	0.47619	0.57619	1.736
3	1.331	0.7513	3.310	0.30211	0.40211	2.487
4	1.464	0.6830	4.641	0.21547	0.31547	3.170
5	1.611	0.6209	6.105	0.16380	0.26380	3.791
6	1.772	0.5645	7.716	0.12961	0.22961	4.355
7	1.949	0.5132	9.487	0.10541	0.20541	4.868
8	2.144	0.4665	11.436	0.08744	0.18744	5.335
9	2.358	0.4241	13.579	0.07364	0.17364	5.759
10	2.594	0.3855	15.937	0.06275	0.16275	6.144
11	2.853	0.3505	18.531	0.05396	0.15396	6.495
12	3.138	0.3186	21.384	0.04676	0.14676	6.814
13	3.452	0.2897	24.523	0.04078	0.14078	7.103
14	3.797	0.2633	27.975	0.03575	0.13575	7.367
15	4.177	0.2394	31.772	0.03147	0.13147	7.606
16	4.595	0.2176	35.950	0.02782	0.12782	7.824
17	5.054	0.1978	40.545	0.02466	0.12466	8.022
18	5.560	0.1799	45.599	0.02193	0.12193	8.201
19	6.116	0.1635	51.159	0.01955	0.11955	8.365
20	6.727	0.1486	57.275	0.01746	0.11746	8.514
21	7.400	0.1351	64.002	0.01562	0.11562	8.649
22	8.140	0.1228	71.403	0.01401	0.11401	8.772
23	8.954	0.1117	79.543	0.01257	0.11257	8.883
24	9.850	0.1015	88.497	0.01130	0.11130	8.985
25	10.835	0.0923	98.347	0.01017	0.11017	9.077
26	11.918	0.0839	109.182	0.00916	0.10916	9.161
27	13.110	0.0763	121.100	0.00826	0.10826	9.237
28	14.421	0.0693	134.210	0.00745	0.10745	9.307
29	15.863	0.0630	148.631	0.00673	0.10673	9.370
30	17.449	0.0573	164.494	0.00608	0.10608	9.427
35	28.102	0.0356	271.024	0.00369	0.10369	9.644
40	45.259	0.0221	442.593	0.00226	0.10226	9.779
45	72.890	0.0137	718.905	0.00139	0.10139	9.863
50	117.391	0.0085	1163.909	0.00086	0.10086	9.915
55	189.059	0.0053	1880.591	0.00053	0.10053	9.947
60	304.482	0.0033	3034.816	0.00033	0.10033	9.967
65	490.371	0.0020	4893.707	0.00020	0.10020	9.980
70	789.747	0.0013	7887.470	0.00013	0.10013	9.987
75	1271.895	0.0008	12708.954	0.00008	0.10008	9.992
80	2048.400	0.0005	20474.002	0.00005	0.10005	9.995
85	3298.969	0.0003	32979.690	0.00003	0.10003	9.997
90	5313.023	0.0002	53120.226	0.00002	0.10002	9.998
95	8556.676	0.0001	85556.760	0.00001	0.10001	9.999

Source: From *Economic Analysis for Engineering and Managerial Decision Making* by Norman M. Barish. Copyright © 1962 by McGraw-Hill, Inc. Used with permission of McGraw-Hill Book Company.

TABLE 2-2 12% Interest Factors.

Period n	Single-payment compound-amount (SPCA)	Single-payment present-worth (SPPW)	Uniform-series compound-amount (USCA)	Sinking-fund payment (SFP)	Capital recovery (CR)	Uniform-series present-worth (USPW)
	Future value of $1 $(1+i)^n$	Present value of $1 $\frac{1}{(1+i)^n}$	Future value of uniform series of $1 $\frac{(1+i)^n-1}{i}$	Uniform series whose future value is $1 $\frac{i}{(1+i)^n-1}$	Uniform series with present value of $1 $\frac{i(1+i)^n}{(1+i)^n-1}$	Present value of uniform series of $1 $\frac{(1+i)^n-1}{i(1+i)^n}$
1	1.120	0.8929	1.000	1.00000	1.12000	0.893
2	1.254	0.7972	2.120	0.47170	0.59170	1.690
3	1.405	0.7118	3.374	0.29635	0.41635	2.402
4	1.574	0.6355	4.779	0.20923	0.32923	3.037
5	1.762	0.5674	6.353	0.15741	0.27741	3.605
6	1.974	0.5066	8.115	0.12323	0.24323	4.111
7	2.211	0.4523	10.089	0.09912	0.21912	4.564
8	2.476	0.4039	12.300	0.08130	0.20130	4.968
9	2.773	0.3606	14.776	0.06768	0.18768	5.328
10	3.106	0.3220	17.549	0.05698	0.17698	5.650
11	3.479	0.2875	20.655	0.04842	0.16842	5.938
12	3.896	0.2567	24.133	0.04144	0.16144	6.194
13	4.363	0.2292	28.029	0.03568	0.15568	6.424
14	4.887	0.2046	32.393	0.03087	0.15087	6.628
15	5.474	0.1827	37.280	0.02682	0.14682	6.811
16	6.130	0.1631	42.753	0.02339	0.14339	6.974
17	6.866	0.1456	48.884	0.02046	0.14046	7.120
18	7.690	0.1300	55.750	0.01794	0.13794	7.250
19	8.613	0.1161	63.440	0.01576	0.13576	7.366
20	9.646	0.1037	72.052	0.01388	0.13388	7.469
21	10.804	0.0926	81.699	0.01224	0.13224	7.562
22	12.100	0.0826	92.503	0.01081	0.13081	7.645
23	13.552	0.0738	104.603	0.00956	0.12956	7.718
24	15.179	0.0659	118.155	0.00846	0.12846	7.784
25	17.000	0.0588	133.334	0.00750	0.12750	7.843
26	19.040	0.0525	150.334	0.00665	0.12665	7.896
27	21.325	0.0469	169.374	0.00590	0.12590	7.943
28	23.884	0.0419	190.699	0.00524	0.12524	7.984
29	26.750	0.0374	214.583	0.00466	0.12466	8.022
30	29.960	0.0334	241.333	0.00414	0.12414	8.055
35	52.800	0.0189	431.663	0.00232	0.12232	8.176
40	93.051	0.0107	767.091	0.00130	0.12130	8.244
45	163.988	0.0061	1358.230	0.00074	0.12074	8.283
50	289.002	0.0035	2400.018	0.00042	0.12042	8.304
55	509.321	0.0020	4236.005	0.00024	0.12024	8.317
60	897.597	0.0011	7471.641	0.00013	0.12013	8.324
65	1581.872	0.0006	13173.937	0.00008	0.12008	8.328
70	2787.800	0.0004	23223.332	0.00004	0.12004	8.330
75	4913.056	0.0002	40933.799	0.00002	0.12002	8.332
80	8658.483	0.0001	72145.692	0.00001	0.12001	8.332

Source: From *Economic Analysis for Engineering and Managerial Decision Making* by Norman M. Barish.

TABLE 2-3 15% Interest Factors.

Period n	Single-payment compound-amount (SPCA)	Single-payment present-worth (SPPW)	Uniform-series compound-amount (USCA)	Sinking-fund payment (SFP)	Capital recovery (CR)	Uniform-series present-worth (USPW)
	Future value of $1 $(1+i)^n$	Present value of $1 $\frac{1}{(1+i)^n}$	Future value of uniform series of $1 $\frac{(1+i)^n-1}{i}$	Uniform series whose future value is $1 $\frac{i}{(1+i)^n-1}$	Uniform series with present value of $1 $\frac{i(1+i)^n}{(1+i)^n-1}$	Present value of uniform series of $1 $\frac{(1+i)^n-1}{i(1+i)^n}$
1	1.150	0.8696	1.000	1.00000	1.15000	0.870
2	1.322	0.7561	2.150	0.46512	0.61512	1.626
3	1.521	0.6575	3.472	0.28798	0.43798	2.283
4	1.749	0.5718	4.993	0.20027	0.35027	2.855
5	2.011	0.4972	6.742	0.14832	0.29832	3.352
6	2.313	0.4323	8.754	0.11424	0.26424	3.784
7	2.660	0.3759	11.067	0.09036	0.24036	4.160
8	3.059	0.3269	13.727	0.07285	0.22285	4.487
9	3.518	0.2843	16.786	0.05957	0.20957	4.772
10	4.046	0.2472	20.304	0.04925	0.19925	5.019
11	4.652	0.2149	24.349	0.04107	0.19107	5.234
12	5.350	0.1869	29.002	0.03448	0.18448	5.421
13	6.153	0.1625	34.352	0.02911	0.17911	5.583
14	7.076	0.1413	40.505	0.02469	0.17469	5.724
15	8.137	0.1229	47.580	0.02102	0.17102	5.847
16	9.358	0.1069	55.717	0.01795	0.16795	5.954
17	10.761	0.0929	65.075	0.01537	0.16537	6.047
18	12.375	0.0808	75.836	0.01319	0.16319	6.128
19	14.232	0.0703	88.212	0.01134	0.16134	6.198
20	16.367	0.0611	102.444	0.00976	0.15976	6.259
21	18.822	0.0531	118.810	0.00842	0.15842	6.312
22	21.645	0.0462	137.632	0.00727	0.15727	6.359
23	24.891	0.0402	159.276	0.00628	0.15628	6.399
24	28.625	0.0349	184.168	0.00543	0.15543	6.434
25	32.919	0.0304	212.793	0.00470	0.15470	6.464
26	37.857	0.0264	245.712	0.00407	0.15407	6.491
27	43.535	0.0230	283.569	0.00353	0.15353	6.514
28	50.066	0.0200	327.104	0.00306	0.15306	6.534
29	57.575	0.0174	377.170	0.00265	0.15265	6.551
30	66.212	0.0151	434.745	0.00230	0.15230	6.566
35	133.176	0.0075	881.170	0.00113	0.15113	6.617
40	267.864	0.0037	1779.090	0.00056	0.15056	6.642
45	538.769	0.0019	3585.128	0.00028	0.15028	6.654
50	1083.657	0.0009	7217.716	0.00014	0.15014	6.661
55	2179.622	0.0005	14524.148	0.00007	0.15007	6.664
60	4383.999	0.0002	29219.992	0.00003	0.15003	6.665
65	8817.787	0.0001	58778.583	0.00002	0.15002	6.666

Source: From *Economic Analysis for Engineering and Managerial Decision Making* by Norman M. Barish. Copyright © 1962 by McGraw-Hill, Inc. Used with permission of McGraw-Hill Book Company.

TABLE 2-4 20% Interest Factors.

Period n	Single-payment compound-amount (SPCA)	Single-payment present-worth (SPPW)	Uniform-series compound-amount (USCA)	Sinking-fund payment (SFP)	Capital recovery (CR)	Uniform-series present-worth (USPW)
	Future value of \$1 $(1+i)^n$	Present value of \$1 $\frac{1}{(1+i)^n}$	Future value of uniform series of \$1 $\frac{(1+i)^n-1}{i}$	Uniform series whose future value is \$1 $\frac{i}{(1+i)^n-1}$	Uniform series with present value of \$1 $\frac{i(1+i)^n}{(1+i)^n-1}$	Present value of uniform series of \$1 $\frac{(1+i)^n-1}{i(1+i)^n}$
1	1.200	0.8333	1.000	1.00000	1.20000	0.833
2	1.440	0.6944	2.200	0.45455	0.65455	1.528
3	1.728	0.5787	3.640	0.27473	0.47473	2.106
4	2.074	0.4823	5.368	0.18629	0.38629	2.589
5	2.488	0.4019	7.442	0.13438	0.33438	2.991
6	2.986	0.3349	9.930	0.10071	0.30071	3.326
7	3.583	0.2791	12.916	0.07742	0.27742	3.605
8	4.300	0.2326	16.499	0.06061	0.26061	3.837
9	5.160	0.1938	20.799	0.04808	0.24808	4.031
10	6.192	0.1615	25.959	0.03852	0.23852	4.192
11	7.430	0.1346	32.150	0.03110	0.23110	4.327
12	8.916	0.1122	39.581	0.02526	0.22526	4.439
13	10.699	0.0935	48.497	0.02062	0.22062	4.533
14	12.839	0.0779	59.196	0.01689	0.21689	4.611
15	15.407	0.0649	72.035	0.01388	0.21388	4.675
16	18.488	0.0541	87.442	0.01144	0.21144	4.730
17	22.186	0.0451	105.931	0.00944	0.20944	4.775
18	26.623	0.0376	128.117	0.00781	0.20781	4.812
19	31.948	0.0313	154.740	0.00646	0.20646	4.843
20	38.338	0.0261	186.688	0.00536	0.20536	4.870
21	46.005	0.0217	225.026	0.00444	0.20444	4.891
22	55.206	0.0181	271.031	0.00369	0.20369	4.909
23	66.247	0.0151	326.237	0.00307	0.20307	4.925
24	79.497	0.0126	392.484	0.00255	0.20255	4.937
25	95.396	0.0105	471.981	0.00212	0.20212	4.948
26	114.475	0.0087	567.377	0.00176	0.20176	4.956
27	137.371	0.0073	681.853	0.00147	0.20147	4.964
28	164.845	0.0061	819.223	0.00122	0.20122	4.970
29	197.814	0.0051	984.068	0.00102	0.20102	4.975
30	237.376	0.0042	1181.882	0.00085	0.20085	4.979
35	590.668	0.0017	2948.341	0.00034	0.20034	4.992
40	1469.772	0.0007	7343.858	0.00014	0.20014	4.997
45	3657.262	0.0003	18281.310	0.00005	0.20005	4.999
50	9100.438	0.0001	45497.191	0.00002	0.20002	4.999

Source: From *Economic Analysis for Engineering and Managerial Decision Making* by Norman M. Barish. Copyright © 1962 by McGraw-Hill, Inc. Used with permission of McGraw-Hill Book Company.

TABLE 2-5 25% Interest Factors.

Period n	Single-payment compound-amount (SPCA)	Single-payment present-worth (SPPW)	Uniform-series compound amount (USCA)	Sinking-fund payment (SFP)	Capital recovery (CR)	Uniform-series present-worth (USPW)
	Future value of $1 $(1+i)^n$	Present value of $1 $\frac{1}{(1+i)^n}$	Future value of uniform series of $1 $\frac{(1+i)^n-1}{i}$	Uniform series whose future value is $1 $\frac{i}{(1+i)^n-1}$	Uniform series with present value of $1 $\frac{i(1+i)^n}{(1+i)^n-1}$	Present value of uniform series of $1 $\frac{(1+i)^n-1}{i(1+i)^n}$
1	1.250	0.8000	1.000	1.00000	1.25000	0.800
2	1.562	0.6400	2.250	0.44444	0.69444	1.440
3	1.953	0.5120	3.812	0.26230	0.51230	1.952
4	2.441	0.4096	5.766	0.17344	0.42344	2.362
5	3.052	0.3277	8.207	0.12185	0.37185	2.689
6	3.815	0.2621	11.259	0.08882	0.33882	2.951
7	4.768	0.2097	15.073	0.06634	0.31634	3.161
8	5.960	0.1678	19.842	0.05040	0.30040	3.329
9	7.451	0.1342	25.802	0.03876	0.28876	3.463
10	9.313	0.1074	33.253	0.03007	0.28007	3.571
11	11.642	0.0859	42.566	0.02349	0.27349	3.656
12	14.552	0.0687	54.208	0.01845	0.26845	3.725
13	18.190	0.0550	68.760	0.01454	0.26454	3.780
14	22.737	0.0440	86.949	0.01150	0.26150	3.824
15	28.422	0.0352	109.687	0.00912	0.25912	3.859
16	35.527	0.0281	138.109	0.00724	0.25724	3.887
17	44.409	0.0225	173.636	0.00576	0.25576	3.910
18	55.511	0.0180	218.045	0.00459	0.25459	3.928
19	69.389	0.0144	273.556	0.00366	0.25366	3.942
20	86.736	0.0115	342.945	0.00292	0.25292	3.954
21	108.420	0.0092	429.681	0.00233	0.25233	3.963
22	135.525	0.0074	538.101	0.00186	0.25186	3.970
23	169.407	0.0059	673.626	0.00148	0.25148	3.976
24	211.758	0.0047	843.033	0.00119	0.25119	3.981
25	264.698	0.0038	1054.791	0.00095	0.25095	3.985
26	330.872	0.0030	1319.489	0.00076	0.25076	3.988
27	413.590	0.0024	1650.361	0.00061	0.25061	3.990
28	516.988	0.0019	2063.952	0.00048	0.25048	3.992
29	646.235	0.0015	2580.939	0.00039	0.25039	3.994
30	807.794	0.0012	3227.174	0.00031	0.25031	3.995
35	2465.190	0.0004	9856.761	0.00010	0.25010	3.998
40	7523.164	0.0001	30088.655	0.00003	0.25003	3.999

Source: From *Economic Analysis for Engineering and Managerial Decision Making* by Norman M. Barish. Copyright © 1962 by McGraw-Hill, Inc. Used with permission of McGraw-Hill Book Company.

TABLE 2-6 30% Interest Factors.

Period n	Single-payment compound-amount (SPCA) Future value of \$1 $(1+i)^n$	Single-payment present-worth (SPPW) Present value of \$1 $\frac{1}{(1+i)^n}$	Uniform-series compound-amount (USCA) Future value of uniform series of \$1 $\frac{(1+i)^n-1}{i}$	Sinking-fund payment (SFP) Uniform series whose future value is \$1 $\frac{i}{(1+i)^n-1}$	Capital recovery (CR) Uniform series with present value of \$1 $\frac{i(1+i)^n}{(1+i)^n-1}$	Uniform-series present-worth (USPW) Present value of uniform series of \$1 $\frac{(1+i)^n-1}{i(1+i)^n}$
1	1.300	0.7692	1.000	1.00000	1.30000	0.769
2	1.690	0.5917	2.300	0.43478	0.73478	1.361
3	2.197	0.4552	3.990	0.25063	0.55063	1.816
4	2.856	0.3501	6.187	0.16163	0.46163	2.166
5	3.713	0.2693	9.043	0.11058	0.41058	2.436
6	4.827	0.2072	12.756	0.07839	0.37839	2.643
7	6.275	0.1594	17.583	0.05687	0.35687	2.802
8	8.157	0.1226	23.858	0.04192	0.34192	2.925
9	10.604	0.0943	32.015	0.03124	0.33124	3.019
10	13.786	0.0725	42.619	0.02346	0.32346	3.092
11	17.922	0.0558	56.405	0.01773	0.31773	3.147
12	23.298	0.0429	74.327	0.01345	0.31345	3.190
13	30.288	0.0330	97.625	0.01024	0.31024	3.223
14	39.374	0.0254	127.913	0.00782	0.30782	3.249
15	51.186	0.0195	167.286	0.00598	0.30598	3.268
16	66.542	0.0150	218.472	0.00458	0.30458	3.283
17	86.504	0.0116	285.014	0.00351	0.30351	3.295
18	112.455	0.0089	371.518	0.00269	0.30269	3.304
19	146.192	0.0068	483.973	0.00207	0.30207	3.311
20	190.050	0.0053	630.165	0.00159	0.30159	3.316
21	247.065	0.0040	820.215	0.00122	0.30122	3.320
22	321.184	0.0031	1067.280	0.00094	0.30094	3.323
23	417.539	0.0024	1388.464	0.00072	0.30072	3.325
24	542.801	0.0018	1806.003	0.00055	0.30055	3.327
25	705.641	0.0014	2348.803	0.00043	0.30043	3.329
26	917.333	0.0011	3054.444	0.00033	0.30033	3.330
27	1192.533	0.0008	3971.778	0.00025	0.30025	3.331
28	1550.293	0.0006	5164.311	0.00019	0.30019	3.331
29	2015.381	0.0005	6714.604	0.00015	0.30015	3.332
30	2619.996	0.0004	8729.985	0.00011	0.30011	3.332
35	9727.860	0.0001	32422.868	0.00003	0.30003	3.333

Source: From *Economic Analysis for Engineering and Managerial Decision Making* by Norman M. Barish. Copyright © 1962 by McGraw-Hill, Inc. Used with permission of McGraw-Hill Book Company.

TABLE 2-7 40% Interest Factors.

Period n	Single-payment compound-amount (SPCA)	Single-payment present-worth (SPPW)	Uniform-series compound-amount (USCA)	Sinking-fund payment (SFP)	Capital recovery (CR)	Uniform-series present-worth (USPW)
	Future value of \$1 $(1+i)^n$	Present value of \$1 $\frac{1}{(1+i)^n}$	Future value of uniform series of \$1 $\frac{(1+i)^n-1}{i}$	Uniform series whose future value is \$1 $\frac{i}{(1+i)^n-1}$	Uniform series with present value of \$1 $\frac{i(1+i)^n}{(1+i)^n-1}$	Present value of uniform series of \$1 $\frac{(1+i)^n-1}{i(1+i)^n}$
1	1.400	0.7143	1.000	1.00000	1.40000	0.714
2	1.960	0.5102	2.400	0.41667	0.81667	1.224
3	2.744	0.3644	4.360	0.22936	0.62936	1.589
4	3.842	0.2603	7.104	0.14077	0.54077	1.849
5	5.378	0.1859	10.946	0.09136	0.49136	2.035
6	7.530	0.1328	16.324	0.06126	0.46126	2.168
7	10.541	0.0949	23.853	0.04192	0.44192	2.263
8	14.758	0.0678	34.395	0.02907	0.42907	2.331
9	20.661	0.0484	49.153	0.02034	0.42034	2.379
10	28.925	0.0346	69.814	0.01432	0.41432	2.414
11	40.496	0.0247	98.739	0.01013	0.41013	2.438
12	56.694	0.0176	139.235	0.00718	0.40718	2.456
13	79.371	0.0126	195.929	0.00510	0.40510	2.469
14	111.120	0.0090	275.300	0.00363	0.40363	2.478
15	155.568	0.0064	386.420	0.00259	0.40259	2.484
16	217.795	0.0046	541.988	0.00185	0.40185	2.489
17	304.913	0.0033	759.784	0.00132	0.40132	2.492
18	426.879	0.0023	1064.697	0.00094	0.40094	2.494
19	597.630	0.0017	1491.576	0.00067	0.40067	2.496
20	836.683	0.0012	2089.206	0.00048	0.40048	2.497
21	1171.356	0.0009	2925.889	0.00034	0.40034	2.498
22	1639.898	0.0006	4097.245	0.00024	0.40024	2.498
23	2295.857	0.0004	5737.142	0.00017	0.40017	2.499
24	3214.200	0.0003	8032.999	0.00012	0.40012	2.499
25	4499.880	0.0002	11247.199	0.00009	0.40009	2.499
26	6299.831	0.0002	15747.079	0.00006	0.40006	2.500
27	8819.764	0.0001	22046.910	0.00005	0.40005	2.500

Source: From *Economic Analysis for Engineering and Managerial Decision Making* by Norman M. Barish. Copyright © 1962 by McGraw-Hill, Inc. Used with permission of McGraw-Hill Book Company.

TABLE 2-8 50% Interest Factors.

Period n	Single-payment compound-amount (SPCA)	Single-payment present-worth (SPPW)	Uniform-series compound-amount (USCA)	Sinking-fund payment (SFP)	Capital recovery (CR)	Uniform-series present-worth (USPW)
	Future value of $1 $(1+i)^n$	Present value of $1 $\frac{1}{(1+i)^n}$	Future value of uniform series of $1 $\frac{(1+i)^n-1}{i}$	Uniform series whose future value is $1 $\frac{i}{(1+i)^n-1}$	Uniform series with present value of $1 $\frac{i(1+i)^n}{(1+i)^n-1}$	Present value of uniform series of $1 $\frac{(1+i)^n-1}{i(1+i)^n}$
1	1.500	0.6667	1.000	1.00000	1.50000	0.667
2	2.250	0.4444	2.500	0.40000	0.90000	1.111
3	3.375	0.2963	4.750	0.21053	0.71053	1.407
4	5.062	0.1975	8.125	0.12308	0.62308	1.605
5	7.594	0.1317	13.188	0.07583	0.57583	1.737
6	11.391	0.0878	20.781	0.04812	0.54812	1.824
7	17.086	0.0585	32.172	0.03108	0.53108	1.883
8	25.629	0.0390	49.258	0.02030	0.52030	1.922
9	38.443	0.0260	74.887	0.01335	0.51335	1.948
10	57.665	0.0173	113.330	0.00882	0.50882	1.965
11	86.498	0.0116	170.995	0.00585	0.50585	1.977
12	129.746	0.0077	257.493	0.00388	0.50388	1.985
13	194.620	0.0051	387.239	0.00258	0.50258	1.990
14	291.929	0.0034	581.859	0.00172	0.50172	1.993
15	437.894	0.0023	873.788	0.00114	0.50114	1.995
16	656.841	0.0015	1311.682	0.00076	0.50076	1.997
17	985.261	0.0010	1968.523	0.00051	0.50051	1.998
18	1477.892	0.0007	2953.784	0.00034	0.50034	1.999
19	2216.838	0.0005	4431.676	0.00023	0.50023	1.999
20	3325.257	0.0003	6648.513	0.00015	0.50015	1.999
21	4987.885	0.0002	9973.770	0.00010	0.50010	2.000
22	7481.828	0.0001	14961.655	0.00007	0.50007	2.000

Source: From *Economic Analysis for Engineering and Managerial Decision Making* by Norman M. Barish.

TABLE 2-9 Five-Year Escalation Table.

Present Worth of a Series of Escalating Payments Compounded Annually
Discount-Escalation Factors for n = 5 Years

Discount Rate	Annual Escalation Rate					
	0.10	0.12	0.14	0.16	0.18	0.20
0.10	5.000000	5.279234	5.572605	5.880105	6.202627	6.540569
0.11	4.866862	5.136200	5.420152	5.717603	6.029313	6.355882
0.12	4.738562	5.000000	5.274242	5.561868	5.863289	6.179066
0.13	4.615647	4.869164	5.133876	5.412404	5.704137	6.009541
0.14	4.497670	4.742953	5.000000	5.269208	5.551563	5.847029
0.15	4.384494	4.622149	4.871228	5.131703	5.404955	5.691165
0.16	4.275647	4.505953	4.747390	5.000000	5.264441	5.541511
0.17	4.171042	4.394428	4.628438	4.873699	5.129353	5.397964
0.18	4.070432	4.287089	4.513947	4.751566	5.000000	5.259749
0.19	3.973684	4.183921	4.403996	4.634350	4.875619	5.126925
0.20	3.880510	4.084577	4.298207	4.521178	4.755725	5.000000
0.21	3.790801	3.989001	4.196400	4.413341	4.640260	4.877689
0.22	4.704368	3.896891	4.098287	4.308947	4.529298	4.759649
0.23	3.621094	3.808179	4.003835	4.208479	4.422339	4.645864
0.24	3.540773	3.722628	3.912807	4.111612	4.319417	4.536517
0.25	3.463301	3.640161	3.825008	4.018249	4.220158	4.431144
0.26	3.388553	3.560586	3.740376	3.928286	4.124553	4.329514
0.27	3.316408	3.483803	3.658706	3.841442	4.032275	4.231583
0.28	3.246718	3.409649	3.579870	3.757639	3.943295	4.137057
0.29	3.179393	3.338051	3.503722	3.676771	3.857370	4.045902
0.30	3.114338	3.268861	3.430201	3.598653	3.774459	3.957921
0.31	3.051452	3.201978	3.359143	3.523171	3.694328	3.872901
0.32	2.990618	3.137327	3.290436	3.450224	3.616936	3.790808
0.33	2.931764	3.074780	3.224015	3.379722	3.542100	3.711472
0.34	2.874812	3.014281	3.159770	3.311524	3.469775	3.634758

TABLE 2-10 Ten-Year Escalation Table.

Present Worth of a Series of Escalating Payments Compounded Annually
Discount-Escalation Factors for n = 10 Years

Discount Rate	Annual Escalation Rate 0.10	0.12	0.14	0.16	0.18	0.20
0.10	10.000000	11.056250	12.234870	13.548650	15.013550	16.646080
0.11	9.518405	10.508020	11.613440	12.844310	14.215140	15.741560
0.12	9.068870	10.000000	11.036530	12.190470	13.474590	14.903510
0.13	8.650280	9.526666	10.498990	11.582430	12.786980	14.125780
0.14	8.259741	9.084209	10.000000	11.017130	12.147890	13.403480
0.15	7.895187	8.672058	9.534301	10.490510	11.552670	12.731900
0.16	7.554141	8.286779	9.099380	10.000000	10.998720	12.106600
0.17	7.234974	7.926784	8.693151	9.542653	10.481740	11.524400
0.18	6.935890	7.589595	8.312960	9.113885	10.000000	10.980620
0.19	6.655455	7.273785	7.957330	8.713262	9.549790	10.472990
0.20	6.392080	6.977461	7.624072	8.338518	9.128122	10.000000
0.21	6.144593	6.699373	7.311519	7.987156	8.733109	9.557141
0.22	5.911755	6.437922	7.017915	7.657542	8.363208	9.141752
0.23	5.692557	6.192047	6.742093	7.348193	8.015993	8.752133
0.24	5.485921	5.960481	6.482632	7.057347	7.690163	8.387045
0.25	5.290990	5.742294	6.238276	6.783767	7.383800	8.044173
0.26	5.106956	5.536463	6.008083	6.526298	7.095769	7.721807
0.27	4.933045	5.342146	5.790929	6.283557	6.824442	7.418647
0.28	4.768518	5.158489	5.585917	6.054608	6.568835	7.133100
0.29	4.612762	4.984826	5.392166	5.838531	6.327682	6.864109
0.30	4.465205	4.820429	5.209000	5.634354	6.100129	6.610435
0.31	4.325286	4.664669	5.035615	5.441257	5.885058	6.370867
0.32	4.192478	4.517015	4.871346	5.258512	5.681746	6.144601
0.33	4.066339	4.376884	4.715648	5.085461	5.489304	5.930659
0.34	3.946452	4.243845	4.567942	4.921409	5.307107	5.728189

TABLE 2-11 Fifteen-Year Escalation Table.

Present Worth of a Series of Escalating Payments Compounded Annually
Discount-Escalation Factors for n = 15 years

Discount Rate	Annual Escalation Rate					
	0.10	0.12	0.14	0.16	0.18	0.20
0.10	15.000000	17.377880	20.199780	23.549540	27.529640	32.259620
0.11	13.964150	16.126230	18.690120	21.727370	25.328490	29.601330
0.12	13.026090	15.000000	17.332040	20.090360	23.355070	27.221890
0.13	12.177030	13.981710	16.105770	18.616160	21.581750	25.087260
0.14	11.406510	13.057790	15.000000	17.287320	19.985530	23.169060
0.15	10.706220	12.220570	13.998120	16.086500	18.545150	21.442230
0.16	10.068030	11.459170	13.088900	15.000000	17.244580	19.884420
0.17	9.485654	10.766180	12.262790	14.015480	16.066830	18.477610
0.18	8.953083	10.133630	11.510270	13.118840	15.000000	17.203010
0.19	8.465335	9.555676	10.824310	12.303300	14.030830	16.047480
0.20	8.017635	9.026333	10.197550	11.560150	13.148090	15.000000
0.21	7.606115	8.540965	9.623969	10.881130	12.343120	14.046400
0.22	7.227109	8.094845	9.097863	10.259820	11.608480	13.176250
0.23	6.877548	7.684317	8.614813	9.690559	10.936240	12.381480
0.24	6.554501	7.305762	8.170423	9.167798	10.320590	11.655310
0.25	6.255518	6.956243	7.760848	8.687104	9.755424	10.990130
0.26	5.978393	6.632936	7.382943	8.244519	9.236152	10.379760
0.27	5.721101	6.333429	7.033547	7.836080	8.757889	9.819020
0.28	5.481814	6.055485	6.710042	7.458700	8.316982	9.302823
0.29	5.258970	5.797236	6.410005	7.109541	7.909701	8.827153
0.30	5.051153	5.556882	6.131433	6.785917	7.533113	8.388091
0.31	4.857052	5.332839	5.872303	6.485500	7.184156	7.982019
0.32	4.675478	5.123753	5.630905	6.206250	6.860492	7.606122
0.33	4.505413	4.928297	5.405771	5.946343	6.559743	7.257569
0.34	4.345926	4.745399	5.195502	5.704048	6.280019	6.933897

TABLE 2-12 Twenty-Year Escalation Table.

Present Worth of a Series of Escalating Payments Compounded Annually
Discount-Escalation Factors for n = 20 Years

Discount Rate	Annual Escalation Rate					
	0.10	0.12	0.14	0.16	0.18	0.20
0.10	20.000000	24.295450	29.722090	36.592170	45.308970	56.383330
0.11	18.213210	22.002090	26.776150	32.799710	40.417480	50.067940
0.12	16.642370	20.000000	24.210030	29.505400	36.181240	44.614710
0.13	15.259850	18.243100	21.964990	26.634490	32.502270	39.891400
0.14	14.038630	16.694830	20.000000	24.127100	29.298170	35.789680
0.15	12.957040	15.329770	18.271200	21.929940	26.498510	32.218060
0.16	11.995640	14.121040	16.746150	20.000000	24.047720	29.098950
0.17	11.138940	13.048560	15.397670	18.300390	21.894660	26.369210
0.18	10.373120	12.093400	14.201180	16.795710	20.000000	23.970940
0.19	9.686791	11.240870	13.137510	15.463070	18.326720	21.860120
0.20	9.069737	10.477430	12.188860	14.279470	16.844020	20.000000
0.21	8.513605	9.792256	11.340570	13.224610	15.527270	18.353210
0.22	8.010912	9.175267	10.579620	12.282120	14.355520	16.890730
0.23	7.555427	8.618459	9.895583	11.438060	13.309280	15.589300
0.24	7.141531	8.114476	9.278916	10.679810	12.373300	14.429370
0.25	6.764528	7.657278	8.721467	9.997057	11.533310	13.392180
0.26	6.420316	7.241402	8.216490	9.380883	10.778020	12.462340
0.27	6.105252	6.862203	7.757722	8.823063	10.096710	11.626890
0.28	5.816151	6.515563	7.339966	8.316995	9.480940	10.874120
0.29	5.550301	6.198027	6.958601	7.856833	8.922847	10.194520
0.30	5.305312	5.906440	6.609778	7.437339	8.416060	9.579437
0.31	5.079039	5.638064	6.289875	7.054007	7.954518	9.021190
0.32	4.869585	5.390575	5.995840	6.702967	7.533406	8.513612
0.33	4.675331	5.161809	5.725066	6.380829	7.148198	8.050965
0.34	4.494838	4.949990	5.475180	6.084525	6.795200	7.628322

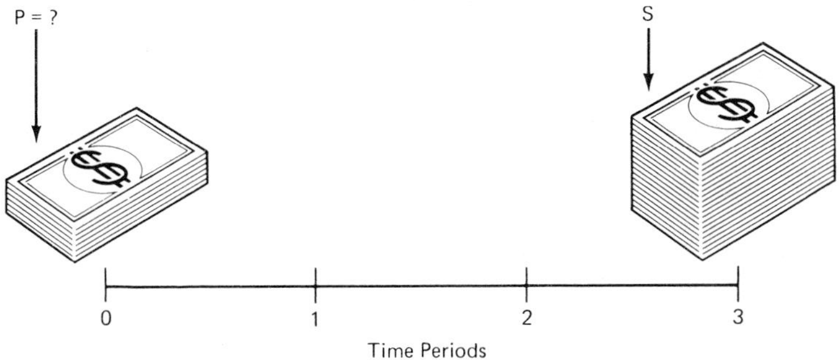

Fig. 2-2 Single payment present worth (SPPW).

Uniform Series Compound Amount—USCA

The USCA factor is used to determine the amount S that an equal annual payment R will accumulate to in n years at i percent interest. If R (uniform annual payment) is known, and S (the future worth of these payments) is required, then Equation 2-6 is used.

$$S = R \times (\text{USCA})i_n \tag{2-6}$$

$$\text{USCA} = \frac{(1 + i)^n - 1}{i} \tag{2-7}$$

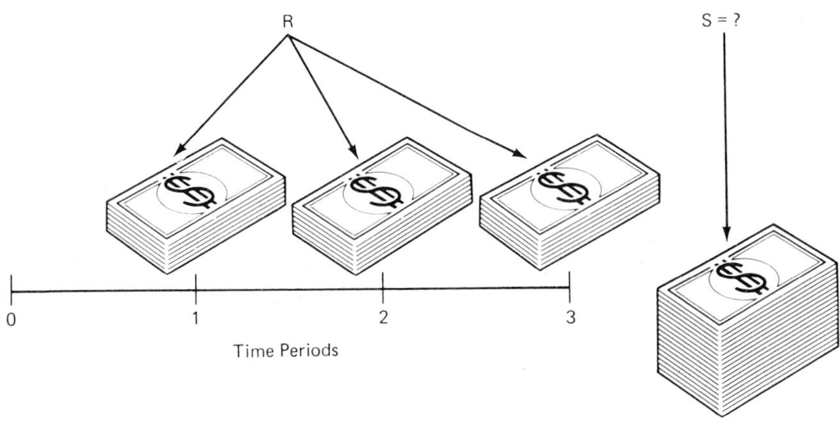

Fig. 2-3 Uniform series compound amount (USCA).

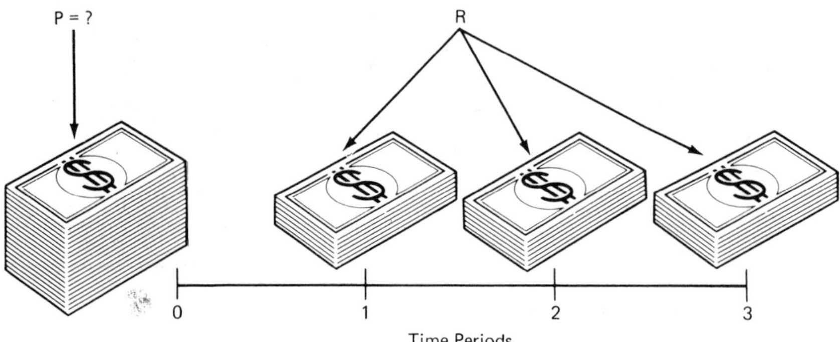

Fig. 2-4 Uniform series present worth (USPW).

Uniform Series Present Worth—(USPW)

The USPW factor is used to determine the present amount P that can be paid by equal payments of R (uniform annual payment) at i percent interest, for n years. If R is known, and P is required, then Equation 2-8 is used.

$$P = R \times (\text{USPW})i_n \tag{2-8}$$

$$\text{USPW} = \frac{(1+i)^n - 1}{i(1+i)^n} \tag{2-9}$$

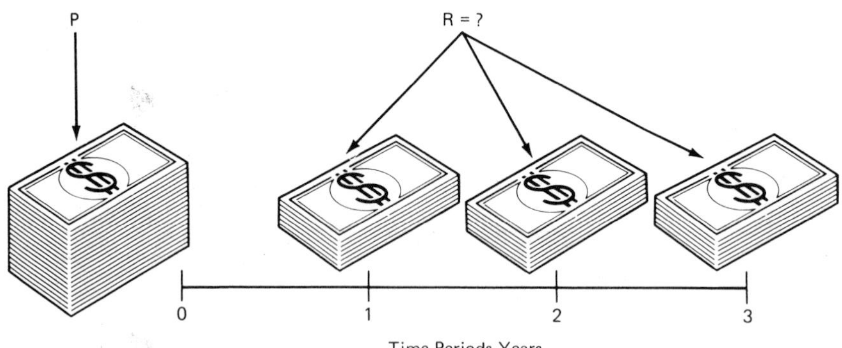

Fig. 2-5 Capital recovery (CR).

Capital Recovery—CR

The CR factor is used to determine an annual payment R required to pay off a present amount P at i percent interest, for n years. If the present sum of money, P, spent today is known, and the uniform payment R needed to pay back P over a stated period of time is required, then Equation 2-10 is used.

$$R = P \times (\mathrm{CR})i_n \qquad (2\text{-}10)$$

$$\mathrm{CR} = \frac{i(1+i)^n}{(1+i)^n - 1} \qquad (2\text{-}11)$$

Sinking Fund Payment—SFP

The SFP factor is used to determine the equal annual amount R that must be invested for n years at i percent interest in order to accumulate a specified future amount. If S (the future worth of a series of annual payments) is known, and R (value of those annual payments) is required, then Equation 2-12 is used.

$$R = S \times (\mathrm{SFP})i_n \qquad (2\text{-}12)$$

$$\mathrm{SFP} = \frac{i}{(1+i)^n - 1} \qquad (2\text{-}13)$$

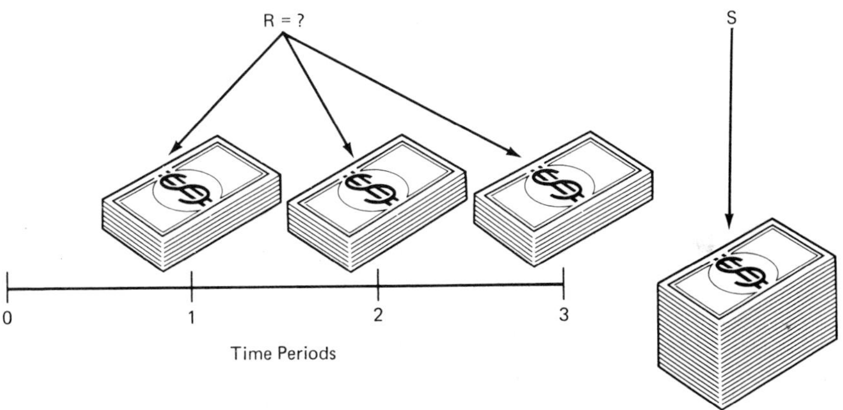

Fig. 2-6 Sinking fund payment (SFP).

Gradient Present Worth—GPW

The GPW factor is used to determine the present amount P that can be paid by annual amounts R which escalate at e percent, at i percent interest, for n years. If R is known, and P is required, then Equation 2-14 is used. The GPW factor is a relatively new term which has gained in importance due to the impact of inflation.

$$P = R \times (\text{GPW})i_n \tag{2-14}$$

$$\text{GPW} = \frac{\dfrac{1+e}{1+i}\left[1 - \left(\dfrac{1+e}{1+i}\right)^n\right]}{1 - \dfrac{1+e}{1+i}} \tag{2-15}$$

The three most commonly used methods in life cycle costing are the annual cost, present worth and rate-of-return analysis.

In the present worth method a minimum rate of return (i) is stipulated. All future expenditures are converted to present values using the interest factors. The alternative with lowest effective first cost is the most desirable.

A similar procedure is implemented in the annual cost method. The difference is that the first cost is converted to an annual expenditure. The alternative with lowest effective annual cost is the most desirable.

In the rate-of-return method, a trial-and-error procedure is usually required. Interpolation from the interest tables can determine what rate of return (i) will give an interest factor which will make the

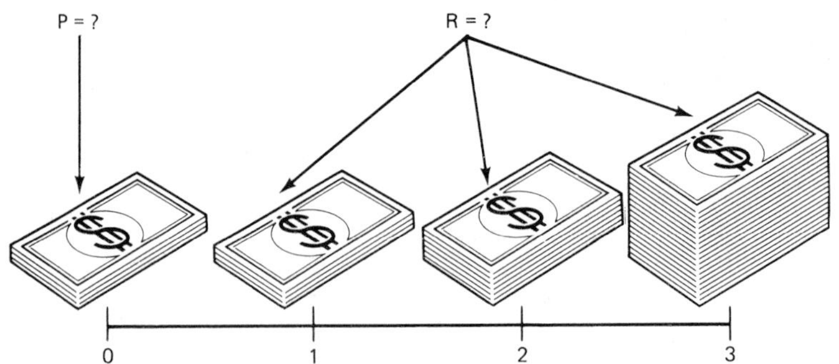

Fig. 2-7 Gradient present worth (CPW).

overall cash flow balance. The rate-of-return analysis gives a good indication of the overall ranking of independent alternates.

The effect of escalation in fuel costs can influence greatly the final decision. When an annual cost grows at a steady rate it may be treated as a gradient and the gradient present worth factor can be used.

Special thanks are given to Rudolph R. Yanuck and Dr. Robert Brown for the use of their specially designed interest and escalation tables used in this text.

When life cycle costing is used to compare several alternatives the differences between costs are important. For example, if one alternate forces additional maintenance or an operating expense to occur, then these factors as well as energy costs need to be included. Remember, what was previously spent for the item to be replaced is irrelevant. The only factor to be considered is whether the new cost can be justified based on projected savings over its useful life.

THE JOB SIMULATION EXPERIENCE

Throughout the text you will experience job situations and problems. Each simulation experience is denoted by SIM. The answer will be given below the problem. Cover the answers, then you can "play the game."

SIM 2-1

An evaluation needs to be made to replace all 40-watt fluorescent lamps with a new lamp that saves 12 percent or 4.8 watts and gives the same output. The cost of each lamp is $2.80.

Assuming a rate of return before taxes of 25 percent is required, can the immediate replacement be justified? Hours of operation are 5800 and the lamp life is two years. Electricity costs 7.0¢/kWh.

ANSWER

$$R = P \times \text{CR}$$

$$R = 5800 \times 4.8 \times 0.070/1000 = \$1.94$$

$$\text{CR} = 1.94/2.80 = .69$$

From Table 2-5 a rate of return of 25 percent is obtained. When analyzing energy conservation measures, never look at what was previously spent or the life remaining. Just determine if the new expenditure will pay for itself.

SIM 2-2

An electrical energy audit indicates electrical motor consumption is 4×10^6 KWH per year. By upgrading the motor spares with high efficiency motors a 10% savings can be realized. The additional cost for these motors is estimated at $80,000. Assuming an 8¢ per KWH energy charge and 20-year life, is the expenditure justified based on a minimum rate of return of 20% before taxes? Solve the problem using the present worth, annual cost, and rate-of-return methods.

Analysis

Present Worth Method

	Alternate 1 *Present Method*	*Alternate 2* *Use High Efficiency Motor Spares*
(1) First Cost *(P)*	–	$80,000
(2) Annual Cost *(R)*	$4 \times 10^6 \times .08$ = $320,000	$.9 \times \$320{,}000$ = $288,000
USPW (Table 2-4)	4.87	4.87
(2) *R* X USPW =	$1,558.400	$1,402,560
Present Worth (1) + (3)	$1,558,400	$1,482,560 ← Choose Alternate with Lowest First Cost

Annual Cost Method

	Alternate 1	*Alternate 2*
(1) First Cost *(P)*	–	$80,000
(2) Annual Cost *(R)*	$320,000	$288,000
CR (Table 16-4)	.2	.2
(3) *P* X CR	–	$16,000
Annual Cost (2) + (3)	$320,000	$304,000 ← Choose Alternate with Lowest First Cost

Rate of Return Method

$$P = R(\text{USPW}) = (\$320{,}000 - \$288{,}000) \times \text{USPW}$$

$$\text{USPW} = \frac{80{,}000}{32{,}000} = 2.5$$

What value of i will make USPW = 2.5? $i = 40\%$ (Table 2-7).

SIM 2-3

Show the effect of 10 percent escalation on the rate of return analysis given the

Energy equipment investment = \$20,000
After tax savings = \$ 2600
Equipment life (n) = 15 years

ANSWER

Without escalation:

$$CR = \frac{R}{P} = \frac{2600}{20{,}000} = 0.13$$

From Table 2-1, the rate of return is 10 percent.

With 10 percent escalation assumed:

$$GPW = \frac{P}{G} = \frac{20{,}000}{2600} = 7.69$$

From Table 2-11, the rate of return is 21 percent.

Thus we see that taking into account a modest escalation rate can dramatically affect the justification of the project.

MAKING DECISIONS FOR ALTERNATE INVESTMENTS

There are several methods for determining which energy conservation alternative is the most economical. Probably the most familiar and trusted method is the annual cost method.[35]

When evaluating replacement of processes or equipment *do not* consider what was previously spent. The decision will be based on whether the new process or equipment proves to save substantially enough in operating costs to justify the expenditure.

Equation 2-16 is used to convert the lump sum investment P into the annual cost. In the case where the asset has a value after the end of its useful life, the annual cost becomes:

$$AC = (P - L)\,CR + iL \tag{2-16}$$

where

AC is the annual cost

L is the net sum of money that can be realized for a piece of equipment, over and above its removal cost, when it is returned at the end of the service life. L is referred to as the salvage value.

As a practical point, the salvage value is usually small and can be neglected, considering the accuracy of future costs. The annual cost technique can be implemented by using the following format:

	Alternate 1	Alternate 2
1. First cost (P)		
2. Estimated life (n)		
3. Estimated salvage value at end of life (L)		
4. Annual disbursements, including energy costs & maintenance (E)		
5. Minimum acceptable return *before* taxes (i)		
6. CR n, i		
7. $(P - L)$CR		
8. Li		
9. AC = $(P - L)$CR + $Li + E$		

Choose alternate with lowest AC

The alternative with the lowest annual cost is the desired choice.

SIM 2-4

A new water line must be constructed from an existing pumping station to a reservoir. Estimates of construction and pumping costs for each pipe size have been made.

Pipe Size	Estimated Construction Costs	Cost/Hour for Pumping
8″	$ 80,000	$4.00
10″	$100,000	$3.00
12″	$160,000	$1.50

The annual cost is based on a 16-year life and a desired return on investment before taxes of 10 percent. Which is the most economical pipe size for pumping 4000 hours/year?

ANSWER

	8″ Pipe	10″ Pipe	12″ Pipe
P	\$80,000	\$100,000	\$160,000
n	16	16	16
L	—	—	—
E	16,000	12,000	6000
i	10%	10%	10%
$CR = 0.127$	—	—	—
$(P - L)CR$	10,160	12,700	20,320
Li	—	—	—
AC	\$26,160	\$24,700 (*Choice*)	\$26,320

DEPRECIATION, TAXES, AND THE TAX CREDIT

Depreciation

Depreciation affects the "accounting procedure" for determining profits and losses and the income tax of a company. In other words, for tax purposes the expenditure for an asset such as a pump or motor cannot be fully expensed in its first year. The original investment must be charged off for tax purposes over the useful life of the asset. A company usually wishes to expense an item as quickly as possible.

The Internal Revenue Service allows several methods for determining the annual depreciation rate.

Straight-Line Depreciation. The simplest method is referred to as a straight-line depreciation and is defined as:

$$D = \frac{P - L}{n} \tag{2-17}$$

where

- D is the annual depreciation rate
- L is the value of equipment at the end of its useful life, commonly referred to as salvage value
- n is the life of the equipment, which is determined by Internal Revenue Service guidelines
- P is the initial expenditure.

Sum-of-Years Digits. Another method is referred to as the sum-of-years digits. In this method the depreciation rate is determined by finding the sum of digits using the following formula,

$$N = n\frac{(n + 1)}{2} \tag{2-18}$$

where n is the life of equipment.

Each year's depreciation rate is determined as follows.

First year $$D = \frac{n}{N}(P - L) \tag{2-19}$$

Second year $$D = \frac{n - 1}{N}(P - L) \tag{2-20}$$

n year $$D = \frac{1}{N}(P - L) \tag{2-21}$$

Declining-Balance Depreciation. The declining-balance method allows for larger depreciation charges in the early years which is sometimes referred to as fast write-off.

The rate is calculated by taking a constant percentage of the declining undepreciated balance. The most common method used to calculate the declining balance is to predetermine the depreciation rate. Under certain circumstances a rate equal to 200 percent of the straight-line depreciation rate may be used. Under other circumstances the rate is limited to $1\frac{1}{2}$ or $\frac{1}{4}$ times as great as straight-line depreciation. In this method the salvage value or undepreciated book value is established once the depreciation rate is preestablished.

To calculate the undepreciated book value, Equation 2-22 used.

$$D = 1 - \left(\frac{L}{P}\right)^{1/N} \tag{2-22}$$

where

D is the annual depreciation rate
L is the salvage value
P is the first cost.

The Tax Reform Act of 1986 (hereafter referred to as the "Act") represented true tax reform, as it made sweeping changes in many basic federal tax code provisions for both individuals and corporations. The Act has had significant impact on financing for cogeneration, alternative energy and energy efficiency transactions, due to substantial modifications in provisions concerning depreciation, investment and energy tax credits, tax-exempt financing, tax rates, the corporate minimum tax and tax shelters generally.

The Act lengthened the recovery periods for most depreciable assets. The Act also repealed the 10 percent investment tax credit ("ITC") for property placed in service on or after January 1, 1986, subject to the transition rules.

Tax Considerations

Tax-deductible expenses such as maintenance, energy, operating costs, insurance, and property taxes reduce the income subject to taxes.

For the after-tax life cycle analysis and payback analysis the actual incurred and annual savings is given as follows.

$$AS = (1 - I)\, E + ID \tag{2-23}$$

where

AS	is the yearly annual after-tax savings (excluding effect of tax credit)
E	is the yearly annual energy savings (difference between original expenses and expenses after modification)
D	is the annual depreciation rate
I	is the income tax bracket.

Equation 2-23 takes into account that the yearly annual energy savings is partially offset by additional taxes which must be paid due to reduced operating expenses. On the other hand, the depreciation allowance reduces taxes directly.

After-Tax Analysis

To compute a rate of return which accounts for taxes, depreciation, escalation, and tax credits, a cash-flow analysis is usually required. This method analyzes all transactions including first and operating costs. To determine the after-tax rate of return a trial and error or computer analysis is required.

All money is converted to the present assuming an interest rate. The summation of all present dollars should equal zero when the correct interest rate is selected, as illustrated in Fig. 2-8.

This analysis can be made assuming a fuel escalation rate by using the gradient present worth interest of the present worth factor.

Year	1 Investment	2 Tax Credit	3 After Tax Savings (AS)	4 Single Payment Present Worth Factor	(2 + 3) X 4 Present Worth
0	$-P$				$-P$
1		$+TC$	AS	$SPPW_1$	$+P_1$
2			AS	$SPPW_2$	P_2
3			AS	$SPPW_3$	P_3
4			AS	$SPPW_4$	P_4
Total					ΣP

$$AS = (1 - I)E + ID$$

Trial and Error Solution:
Correct i when $\Sigma P = 0$

Fig. 2-8 Cash flow rate of return analysis.

SIM 2-5

Develop a set of curves that indicate the capital that can be invested to give a rate of return of 15 percent after taxes for each \$1000 saved for the following conditions.

1. The effect of escalation is not considered.
2. A 5 percent fuel escalation is considered.
3. A 10 percent fuel escalation is considered.
4. A 14 percent fuel escalation is considered.
5. A 20 percent fuel escalation is considered.

Calculate for 5-, 10-, 15-, 20-year life.

Assume straight-line depreciation over useful life, 34 percent income tax bracket, and no tax credit.

ANSWER

$$AS = (1 - I)E + ID$$

$$I = 0.34, \qquad E = \$1000$$

$$AS = 660 + \frac{0.34P}{N}$$

Thus, the after-tax savings (AS) are comprised of two components. The first component is a uniform series of \$660 escalating at e percent/year. The second component is a uniform series of $0.34P/N$.

Each component is treated individually and converted to present day values using the GPW factor and the USPW factor, respectively. The sum of these two present worth factors must equal P. In the case of no escalation, the formula is:

$$P = 660 \text{ USPW} + \frac{0.34\,P}{N} \text{ USPW.}$$

In the case of escalation:

$$P = 660 \text{ GPW} + \frac{0.34P}{N} \text{ USPW.}$$

Since there is only one unknown, the formulas can be readily solved. The results are indicated below.

	$N = 5$ \$$P$	$N = 10$ \$$P$	$N = 15$ \$$P$	$N = 20$ \$$P$
$e = 0$	2869	4000	4459	4648
$e = 10\%$	3753	6292	8165	9618
$e = 14\%$	4170	7598	10,676	13,567
$e = 20\%$	4871	10,146	16,353	23,918

Figure 2-9 illustrates the effects of escalation. This figure can be used as a quick way to determine after-tax economics of energy utilization expenditures.

SIM 2-6

It is desired to have an after-tax savings of 15 percent. Comment on the investment that can be justified if it is assumed that the fuel rate escalation should not be considered and the annual energy savings is $2000 with an equipment economic life of 15 years.

Comment on the above, assuming a 14 percent fuel escalation.

ANSWER

From Fig. 2-9, for each $1000 energy savings, an investment of $4400 is justified or $8800 for a $2000 savings when no fuel increase is accounted for.

With a 14 percent fuel escalation rate an investment of $10,600 is justified for each $1000 energy savings, thus $21,200 can be justified for $2000 savings. Thus, a much higher expenditure is economically justifiable and will yield the same after-tax rate of return of 15 percent when a fuel escalation of 14 percent is considered.

Figure 2-9. Effects of Escalation On Investment Requirements

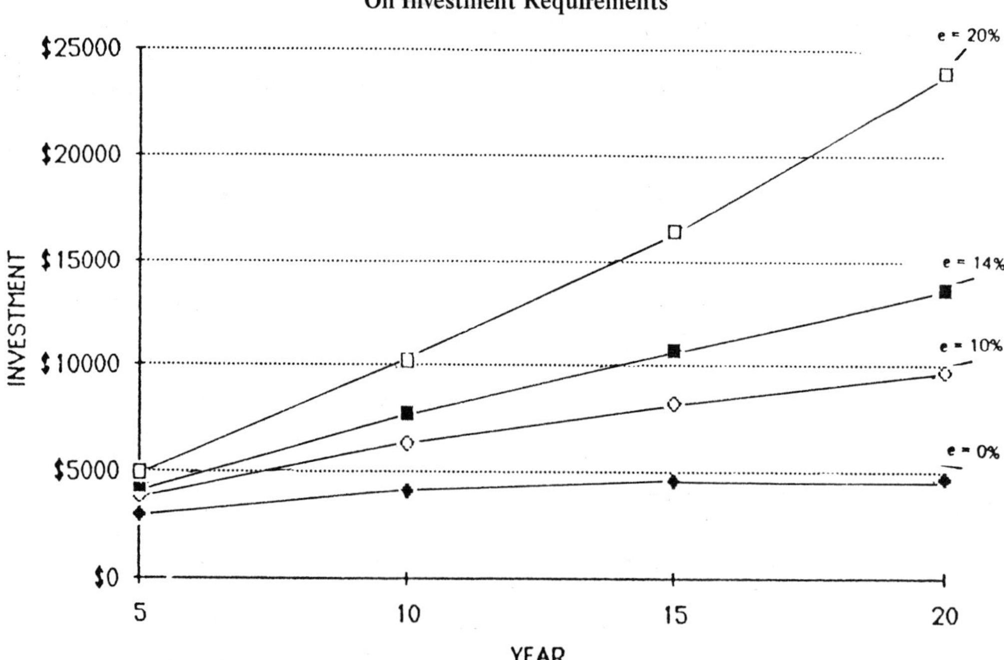

Note: Maximum investment in order to attain a 15% after-tax rate of return on investment for annual savings of $1000.

IMPACT OF FUEL INFLATION ON LIFE CYCLE COSTING

As illustrated by problem 2-5 a modest estimate of fuel inflation as a major impact on improving the rate of return on investment of the project. The problem facing the energy engineer is how to forecast what the future of energy costs will be. All too often no fuel inflation is considered because of the difficulty of projecting the future. In making projections the following guidelines may be helpful:

- Is there a rate increase that can be forecast based on new nuclear generating capacity? In locations such as Georgia, California, and Arizona electric rates will rise at a faster rate due to commissioning of new nuclear plants and rate increases approved by the Public Service Commission of that state.
- What has been the historical rate increase for the facility? Even with fluctuations there are likely to be trends to follow.
- What events on a national or international level would impact on your costs? New state taxes, new production quotas by OPEC and other factors affecting your fuel prices.
- What do the experts say? Energy economists, forecasting services, and your local utility projections all should be taken into account.

COMPUTER ANALYSIS

The Alliance to Save Energy, 1925 K Street, NW, Suite 206, Washington, DC 20006, has introduced an investment analysis software package, ENVEST, which costs only \$75 for 5½-inch disks and \$85 for 3½-inch hard drive disks, and includes a 170-page user manual and 30 days of telephone support. The program can be run on an IBM PC, PCXT, PCAT with 256K ram. The program enables the user to:

- Generate spreadsheets and graphs showing the yearly cash flows from any energy-related investment.
- Compute payback, internal rate of return, and other important investment measures.

- Experiment with differing energy price projections.
- Perform sensitivity analysis on key assumptions.
- Compare alternative financing options, including loans, leases, and shared savings.
- Store data on over 100 energy efficiency investments.

3
The Facility Survey

The survey of the facility is considered a very important part of the industrial energy audit. Chapter 1 indicated that there are many types of surveys, from a simple walk-through to a complete quantification of uses and losses. This chapter will illustrate various types of instruments that can aid in the industrial audit.

COMPARING CATALOGUE DATA WITH ACTUAL PERFORMANCE

Many energy managers are surprised when they record actual performance data of equipment and compare it with catalogue information. There is usually a great disparity between the two.

Each manufacturer has design tolerance for its equipment. For critical equipment, performance guarantees or tests should be incorporated into the initial specifications.

As part of the facility survey, nameplate data of pumps, motors, chillers, fans, etc. should be taken. The nameplate data should also be compared to actual running conditions.

The initial survey can detect motors that were sized too big. By replacing the motor with a smaller one, energy savings can be realized.

INFRARED EQUIPMENT

Some companies may have the wrong impression that infrared equipment can meet most of their instrumentation needs.

The primary use of infrared equipment in an energy utilization program is to detect building or equipment losses. Thus it is just one of the many options available.

Several energy managers find infrared in use in their plant prior to the energy utilization program. Infrared equipment, in many instances, was purchased by the electrical department and used to detect electrical hot spots.

Infrared energy is an invisible part of the electromagnetic spectrum. It exists naturally and can be measured by remote heat-sensing equipment. Within the last four years lightweight portable infrared systems became available to help determine energy losses. Differences in the infrared emissions from the surface of objects cause color variations to appear on the scanner. The hotter the object, the more infrared radiated. With the aid of an isotherm circuit the intensity of these radiation levels can be accurately measured and quantified. In essence the infrared scanning device is a diagnostic tool which can be used to determine building heat losses. Equipment costs range from \$400 to \$50,000.

An overview energy scan of the plant can be made through an aerial survey using infrared equipment. Several companies offer aerial scan services starting at \$1500. Aerial scans can determine underground stream pipe leaks, hot gas discharges, leaks, etc.

Since IR detection and measurement equipment have gained increased importance in the energy audit process, a summary of the fundamentals are reviewed in this section.

The visible portion of the spectrum runs from .4 to .75 micrometers (μm). The infrared or thermal radiation begins at this point and extends to approximately 1000 μm. Objects such as people, plants, or buildings will emit radiation with wavelengths around 10 μm.

Infrared instruments are required to detect and measure the thermal radiation. To calibrate the instrument a special "black body" radiator is used. A black body radiator absorbs all the radiation that impinges on it and has an absorbing efficiency or emissivity of 1.

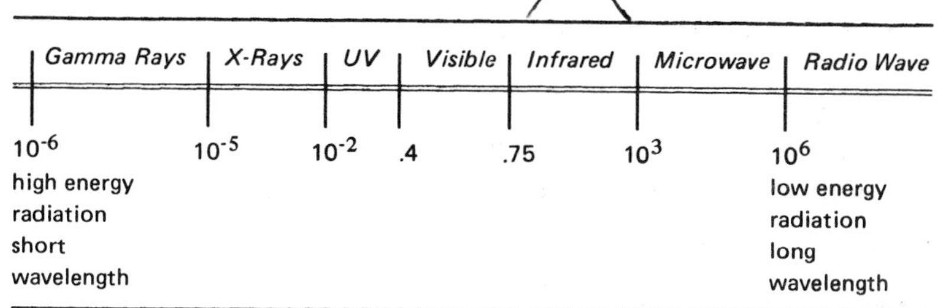

Fig. 3-1 Electromagnetic spectrum.

The accuracy of temperature measurements by infrared instruments depends on the three processes which are responsible for an object acting like a black body. These processes—absorbed, reflected, and transmitted radiation—are responsible for the total radiation reaching an infrared scanner.

The real temperature of the object is dependent only upon its emitted radiation.

Corrections to apparent temperatures are made by knowing the emissivity of an object at a specified temperature.

The heart of the infrared instrument is the infrared detector. The detector absorbs infrared energy and converts it into electrical voltage or current. The two principal types of detectors are the thermal and photo type. The thermal detector generally requires a given period of time to develop an image on photographic film. The photo detectors are more sensitive and have a higher response time. Television-like displays on a cathode ray tube permit studies of dynamic thermal events on moving objects in real time.

There are various ways of displaying signals produced by infrared detectors. One way is by use of an isotherm contour. The lightest areas of the picture represent the warmest areas of the subject and the darkest areas represent the coolest portions. These instruments can show thermal variations of less than 0.1°C and can cover a range of -30°C to over 2000°C.

The isotherm can be calibrated by means of a black body radiator so that a specific temperature is known. The scanner can then be moved and the temperatures of the various parts of the subject can be made.

MEASURING ELECTRICAL SYSTEM PERFORMANCE

The ammeter, voltmeter, wattmeter, power factor meter, and foot-candle meter are usually required to do an electrical survey. These instruments are described below.

Ammeter and Voltmeter

To measure electrical currents, ammeters are used. For most audits, alternating currents are measured. Ammeters used in audits are portable and are designed to be easily attached and removed.

There are many brands and styles of snap-on ammeters commonly available that can read up to 1000 amperes continuously. This range can be extended to 4000 amperes continuously for some models with an accessory step-down current transformer.

The snap-on ammeters can be either indicating or recording with a printout. After attachment, the recording ammeter can keep recording current variations for as long as a full month on one roll of recording paper. This allows studying current variations in a conductor for extended periods without constant operator attention.

The ammeter supplies a direct measurement of electrical current which is one of the parameters needed to calculate electrical energy. The second parameter required to calculate energy is voltage, and it is measured by a voltmeter.

Several types of electrical meters can read the voltage or current. A voltmeter measures the difference in electrical potential between two points in an electrical circuit.

In series with the probes are the galvanometer and a fixed resistance (which determine the voltage scale). The current through this fixed resistance circuit is then proportional to the voltage and the galvanometer deflects in proportion to the voltage.

The voltage drops measured in many instances are fairly constant and need only be performed once. If there are appreciable fluctuations, additional readings or the use of a recording voltmeter may be indicated.

Most voltages measured in practice are under 600 volts and there are many portable voltmeter/ammeter clamp-ons available for this and lower ranges.

Wattmeter and Power Factor Meter

The portable wattmeter can be used to indicate by direct reading electrical energy in watts. It can also be calculated by measuring voltage, current and the angle between them (power factor angle).

The basic wattmeter consists of three voltage probes and a snap-on current coil which feeds the wattmeter movement.

The typical operating limits are 300 kilowatts, 650 volts, and 600 amperes. It can be used on both one- and three-phase circuits.

The portable power factor meter is primarily a three-phase instrument. One of its three voltage probes is attached to each conductor phase and a snap-on jaw is placed about one of the phases. By disconnecting the wattmeter circuitry, it will directly read the power factor of the circuit to which it is attached.

It can measure power factor over a range of 1.0 leading to 1.0 lagging with "ampacities" up to 1500 amperes at 600 volts. This range covers the large bulk of the applications found in light industry and commerce.

The power factor is a basic parameter whose value must be known to calculate electric energy usage. Diagnostically it is a useful instrument to determine the sources of poor power factor in a facility.

Portable digital KWH and KW demand units are now available.

Digital read-outs of energy usage in both KWH and KW demand or in dollars and cents, including instantaneous usage, accumulated usage, projected usage for a particular billing period, alarms when over-target levels are desired for usage, and control-outputs for load-shedding and cycling are possible.

Continuous displays or intermittent alternating displays are available at the touch of a button of any information needed such as the cost of operating a production machine for one shift, one hour or one week.

Footcandle Meter

Footcandle meters measure illumination in units of footcandles through light-sensitive barrier layer of cells contained within them. They are usually pocket size and portable and are meant to be used as field instruments to survey levels of illumination. Footcandle meters differ from conventional photographic lightmeters in that they are color and cosine corrected.

TEMPERATURE MEASUREMENTS

To maximize system performance, knowledge of the temperature of a fluid, surface, etc. is essential. Several types of temperature devices are described in this section.

Thermometer

There are many types of thermometers that can be used in an energy audit. The choice of what to use is usually dictated by cost, durability, and application.

For air-conditioning, ventilation and hot-water service applications (temperature ranges 50°F to 250°F) a multipurpose portable battery-operated thermometer is used. Three separate probes are usually provided to measure liquid, air or surface temperatures.

For boiler and oven stacks (1000°F) a dial thermometer is used. Thermocouples are used for measurements above 1000°F.

Surface Pyrometer

Surface pyrometers are instruments which measure the temperature of surfaces. They are somewhat more complex than other temperature instruments because their probe must make intimate contact with the surface being measured.

Surface pyrometers are of imense help in assessing heat losses through walls and also for testing steam traps.

They may be divided into two classes: low-temperature (up to 250°F) and high-temperature (up to 600°F to 700°F). The low-temperature unit is usually part of the multipurpose thermometer kit. The high-temperature unit is more specialized, but needed for evaluating fired units and general steam service.

There are also noncontact surface pyrometers which measure infrared radiation from surfaces in terms of temperature. These are suitable for general work and also for measuring surfaces which are visually but not physically accessible.

A more specialized instrument is the optical pyrometer. This is for high-temperature work (above 1500°F) because it measures the temperature of bodies which are incandescent because of their temperature.

Psychrometer

A psychrometer is an instrument which measures relative humidity based on the relation of the dry-bulb temperature and the wet-bulb temperature.

Relative humidity is of prime importance in HVAC and drying operations. Recording psychrometers are also available. Above 200°F humidity studies constitute a specialized field of endeavor.

Portable Electronic Thermometer

The portable electronic thermometer is an adaptable temperature measurement tool. The battery-powered basic instrument, when housed in a carrying case, is suitable for laboratory or industrial use.

A pocket-size digital, battery-operated thermometer is especially convenient for spot checks or where a number of rapid readings of process temperatures need to be taken.

Thermocouple Probe

No matter what sort of indicating instrument is employed, the thermocouple used should be carefully selected to match the application and properly positioned if a representative temperature is to be measured. The same care is needed for all sensing devices—thermocouple, bimetals, resistance elements, fluid expansion, and vapour pressure bulbs.

Suction Pyrometer

Errors arise if a normal sheathed thermocouple is used to measure gas temperatures, especially high ones. The suction pyrometer overcomes these by shielding the thermocouple from wall radiation and drawing gases over it at high velocity to ensure good convective heat transfer. The thermocouple thus produces a reading which approaches the true temperature at the sampling point rather than a temperature between that of the walls and the gases.

MEASURING COMBUSTION SYSTEMS

To maximize combustion efficiency it is necessary to know the composition of the flue gas. By obtaining a good air-fuel ratio substantial energy will be saved.

Combustion Tester

Combustion testing consists of determining the concentrations of the products of combustion in a stack gas. The products of combustion usually considered are carbon dioxide and carbon monoxide. Oxygen is tested to assure proper excess air levels.

The definitive test for these constituents is an Orsat apparatus. This test consists of taking a measured volume of stack gas and measuring successive volumes after intimate contact with selective absorbing solutions. The reduction in volume after each absorption is the measure of each constituent.

The Orsat has a number of disadvantages. The main ones are that it requires considerable time to set up and use and its operator must have a good degree of dexterity and be in constant practice.

Instead of an Orsat, there are portable and easy to use absorbing instruments which can easily determine the concentrations of the constituents of interest on an individual basis. Setup and operating times are minimal and just about anyone can learn to use them.

The typical range of concentrations are CO_2 : 0–20%, O_2 : 0–21%, and CO: 0–0.5%. The CO_2 or O_2 content, along with knowledge of flue gas temperature and fuel type, allows the flue gas loss to be determined off standard charts.

Boiler Test Kit

The boiler test kit contains the following:

CO_2 Gas analyzer

O_2 Gas analyzer
Inclined monometer
CO Gas analyzer.

The purpose of the components of the kit is to help evaluate fireside boiler operation. Good combustion usually means high carbon dioxide (CO_2), low oxygen (O_2), and little or no trace of carbon monoxide (CO).

Gas Analyzers

The gas analyzers are usually of the Fyrite type. The Fyrite type differs from the Orsat apparatus in that it is more limited in application and less accurate. The chief advantages of the Fyrite are that it is simple and easy to use and is inexpensive. This device is many times used in an energy audit. Three readings using the Fyrite analyzer should be made and the results averaged.

Draft Gauge

The draft gauge is used to measure pressure. It can be the pocket type, or the inclined monometer type.

Smoke Tester

To measure combustion completeness the smoke detector is used. Smoke is unburned carbon, which wastes fuel, causes air pollution, and fouls heat-exchanger surfaces. To use the instrument, a measured volume of flue gas is drawn through filter paper with the probe. The smoke spot is compared visually with a standard scale and a measure of smoke density is determined.

Combustion Analyzer

The combustion electronic analyzer permits fast, close adjustments. The unit contains digital displays. A standard sampler assembly with probe allowsfor stack measurements through a single stack or breaching hole.

MEASURING HEATING, VENTILATION AND AIR-CONDITIONING (HVAC) SYSTEM PERFORMANCE

Air Velocity Measurement

The following suggests the preference, suitability, and approximate costs of particular equipment.

- *Smoke pellets*—limited use but very low cost. Considered to be useful if engineering staff has experience in handling.
- *Anemometer* (deflecting vane)—good indication of air movement with acceptable order of accuracy. Considered useful (approximately $50).
- *Anemometer* (revolving vane)—good indicator of air movement with acceptable accuracy. However, easily subject to damage. Considered useful (approximately $100).
- *Pitot tube*—a standard air measurement device with good levels of accuracy. Considered essential. Can be purchased in various lengths—12″ about $20, 48″ about $35. Must be used with a monometer. These vary considerably in cost, but could be on the order of $20 to $60.
- *Impact tube*—usually packaged air flow meter kits, complete with various jets for testing ducts, grills, open areas, etc. These units are convenient to use and of sufficient accuracy. The costs vary around $150 to $300, and therefore this order of cost could only be justified for a large system.
- *Heated thermocouple*—these units are sensitive and accurate but costly. A typical cost would be about $500 and can only be justified for regular use in a large plant.
- *Hot wire anemometer*—not recommended. Too costly and too complex.

Temperature Measurement

The temperature devices most commonly used are as follows.

- *Glass thermometers*—considered to be the most useful to temperature measuring instruments—accurate and convenient but fragile. Cost runs from $5 each for 12″ long mercury in glass. Engineers should have a selection of various ranges.

- *Resistance thermometers*—considered to be very useful for A/C testing. Accuracy is good and they are reliable and convenient to use. Suitable units can be purchased from $150 up, some with a selection of several temperature ranges.
- *Thermocouples*—similar to resistance thermocouple, but do not require battery power source. Chrome-Alum or iron types are the most useful and have satisfactory accuracy and repeatability. Costs start from $50 and go up.
- *Bimetallic thermometers*—considered unsuitable.
- *Pressure bulb thermometers*—more suitable for permanent installation. Accurate and reasonable in cost—$40 up.
- *Optical pyrometers*—only suitable for furnace settings and therefore limited in use. Cost from $300 up.
- *Radiation pyrometers*—limited in use for A/C work and costs from $500 up.
- *Indicating crayons*—limited in use and not considered suitable for A/C testing—costs around $2/crayon.
- *Thermographs*—use for recording room or space temperature and gives a chart indicating variations over a 12- or 168-hour period. Reasonably accurate. Low cost at around $30 to $60. (Spring-wound drive.)

Pressure Measurement (Absolute and Differential)

Common devices used for measuring pressure in HVAC applications (accuracy, range, application, and limitations are discussed in relation to HVAC work) are as follows.

- *Absolute pressure manometer*—not really suited to HVAC test work.
- *Diaphragm*—not really suited to HVAC test work.
- *Barometer (Hg manometer)*—not really suited to HVAC test work.
- *Micromanometer*—not usually portable, but suitable for fixed measurement of pressure differentials across filter, coils, etc. Cost around $30 and up.
- *Draft gauges*—can be portable and used for either direct pressure or pressure differential. From $30 up.
- *Manometers*—can be portable. Used for direct pressure reading

and with pitot tubes for air flows. Very useful. Costs from $20 up.

- *Swing Vane gauges*—can be portable. Usually used for air flow. Costs about $30.
- *Bourdon tube gauges*—very useful for measuring all forms of system fluid pressures from 5 psi up. Costs vary greatly, from $10 up. Special types for refrigeration plants.

Humidity Measurement

The data given below indicate the type of instruments available for humidity measurement. The following indicates equipment suitable for HVAC applications.

- *Psychrometers*—basically these are wet and dry bulb thermometers. They can be fixed on a portable stand or mounted in a frame with a handle for revolving in air. Costs are low ($10 to $30) and are convenient to use.
- *Dewpoint hygrometers*—not considered suitable for HVAC test work.
- *Dimensional change*—device usually consists of a "hair," which changes in length proportionally with humidity changes. Not usually portable, fragile, and only suitable for limited temperature and humidity ranges.
- *Electrical conductivity*—can be compact and portable but of a higher cost (from $200 up). Very convenient to use.
- *Electrolytic*—as above. But for very low temperature ranges. Therefore unsuitable for HVAC test work.
- *Gravemeter*—not suitable.

4

Electrical System Optimization

APPLYING PROVEN TECHNIQUES TO REDUCE THE ELECTRICAL BILL

Electrical bills can be reduced by up to 30 percent by knowing the utility rate structure, by improving the plant power factor, by reducing peak loads, and by the efficient use of lighting. This chapter will illustrate these aspects as they apply to the energy utilization program.

WHY THE PLANT MANAGER SHOULD UNDERSTAND THE ELECTRIC RATE STRUCTURE

Each plant manager should understand how the plant is billed. Utility companies usually have several rate structures offered to customers. By understanding the electrical characteristics of the plant, the best rate structure for the plant is determined. Understanding the rate structure also enables the plant manager to avoid the penalties the utility company incorporates into their rates.

As a result of the National Energy Plan, public service commissions are conducting hearings to evaluate proposed rate tariff changes. Industry needs to actively participate in these hearings to make sure their interests are represented.

Some "rate reform" proposals in effect subsidize lower residential electricity prices by raising industrial electricity prices. Two of the reform proposals are as follows.

"Lifeline" rates—a plan to set low, below-cost prices for the first 300 to 500 KWH of electricity purchased by a household in a month. Utility revenues lost as a result of providing below-cost electricity would be made up through higher prices to other consumers, notably industry.

"Uniform" or "flat" rates—aimed at having all consumers pay the same price for each KWH of electricity. Since industry buys large volumes of electricity at prices lower than charged to other consumers (the cost of serving industry being lower to the utility), the result would be higher prices to industry and lower prices to other consumers.

ELECTRICAL RATE TARIFF

The basic electrical rate charges contain the following elements.

Billing demand—the maximum KW requirement over a 15-, 30-, or 60-minute interval.

Load factor—the ratio of the average load over a designated period to the peak demand load occurring in that period.

Power factor—the ratio of resistive power to apparent power.

Traditionally, electrical rate tariffs have a decreasing KWH charge with usage. This practice is likely to gradually phase out. New tariffs are containing the following elements.

Time of day—discounts are allowed for electrical usage during off-peak hours.

Ratchet rate—the billing demand is based on 80 to 90 percent of peak demand for any one month. The billing demand will remain at that ratchet for 12 months even though the actual demand for the succeeding months may be less.

The effect of changes in electrical rate tariffs can be significant as illustrated in SIM 4-1.

SIM 4-1

The existing rate structure is as follows.

Demand charge:

First 25 KW of billing demand	$4.00/KW/month
Next 475 KW of billing demand	$3.50/KW/month
Next 1000 KW	$3.25/KW/month

Energy charge:

First 2,000 KWH/month	4¢/KWH
Next 18,000 KWH/month	3¢/KWH
Next 180,000 KWH/month	2.2¢/KWH
Etc.	

The new proposed schedule deletes price breaks for usage.

	Billing Months June–September	*Billing Months October–May*
Demand charge	$13.00/KW/month	$5.00/KW/month
Energy charge	2.5¢/KWH	1.5¢/KWH

Demand charge based on greatest billing demand month.

Comment on the proposed billing as it would affect an industrial customer who uses 475 KW per month, 330 hours, 900 KW winter (8 months) demand, 1200 KW summer (4 months) demand.

ANSWER

The proposed rate schedule has to major changes. First, billing demand is on a ratchet basis and discourages peak demand during summer months. The high demand charge encourages the plant to improve the overall load factor. The increased demand charge is partially offset with a lower energy usage rate.

Original billing

Winter: First 25 KW $ 100
Demand Next 475 KW 1660
Next 400 KW 1300
$3060
Summer: $4035
Total demand: 8 X 3060 + 4 X 4035 = $40,620
Usage charge: 475 KW X 330 Hours = 156,750 KWH
First 2,000 KWH @ 4¢ = $ 80
Next 18,000 KWH @ 3¢ = 540
Next 136,750 KWH @ 2.2¢ = 3,008
$3,628
Total usage: 3628 X 12 + $43,536
Total charge = $84,156 or 4.47¢/KWH

Proposed billing

Demand: 1200 X $13.00 X 4 months = $ 62,400
1200 X $ 5.00 X 8 months = $ 48,000
Total demand: $110,400
Usage: 475 KW X 330 X 2.5¢ X 4 = $ 15,675
475 KW X 330 X 1.5¢ X 8 = $ 18,810
Total usage: $ 34,485
Total charge = $144,885 or 7.7¢/KWH or a 72 percent increase

POWER BASICS—THE KEY TO ELECTRICAL ENERGY REDUCTION

By understanding power basics, one can reduce the electrical bill. The total power requirement is comprised of two components, as illustrated in the power triangle, Fig. 4-1. This diagram shows the resistive portion or kilowatt (KW), ninety degrees out of phase with the reactive portion, kilovolt ampere reactive (Kvar). The reactive current is necessary to build up the flux for the magnetic field of inductive devices, but otherwise it is non-usable. The resistive portion is also known as the active power which is directly converted to useful work. The hypotenuse of the power triangle is referred to as the kilovolt ampere or apparent power (Kva). The angle between KW and Kva is the power factor angle.

$$\text{KW} = \text{Kva} \cos \theta \tag{4-1a}$$

$$\text{Kva} = \text{KW}/\cos \theta \tag{4-1b}$$

$$\text{Kvar} = \text{Kva} \sin \theta \tag{4-1c}$$

$$\text{P.F.} = \cos \theta \tag{4-1d}$$

where P.F. is referred to as the power factor.

Note: Only power portions in phase with each other can be combined. For example: resistive portions of one load can be added to resistive portions of another. The same will hold for reactive loads.

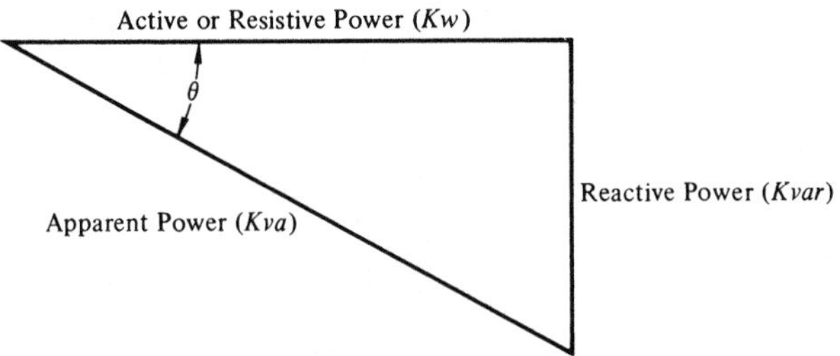

Fig. 4-1 The power triangle.

RELATIONSHIPS BETWEEN POWER, VOLTAGE, AND CURRENT

For a balanced 3-phase load,

$$\text{Power watts} = \sqrt{3}\, V_L\, I_L \cos\theta \qquad (4\text{-}2a)$$

Volt amperes.

For a balanced 1-phase load,

$$\text{Power} = V_L\, I_L \cos\theta \qquad (4\text{-}2b)$$

where

V_L = Voltage between hot legs
I_L = Line current.

MOTOR LOADS

Each electrical load in a system has an inherent power factor. Motor loads are usually specified by horsepower ratings. These may be converted to Kva, by use of Equation 4-3.

$$\text{Kva} = \frac{\text{hp} \times 0.746}{n \times \text{P.F.}} \qquad (4\text{-}3)$$

where

n = Motor efficiency
P.F. = Motor power factor
hp = Motor horsepower.

Most motor manufacturers can supply information on motor efficiencies and power factors. Typical values are illustrated in Table 4-1. From this table, it is evident that smaller motors running partly loaded are the least efficient and have the poorest power factor.

TABLE 4-1 Typical Motor Horsepowers & Efficiencies for 1800 RPM, "T" Frame, Totally Enclosed Fan Cooled (TEFC) Motors.

	1/2 Load		3/4 Load		Full Load	
Motor hp-Range	n	P.F.	n	P.F.	n	P.F.
3–30	83.3	70.1	85.8	79.2	86.2	83.5
40–100	89.2	79.2	90.7	85.4	90.9	87.7

WHAT ARE THE ADVANTAGES OF POWER FACTOR CORRECTION?

Several advantages that usually offset the cost of correcting the power factor are indicated below:

1. The monthly electric bill is lowered due to the utility company power rate structure.
2. The plant system capacity is increased since the transformer load can be increased.
3. Electrical system losses are decreased and voltage regulation is improved.

SIM 4-2

A plant load is comprised of the following:

50 KW lighting at unity power factor
Ten–30 hp motors running at full load
Two–40 hp motors running at full load.

What is the power factor resulting from these loads?

ANSWER

30 hp motors

$$n = 0.862 \qquad \cos\theta = 0.835$$

$$\text{KW} = \frac{\text{hp} \times 0.746}{n} = \frac{10 \times 30 \times 0.746}{0.862} = 259.6$$

$$\theta = 33^\circ$$

$$\tan\theta = 0.65$$

$$\text{Kvar} = \text{KW} \tan\theta = 259.6 \times 0.65 = 168.7 \text{ Kvar}$$

40 hp motors - $\theta = 29^\circ$

$$\text{KW} = \frac{2 \times 40 \times 0.746}{0.90} = 66.3$$

$$\tan\theta = 0.55$$

$$\text{Kvar} = 66.3 \times 0.55 = 36.4$$

$$\text{KW} = 50 + 259.6 + 66.3 = 375.9$$

$$\text{Kvar} = 168.7 + 36.4 = 205.1$$

$$\tan\theta = \frac{\text{Kvar}}{\text{KW}} = \frac{205.1}{375.9} = .546$$

$$\theta = 28^\circ$$

$$\cos\theta = 0.88$$

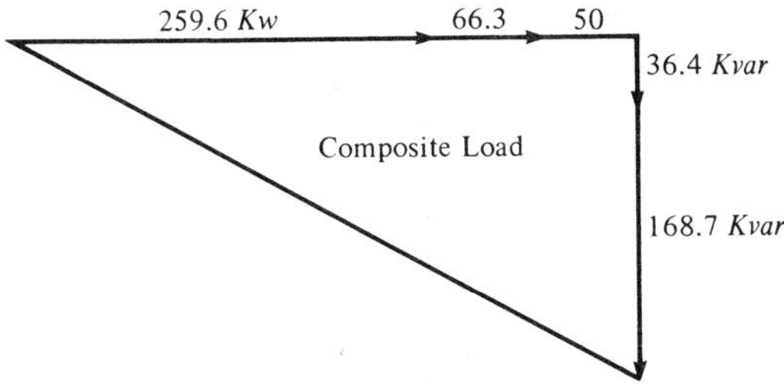

Fig. 4-2 The composite load.

It should be noted that the lighting load at a P.F. of 1 improves the overall power factor. The composite load is illustrated in Fig. 4-2.

HOW TO IMPROVE THE PLANT POWER FACTOR

The plant power factor is improved by:

1. Reducing inefficient loadings; motors running at full load have a significantly better power factor.
2. Providing external capacitors at the motor or at the distribution equipment.
3. Use of energy-efficient motors.
4. Using synchronous motors instead of induction motors. [The main application of synchronous motors are in plants that require new large slow speed motor drives (1200 rpm and below).]

How Capacitors Improve Power Factor

Capacitors supply the reactive kilovars or magnetizing power required for reactive loads. Thus, the kilovars required from the generating source decreases. This is illustrated in Fig. 4-3.

SIM 4-3

Specify the required capacitor kilovars to improve the power factor of SIM 4-2 to 0.95. Indicate the reduction in Kva and line current at 460 volts due to power factor correction.

ANSWER

The KW load of SIM 4-2 is fixed at 375.9 KW.

The desired power factor is 0.95 or $\theta = 18^\circ$.

Induction Motors

A. Partially Loaded, without Power Factor Correction

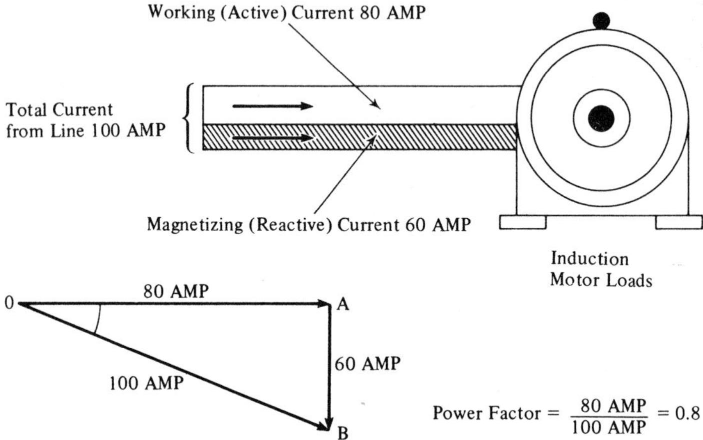

B. Capacitor Installed Near Same Motor to Supply Motor's Magnetizing Current Requirement

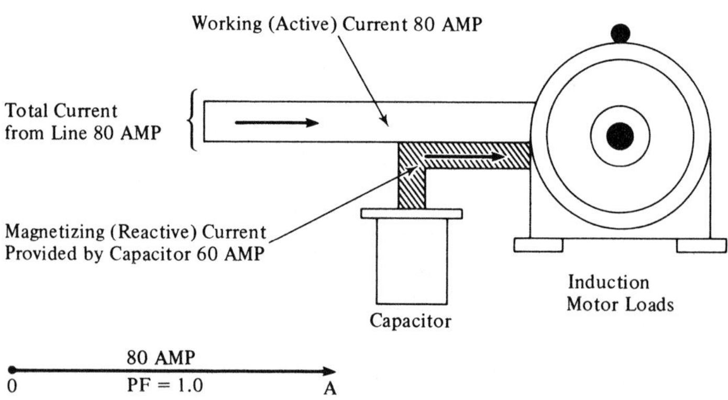

Fig. 4-3 The effect of capacitors. (Reprinted By Permission of *Specifying Engineer*.)

Thus: Kvar = KW tan θ = 375.9 X 0.32 = 120 Kvar. The required capacitor kilovars is 205.1 - 120 = 85.1 Kvar. The load analysis is illustrated in Fig. 4-4.

The Kva is reduced from: Kva_1 = KW/cos θ = 375.9/0.88 = 427.1 Kva, to: Kva_2 = 375.9/0.95 = 395.6. To calculate the line current use Eq. 4-2a.

Thus, I (Before) = $427.1 \times 10^3/\sqrt{3} \times 460$ = 536 amps

I (After) = $395.6 \times 10^3/\sqrt{3} \times 460$ = 497 amps.

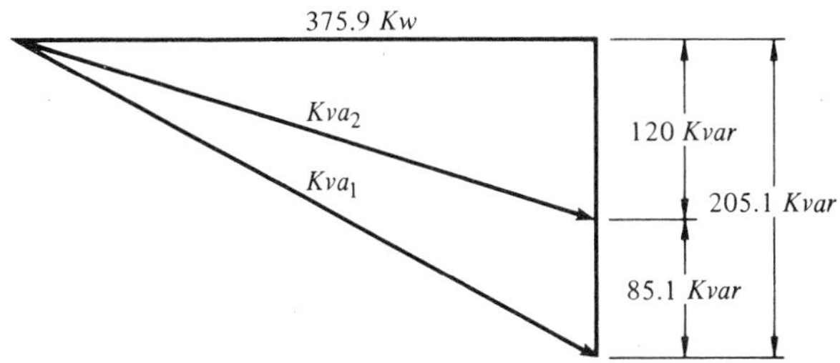

Fig. 4-4 The effect of power factor correction on power triangle.

TABLE 4-2 Shortcut Method—Power Factor Correction.

KW MULTIPLIERS FOR DETERMINING CAPACITOR KILOVARS

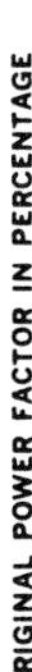

ORIGINAL POWER FACTOR IN PERCENTAGE	DESIRED POWER-FACTOR IN PERCENTAGE																				
	80	**81**	**82**	**83**	**84**	**85**	**86**	**87**	**88**	**89**	**90**	**91**	**92**	**93**	**94**	**95**	**96**	**97**	**98**	**99**	**100**
50	.982	1.008	1.034	1.060	1.086	1.112	1.139	1.165	1.192	1.220	1.248	1.276	1.303	1.337	1.369	1.403	1.441	1.481	1.529	1.590	1.732
51	.936	.962	.988	1.014	1.040	1.066	1.093	1.119	1.146	1.174	1.202	1.230	1.257	1.291	1.323	1.357	1.395	1.435	1.483	1.544	1.688
52	.894	.920	.946	.972	.998	1.024	1.051	1.077	1.104	1.132	1.160	1.188	1.215	1.249	1.281	1.315	1.353	1.393	1.441	1.502	1.644
53	.850	.876	.902	.928	.954	.980	1.007	1.033	1.060	1.088	1.116	1.144	1.171	1.205	1.237	1.271	1.309	1.349	1.397	1.458	1.600
54	.809	.835	.861	.887	.913	.939	.966	.992	1.019	1.047	1.075	1.103	1.130	1.164	1.196	1.230	1.268	1.308	1.356	1.417	1.559
55	.769	.795	.821	.847	.873	.899	.926	.952	.979	1.007	1.035	1.063	1.090	1.124	1.156	1.190	1.228	1.268	1.316	1.377	1.519
56	.730	.756	.782	.808	.834	.860	.887	.913	.940	.968	.996	1.024	1.051	1.085	1.117	1.151	1.189	1.229	1.277	1.338	1.480
57	.692	.718	.744	.770	.796	.822	.849	.875	.902	.930	.958	.986	1.013	1.047	1.079	1.113	1.151	1.191	1.239	1.300	1.442
58	.655	.681	.707	.733	.759	.785	.812	.838	.865	.893	.921	.949	.976	1.010	1.042	1.076	1.114	1.154	1.202	1.263	1.405
59	.618	.644	.670	.696	.722	.748	.775	.801	.828	.856	.884	.912	.939	.973	1.005	1.039	1.077	1.117	1.165	1.226	1.368
60	.584	.610	.636	.662	.688	.714	.741	.767	.794	.822	.849	.878	.905	.939	.971	1.005	1.043	1.083	1.131	1.192	1.334
61	.549	.575	.601	.627	.653	.679	.706	.732	.759	.787	.815	.843	.870	.904	.936	.970	1.008	1.048	1.096	1.157	1.299
62	.515	.541	.567	.593	.619	.645	.672	.698	.725	.753	.781	.809	.836	.870	.902	.936	.974	1.014	1.062	1.123	1.265
63	.483	.509	.535	.561	.587	.613	.640	.666	.693	.721	.749	.777	.804	.838	.870	.904	.942	.982	1.030	1.091	1.233
64	.450	.476	.502	.528	.554	.580	.607	.633	.660	.688	.716	.744	.771	.805	.837	.871	.909	.949	.997	1.058	1.200
65	.419	.445	.471	.497	.523	.549	.576	.602	.629	.657	.685	.713	.740	.774	.806	.840	.878	.918	.966	1.027	1.169
66	.388	.414	.440	.466	.492	.518	.545	.571	.598	.626	.654	.682	.709	.743	.775	.809	.847	.887	.935	.996	1.138
67	.358	.384	.410	.436	.462	.488	.515	.541	.568	.596	.624	.652	.679	.713	.745	.779	.817	.857	.905	.966	1.108
68	.329	.355	.381	.407	.433	.459	.486	.512	.539	.567	.595	.623	.650	.684	.716	.750	.788	.828	.876	.937	1.079
69	.299	.325	.351	.377	.403	.429	.456	.482	.509	.537	.565	.593	.620	.654	.686	.720	.758	.798	.840	.907	1.049
70	.270	.296	.322	.348	.374	.400	.427	.453	.480	.508	.536	.564	.591	.625	.657	.691	.729	.769	.811	.878	1.020
71	.242	.268	.294	.320	.346	.372	.399	.425	.452	.480	.508	.536	.563	.597	.629	.683	.701	.741	.783	.850	.992
72	.213	.239	.265	.291	.317	.343	.370	.396	.423	.451	.479	.507	.534	.568	.600	.634	.672	.712	.754	.821	.963
73	.186	.212	.238	.264	.290	.316	.343	.369	.396	.424	.452	.480	.507	.541	.573	.607	.645	.685	.727	.794	.936
74	.159	.185	.211	.237	.263	.289	.316	.342	.369	.397	.425	.453	.480	.514	.546	.580	.618	.658	.700	.767	.909
75	.132	.158	.184	.210	.236	.262	.289	.315	.342	.370	.398	.426	.453	.487	.519	.553	.591	.631	.673	.740	.882
76	.105	.131	.157	.183	.209	.235	.262	.288	.315	.343	.371	.399	.426	.460	.492	.526	.564	.604	.652	.713	.855
77	.079	.105	.131	.157	.183	.209	.236	.262	.289	.317	.345	.373	.400	.434	.466	.500	.538	.578	.620	.687	.829
78	.053	.079	.105	.131	.157	.183	.210	.236	.263	.291	.319	.347	.374	.408	.440	.474	.512	.552	.594	.661	.803
79	.026	.052	.078	.104	.130	.156	.183	.209	.236	.264	.292	.320	.347	.381	.413	.447	.485	.525	.567	.634	.776
80	.000	.026	.052	.078	.104	.130	.157	.183	.210	.238	.266	.294	.321	.355	.387	.421	.450	.499	.541	.608	.750
81	—	.000	.026	.052	.078	.104	.131	.157	.184	.212	.240	.268	.295	.329	.361	.395	.433	.473	.515	.582	.724
82	—	—	.000	.026	.052	.078	.105	.131	.158	.186	.214	.242	.269	.303	.335	.369	.407	.447	.489	.556	.698
83	—	—	—	.000	.026	.052	.079	.105	.132	.160	.188	.216	.243	.277	.309	.343	.381	.421	.463	.530	.672
84	—	—	—	—	.000	.026	.053	.079	.106	.134	.162	.190	.217	.251	.283	.317	.355	.395	.437	.504	.645
85	—	—	—	—	—	.000	.027	.053	.080	.108	.136	.164	.191	.225	.257	.291	.329	.369	.417	.478	.620

Example: Total kw input of load from wattmeter reading 100 kw at a power factor of 60%. The leading reactive kvar necessary to raise the power factor to 90% is found by multiplying the 100 kw by the factor found in the table, which is .849. Then 100 kw × 0.849 = 84.9 kvar. Use 85 kvar.

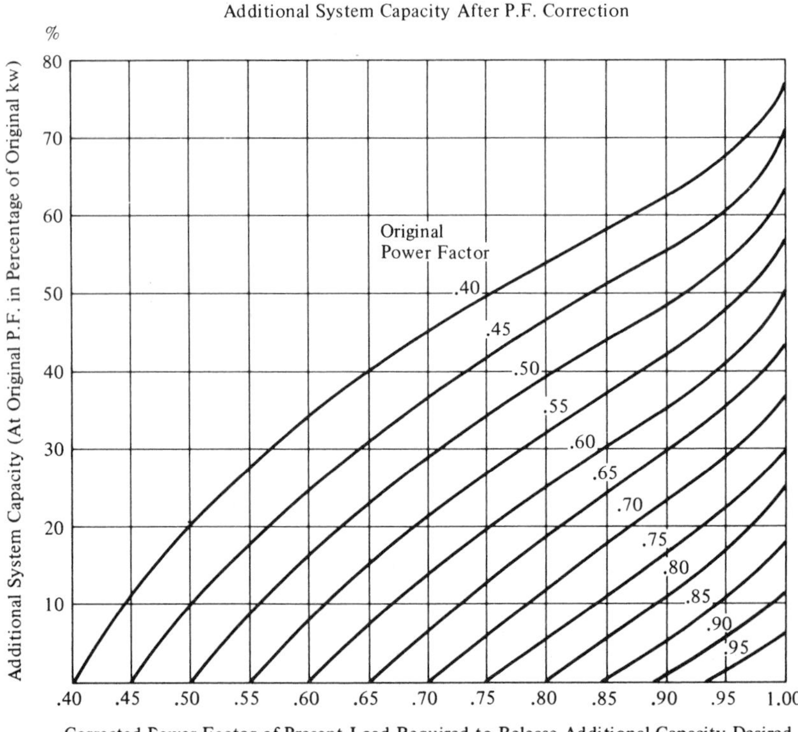

Example: Ten percent of additional capacity is required; the plant power factor is 0.85. What value should the Power Factor be corrected to in order to release the additional capacity?

Answer: From the Figure above, the Power Factor should be corrected to 0.95.

Fig. 4-5 Shortcut method to determine released capacity by power factor correction. Adapted from *Specifying Engineer.*

Shortcut Methods

A handy shortcut table which can be used to find the value of the capacitor required to improve the plant power factor is illustrated by Table 4-2. Figure 4-5 illustrates the additional capacity available when capacitors are used.

WHERE TO LOCATE CAPACITORS

As indicated, the primary purpose of capacitors is to reduce the power consumption. Additional benefits are derived by capacitor

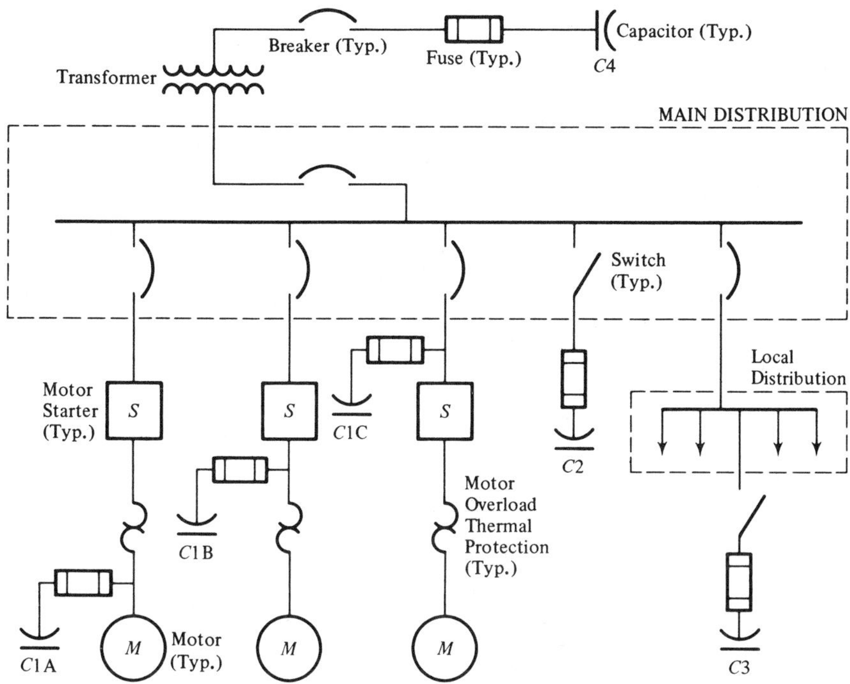

Fig. 4-6 Power distribution diagram illustrating capacitor locations.

location. Fig. 4-6 indicates typical capacitor locations. Maximum benefit of capacitors is derived by locating them as close as possible to the load. At this location, its kilovars are confined to the smallest possible segment, decreasing the load current. This, in turn, will reduce power losses of the system substantially. Power losses are proportional to the square of the current. When power losses are reduced, voltage at the motor increases; thus, motor performance also increases.

Locations *C*1A, *C*1B and *C*1C of Fig. 4-6 indicate three different arrangements at the load. Note that in all three locations, extra switches are not required, since the capacitor is either switched with the motor starter or the breaker before the starter. Case *C*1A is recommended for new installation, since the maximum benefit is derived and the size of the motor thermal protector is reduced. In Case *C*1B, as in Case *C*1A, the capacitor is energized only when the motor is in operation. Case *C*1B is recommended in cases where the installation is existing and the thermal protector does not need to be re-sized. In position $C1C$, the capacitor is permanently connected to the circuit, but does not require a separate switch, since it can be disconnected by the breaker before the starter.

It should be noted that the rating of the capacitor should *not* be greater than the no-load magnetizing Kvar of the motor. If this condition exists, damaging over-voltage or transient torques can occur. This is why most motor manufacturers specify maximum capacitor ratings to be applied to specific motors.

The next preference for capacitor locations as illustrated by Fig. 4-6 is at locations C_2 and C_3. In these locations, a breaker or switch will be required. Location C_4 requires a high voltage breaker. The advantage of locating capacitors at power centers or feeders is that they can be grouped together. When several motors are running intermittently, the capacitors are permitted to be on line all the time, reducing the total power regardless of load. Figures 4-7 and 4-8 illustrate typical capacitor installations.

Fig. 4-7 Installation of capacitors in central area. Machine shop and welding plant. 75 kvar installed capacitors. (*Courtesy of Federal Pacific Electric Company.*)

Fig. 4-8 Installation of capacitors at motor. Black top and gravel plant. 45 kvar installed capacitors. (*Courtesy of Federal Pacific Electric Company.*)

EFFICIENT MOTORS

Another method to improve the plant or building power factor is to use energy efficient motors. Energy efficient motors are available from several manufacturers. Energy efficient motors are approximately 30 percent more expensive than their standard counterpart. Based on the energy cost it can be determined if the added investment is justified. With the emphasis on energy conservation, new lines of energy efficient motors are being introduced. Figures 4-9 and 4-10 illustrate a typical comparison between energy efficient and standard motors.

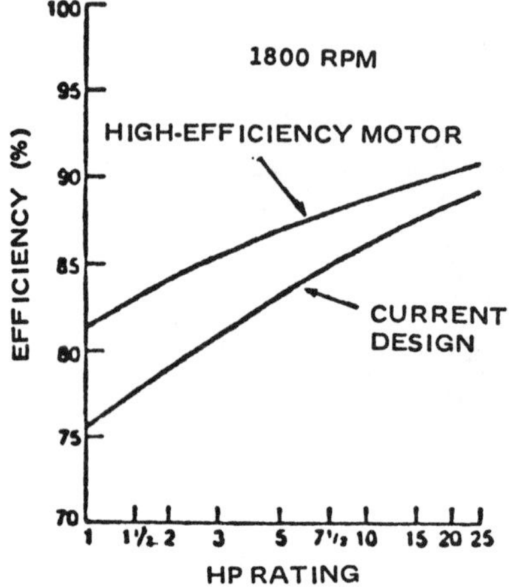

Fig. 4-9 Efficiency vs horsepower rating (dripproof motors).

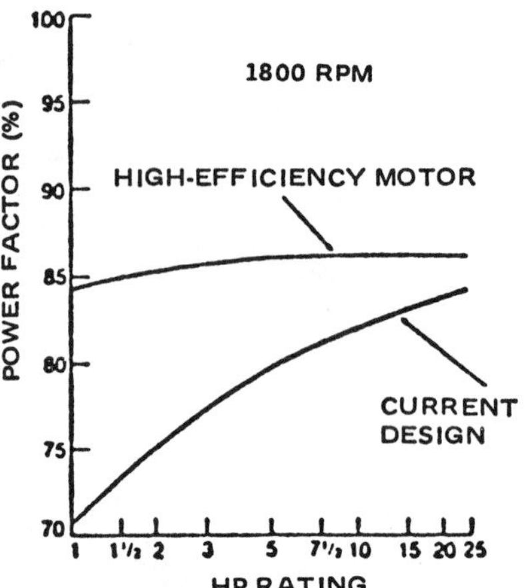

Fig. 4-10 Power factor vs horsepower rating (dripproof motors).

A third method to improve the power factor is to add capacitor banks to lower the total reactive Kvar. The line current will also be reduced, thus the corresponding I^2R loss through cables will also be lowered.

SYNCHRONOUS MOTORS AND POWER FACTOR CORRECTION

Synchronous motors find applications when constant speed operation is essential. A synchronous motor, unlike its induction motor counterpart, requires D.C. power as well as A.C. power. Many synchronous motors are self-excited; thus, the A.C. power to the motor is the only requirement. The D.C. for the field windings is generated intrinsic to the motor. Synchronous motors in ratings above 300 hp and speed below 1200 rpm are often cheaper than induction motors. Another factor for synchronous motor selection is that power factor is improved with their use.

Figure 4-11 illustrates the use of different power factor synchronous motors.[43] The 0.8 power factor synchronous motor delivers leading kilovars similar to that of a capacitor, while the unity power factor synchronous motor only delivers leading kilovars when operat-

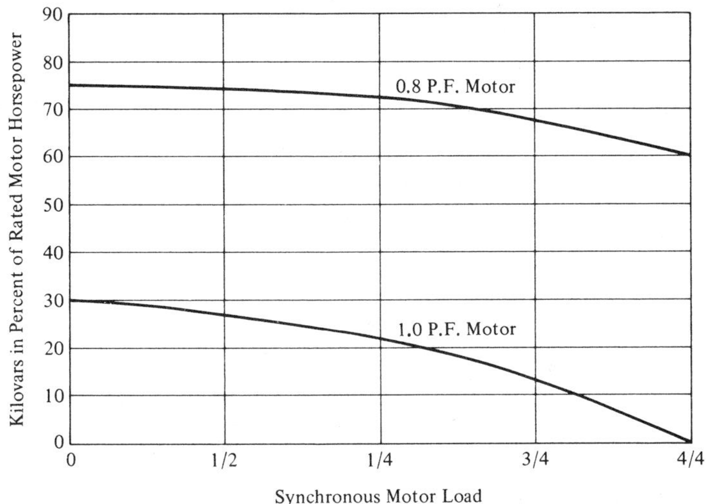

Fig. 4-11 Kilovars supplied by synchronous motors.
Source: Electrical Power Distribution for Industrial Plants Institute of Electrical and Electronic Engineers, 1964.

ing at reduced loads. Both types of synchronous motors improve plant power factor.

WHAT METHOD SHOULD BE USED TO IMPROVE THE PLANT POWER FACTOR?

The use of capacitors is usually the most economical method of improving the plant power factor when the loads consist mainly of groups of small motors. If a very large load operates continuously, synchronous motors should be considered. Synchronous condensers are seldom used.

A synchronous condenser is essentially a synchronous motor which is operated without load. Since synchronous condensers are very large and expensive, they are rarely used in industry.

The improvement of the plant power factor should begin at the source.

Fully loaded motors have good power factors.

The economics of power factor correction indicates that correcting beyond 0.95 is seldom justified.

Since power factor correction using capacitors usually has a payback period of 3 years or less, it should have a high priority in the overall program.

Power factor correction is basic to electrical design. It has been rediscovered due to the increased energy cost. This is one example of rediscovery; if you look hard enough you will find many more.

WHAT IS LOAD MANAGEMENT?

Load management is an umbrella term that describes the methods and technologies a utility can use to control the timing and peak of customer power use. Its objective is to reduce the demand for electricity during peak use periods and increase the demand during off peak periods. The utility rate structure is the vehicle used to meet this objective. The utility rate structure penalizes a customer for peak power demands. Other structures have penalties to discourage the use of electricity during certain times of the day. Load management enables the utility company to use its power generating equipment more efficiently.

WHAT HAVE BEEN SOME OF THE RESULTS OF LOAD MANAGEMENT?

Some of the results of load management programs are as follows:

1. A large glass manufacturer in Tipton Pennsylvania rescheduled the start-up of its tempering ovens to night-time hours–*Result*: Peak demand reduced by 1250 kilowatts.
2. A Flemington, New Jersey plant rescheduled the time for recharging lift truck batteries–*Result:* Peak daytime demand reduced by 100 kilowatts.
3. A tool manufacturing company in Meadville, Pennsylvania installed demand-limiting equipment and started using waste heat from some of its process operations to supplement heat to its building–*Result:* Overall power demand dropped by 1000 kilowatts.
4. In Hamburg, Pennsylvania, a metal coating company rescheduled furnace loading from its first shift to its second shift–*Result:* Demand reduced by 1000 kilowatts during peak use periods. Expected savings $35,000 to $40,000 per year.

From the above, two load management techniques can be seen. One is to re-schedule energy-related activities to non-peak hour times; the other is to automatically shed loads by use of a load demand controller or computer. In either case, the first step is to make a load audit. A load audit indicates the electrical energy characteristics for the billing period. The electric characteristics are compared with plant operations to determine what has caused the peak power usages.

APPLICATION OF AUTOMATIC LOAD SHEDDING

Load shedding should be considered when power usage demands fluctuate substantially and load leveling is feasible because of substantial non-essential or controllable loads. Load shedding has been used widely in the steel industry, but the principles of load shedding can be applied to any large industrial or commercial user. The first step in applying load shedding is to establish a target demand. The target demand is based on actual load readings or on a load analysis. The second step is to identify controllable loads which can be shut off to obtain the desired limit. Examples of controllable loads are electric furnaces, electric boilers, compressors, snow meltrs, air conditioners, heating and ventilating fans, comfort cooling, and non-critical "batch processes." Ideally, if controllable loads matched intermittent peak uses, the plant under load demand control would use a preset amount of kilowatts in a fixed amount of time. Figure 4-12 shows the effect of load shedding on peak power demands.

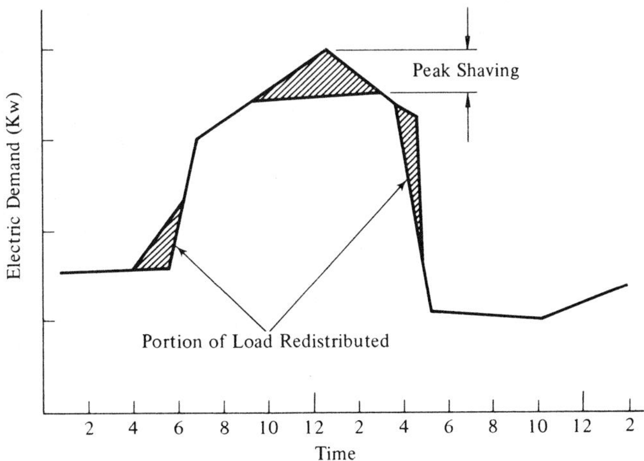

Fig. 4-12 Load shedding.

HOW DOES LOAD DEMAND CONTROL WORK?

Figure 4-13 is the block diagram of a demand control system. The demand controller is essentially a computer. It compares the consumers' actual rate of energy consumption to a predetermined ideal rate of energy consumption, during any demand interval. Let's look at each aspect.

Inputs

The same metering which the utility company uses for billing is used to supply information to the Load Demand control. The Watt Hour Meter supplies information on the kilowatt hours used. The information supplied is in the form of pulses. The Demand Meter supplies information on the end of the demand interval. This period of time is usually 15, 30, or 60 minutes.

Logic

The logic elements compare input data to a predetermined ideal rate. Signals to shed load are activated when the actual usage rate indicates that the present demand will be exceeded. Signals to restore loads are activated whenever a new demand is started. Loads are shed within the last few minutes of the demand interval in order to avoid unnecessary control action. Refer to Fig. 4-14.

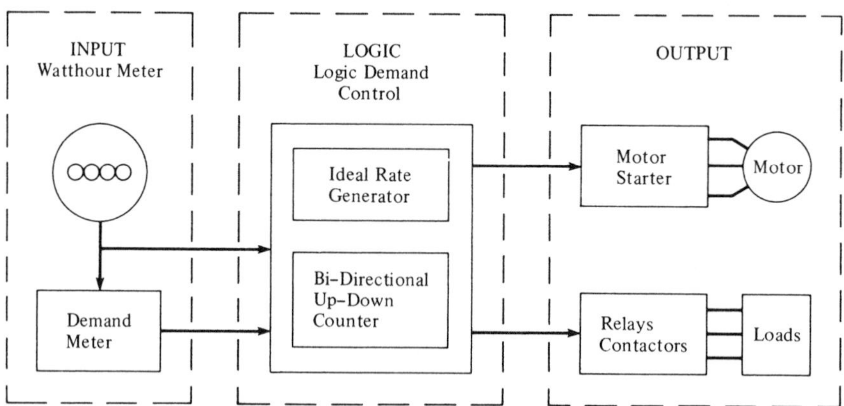

Fig. 4-13 Block diagram of load demand controller. (Reprinted by permission of *Electrical Consultant Magazine.*)

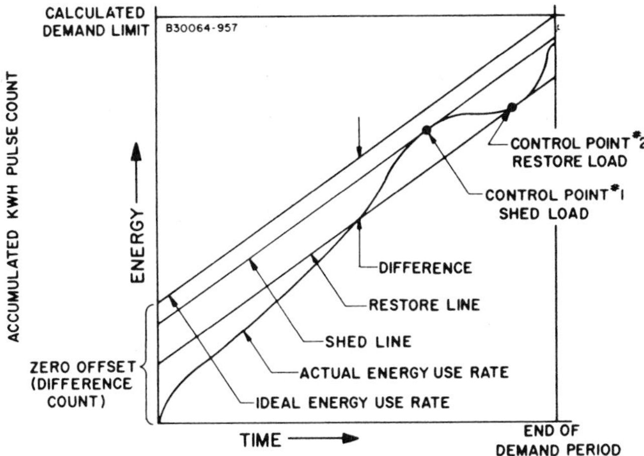

Fig. 4-14 Control curves for load demand controller. (Reprinted by permission of Square D Company.)

Outputs

Signals from the logic elements activate relays, contractors, or motor starters to shed or restore loads.

Caution

In order to accomplish load shedding, be careful to avoid cycling of equipment for short periods of time, i.e., five minutes. For example, turning motors on and off lowers the life of the equipment and could severely damage it. Always check the specifications of the controllables to insure that many starts in a short period of time does not damage the equipment.

Is a Computer Required to Shed Loads?

Before buying a computer to automatically shed loads, consider some of the packaged solid state load demand controllers. These controllers are relatively inexpensive and several models are available to meet most needs. If computer facilities are already available, then computer control of load shedding becomes more attractive. In either case, an analysis of specific electrical usage should be made by an independent consultant in order to determine which system is best.

THE CONFUSION OVER ENERGY MANAGEMENT SYSTEMS

With the myriads of electrical management systems on the market, it is no wonder why the energy manager is confused. First, there is no standardization of equipment specifications. Second, the term *energy management* system itself is confusing.

An energy management system could be one or all of the following:

1. Simple time clock.
2. Simple duty cycle device.
3. Peak load shedding device.
4. Heating, ventilation, and air conditioning controllers (enthalpy controller, temperature switch, etc.).

The main reason the controller is specified in the first place is to reduce peak electrical demand. Once the controller is constructed for that purpose, features such as a time clock and duty cycle control are economically added.

The terminology usually changes from a simple peak load shedding device to an energy management system when the heating and ventilation, control features are added.

The size of the unit to be purchased depends on the number of loads that must be cycled to reduce peak demand.

LIGHTING BASICS—THE KEY TO REDUCING LIGHTING WASTES

By understanding the basics of lighting design, several ways to improve the efficiency of lighting systems will become apparent.

There are two common lighting methods used: One is called the "Lumen" method, while the other is the "Point by Point" method. The Lumen method assumes an equal footcandle level throughout the area. This method is used frequently by lighting designers since it is simplest; however, it wastes energy, since it is the light "at the task" which must be maintained and not the light in the surrounding areas. The "Point by Point" method calculates the lighting requirements for the task in question.

The methods are illustrated by Eqs. 4-4, 4-5, 4-6, and 4-7.

Lumen Method

$$N = \frac{F_1 \times A}{Lu \times L_1 \times L_2 \times Cu} \tag{4-4}$$

where

- N is the number of lamps required.
- F_1 is the required footcandle level at the task. A footcandle is a measure of illumination; one standard candle power measured one foot away.
- A is the area of the room in square feet.
- Lu is the Lumen output per lamp. A Lumen is a measure of lamp intensity: its value is found in the manufacturer's catalogue.
- Cu is the coefficient of utilization. It represents the ratio of the Lumens reaching the working plane to the total Lumens generated by the lamp. The coefficient of utilization makes allowances for light absorbed or reflected by walls, ceilings, and the fixture itself. Its values are found in the manufacturer's catalogue.
- L_1 is the lamp depreciation factor. It takes into account that the lamp Lumen depreciates with time. Its value is found in the manufacturer's catalogue.
- L_2 is the luminaire (fixture) dirt depreciation factor. It takes into account the effect of dirt on a luminaire, and varies with type of luminaire and the atmosphere in which it is operated.

Point by Point Methods

There are three commonly used lighting formulas associated with this method. Eq. 4-5 is used for infinite length illumination sources, such as a row of non-louvered industrial fixtures. Eqs. 4-6 and 4-7 are used for point sources. An incandescent lamp or mercury vapor luminaire is treated as a point source. These equations omit inter reflections; thus the total measured illumination will be greater than the calculated values. Inter reflections can be taken into account by referring to a lighting handbook.

$$F_2 = \frac{0.35 \times CP}{D} \tag{4-5}$$

$$F_3 = \frac{CP \times \cos\theta}{D^2} \tag{4-6}$$

$$F_4 = \frac{CP \times \sin\theta}{D^2} \tag{4-7}$$

where

F_2 is the illumination produced at a point on a plane directly parallel to and directly under the source.
F_3 is the illumination on a horizontal plane.
F_4 is the illumination on a vertical plane.
CP is the candle power of the source in the particular direction. Its value is found in a manufacturer's catalogue.
D is the distance in feet to the point of illumination.
θ is the angle between "D" and the direct component. Refer to Fig. 4-15.

Analysis of Methods

From the two methods of computing lighting requirements, the following conclusions are made:

- The "point by point" method should be used with "at the task" lighting levels.
- Efficient lamps with high lumen and candlepower output should be used.
- Luminaires (fixtures) should be chosen based on a high coefficient of utilization for the application.

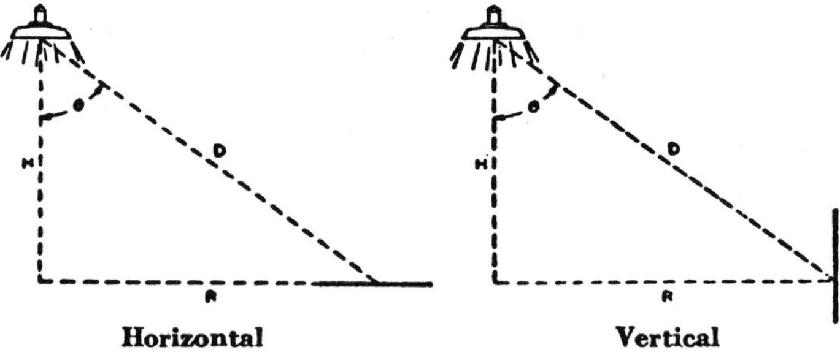

Fig. 4-15 Point by point method of illumination.

- Luminaires should be chosen on the basis of the environment; i.e., in a dirty environment, luminaires should prevent dust build-up.
- Lamps with good lamp lumen depreciation characteristics should be used.

LIGHTING ILLUMINATION REQUIREMENTS

The levels of illumination specified by the Illuminating Engineering Society (IES) in the IES Lighting Handbook are recommended levels at the *Task Surface.* For years, lighting designers have been using IES Task Lighting Values as the criteria for the total space. An option available to lighting designers is to use lighting levels below Task values for areas surrounding the Task location. The level of the surrounding area should be no more than $\frac{1}{3}$ of the weighted average of the footcandle levels for Task areas with a 20 footcandle minimum. (10 footcandle minimum can be used for non-critical general areas.)

THE EFFICIENT USE OF LAMPS

Figure 4-16 illustrates lumen outputs of various types of lamps.

Efficient Types of Incandescents for Limited Use

Attempts to increase the efficiency of incandescent lighting while maintaining good color rendition have led to the manufacture of a number of energy-saving incandescent lamps for limited residential use.

Tungsten Halogen—These lamps vary from the standard incandescent by the addition of halogen gases to the bulb. Halogen gases keep the glass bulb from darkening by preventing the filament from evaporating, and thereby increase liftime up to four times that of a standard bulb. The lumen-per-watt rating is approximately the same for both types of incandescents, but tungsten halogen lamps average 94% efficiency throughout their extended lifetime, offering significant energy and operating cost savings. However, tungsten halogen lamps require special fixtures, and during operation, the surface of the bulb reaches very high temperatures, so they are not commonly used in the home.

Reflector or R-Lamps—Reflector lamps (R-lamps) are incandescents with an interior coating of aluminum that directs the light to

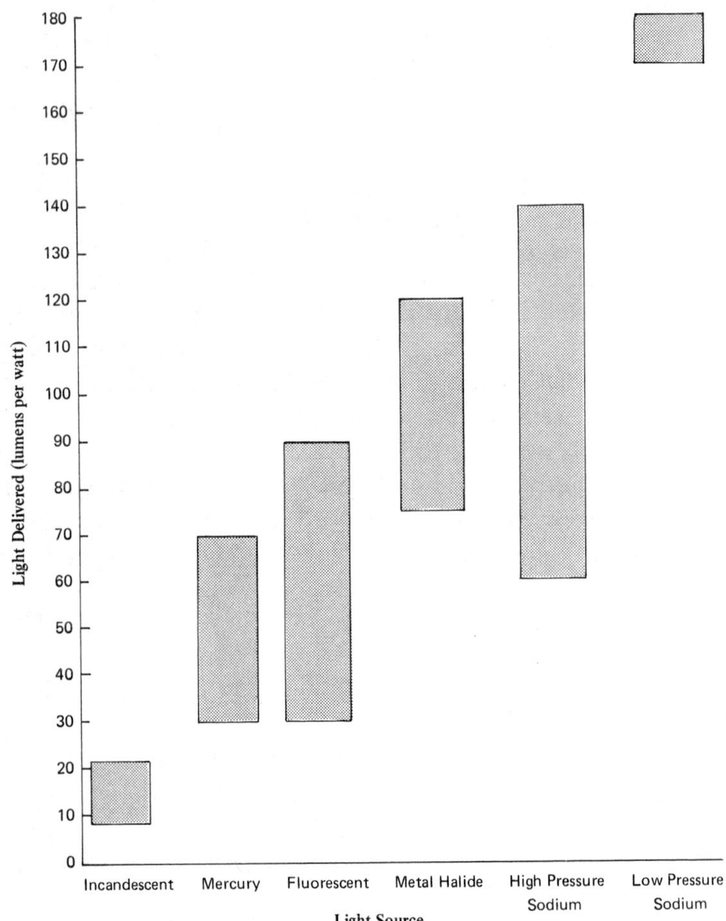

Fig. 4-16 EFFICIENCY OF VARIOUS LIGHT SOURCES

the front of the bulb. Certain incandescent light fixtures, such as recessed or directional fixtures, trap light inside. Reflector lamps project a cone of light out of the fixture and into the room, so that more light is delivered where it is needed. In these fixtures, a 50-watt reflector bulb will provide better lighting and use less energy when substituted for a 100-watt standard incandescent bulb.

Reflector lamps are an appropriate choice for task lighting, because they directly illuminate a work area, and for accent lighting. Reflector lamps are available in 25, 30, 50, 75, and 150 watts. While they have a lower initial efficiency (lumens per watt) than regular incandescents, they direct light more effectively, so that more light is actually delivered than with regular incandescents.

PAR Lamps—Parabolic aluminized reflector (PAR) lamps are reflector lamps with a lens of heavy, durable glass, which makes them an appropriate choice for outdoor flood and spot lighting. They are available in 75, 150, and 250 watts. They have longer lifetimes with less depreciation than standard incandescents.

ER Lamps—Ellipsoidal reflector (ER) lamps are ideally suited for recessed fixtures, because the beam of light produced is focused two inches ahead of the lamp to reduce the amount of light trapped in the fixture. In a directional fixture, a 75-watt ellipsoidal reflector lamp delivers more light than a 150-watt R-lamp.

Mercury vapor lamps find limited use in today's lighting systems because fluorescent and other high intensity discharge (HID) sources have passed them in both lamp efficacy and system efficiency.

Fluorescent lamps have made dramatic advances in the last 10 years. From the introduction of reduced wattage lamps in the mid-1970s, to the marketing of several styles of low wattage, compact lamps recently, there has been a steady parade of new products. The range of colors is more complete than mercury vapor and lamp manufacturers have recently made significant progress in developing fluorescent and metal halide lamps which have much more consistent color rendering properties allowing greater flexibility in mixing these two sources without creating disturbing color mismatches. The recent compact fluorescent lamps open up a whole new market for fluorescent sources. These lamps permit design of much smaller luminaries which can compete with incandescent and mercury vapor in the low cost, square or round fixture market which the incandescent and mercury sources have dominated for so long.

Energy Efficient "Plus" Fluorescents

The "energy efficient plus" fluorescents represent the second generation of improved fluorescent lighting. These bulbs are available for replacement of standard 4-foot, 40-watt bulbs and require only 32 watts of electricity to produce essentially the same light levels. To the authors' knowledge, they are not available for 8-foot fluorescent bulb retrofit. The energy efficient plus fluorescents require a ballast change. The light output is similar to the energy efficient bulbs and the two types may be mixed in the same area if desired.

Energy Efficient Fluorescents-System Change

The third generation of energy efficient fluorescents requires both a ballast and a fixture replacement. Examples of the new generation fluorescent products are:

- General Electric – "Optimizer" and "Maximizer."
- Sylvania – "Octron."

The fixtures and ballasts designed for the third-generation fluorescents are not interchangeable with earlier generations.

Metal halide lamps – fall into a lamp efficacy range of approximately 75-125 lumens per watt. This makes them more energy efficient than mercury vapor but somewhat less so than high pressure sodium. Metal-halide lamps generally have fairly good color rendering qualities. While this lamp displays some very desirable qualities, it also has some distinct drawbacks including relatively short life for an HID lamp, long restrike time to restart after the lamp has been shut off (about 15-20 minutes at 70°F) and a pronounced tendency to shift colors as the lamp ages.

High pressure sodium lamps – introduced a new era of extremely high efficacy in a lamp which operates in fixtures having construction very similar to those used for mercury vapor and metal halide. The 24,000-hour lamp life, good lumen maintenance and high efficacy of these lamps make them ideal sources for industrial and outdoor applications where discrimination of a range of colors is not critical.

The lamp's primary drawback is the rendering of some colors. The lamp produces a high percentage of light in the yellow range of the spectrum. This tends to accentuate colors in the yellow region. Rendering of reds and greens shows a pronounced color shift. In areas where color selection, matching and discrimination are necessary, high pressure sodium should not be used as the only source of light. It is possible to gain quite satisfactory color rendering by mixing high pressure sodium and metal halide in the proper proportions. Since both sources have relatively high efficacies, there is not a significant loss in energy efficiency by making this compromise.

Recently lamp manufacturers have introduced high pressure sodium lamps with improved color rendering qualities. However, as with most things in this world, the improvement in color rendering was not gained without cost–the efficacy of the color-improved lamps is somewhat lower, approximately 90 lumens per watt.

Low pressure sodium lamps – provide the highest efficacy of any of the sources for general lighting with values ranging up to 180 lumens per watt. Low pressure sodium produces an almost pure yellow light with very high efficacy and renders all colors gray except yellow or near yellow. The effect of this is there can be no color discrimination under low pressure sodium lighting and it is suitable for use in a very limited number of applications. It is an acceptable source for warehouse lighting where it is only necessary to read lables but not to choose items by color. This source has application for either indoor or outdoor safety or security lighting, again as long as color rendering is not important.

CONTROL EQUIPMENT

Table 4-3 lists various types of equipment that can be components of a lighting control system, with a description of the predominant characteristic of each type of equipment. Static equipment can alter light levels semipermanently. Dynamic equipment can alter light levels automatically over short intervals to correspond to the activities in a space. Different sets of components can be used to form various lighting control systems in order to accomplish different combinations of control strategies.

Table 4-3. Lighting Control Equipment

System	Remarks
STATIC:	
Delamping	Method for reducing light level 50%.
Impedance Monitors	Method for reducing light level 30, 50%.
DYNAMIC:	
Light Controllers	
Switches/Relays	Method for on-off switching of large banks of lamps.
Voltage/Phase Control	Method for controlling light level continuously 100 to 50%.
Solid-State Dimming Ballasts	Ballasts that operate fluorescent lamps efficiently and can dim them continuously (100 to 10%) with low voltage.
SENSORS:	
Clocks	System to regulate the illumination distribution as a function of time.
Personnel	Sensor that detects whether a space is occupied by sensing the motion of an occupant.
Photocell	Sensor that measures illumination level of a designated area.
COMMUNICATION:	
Computer/Microprocessor	Method for automatically communicating instructions and/or input from sensors to commands to the light controllers.
Power-Line Carrier	Method for carrying information over existing power lines rather than dedicated hard-wired communication lines.

Solid State Ballasts

After more than 10 years of development and 5 years of manufacturing experience, operating fluorescent lamps at high frequency (20 to 30 kHz) with solid-state ballasts has achieved credibility. The fact that all of the major ballast manufacturers offer solid-state ballasts and the major lamp companies have designed new lamps to be operated at high frequency is evidence that the solid-state high frequency ballast is now state-of-the-art.

It has been shown that fluorescent lamps operated at high frequency are 10 to 15% more efficacious than 60 Hz operation. In addition, the solid-state ballast is more efficient than conventional ballasts in conditioning the input power for the lamps, such that the total system efficacy increase is between 20 and 25 percent. That is, for a standard two-lamp 40 watt F40 T-12 rapid start system, overall efficacy is increased from 63 lm/w to over 80 lm/w.

SIM 4-4

The waste treatment plant has an average 2000-kW connected load at a PF of 0.8. Comment on the yearly savings before taxes for installing capacitors to improve the PF to 0.9. The demand charge that accounts in this billing structure for poor PF is $9/Kva/month. The total installation cost of the capacitors is $25/Kvar.

An analysis of the electrical demand indicates that a peak of 2500 kW can be reduced by automatically turning off decorative lighting and lighting for non-essential uses. Comment on adding a package load demand controller to reduce the peak demand to 2000 kW at a PF of 0.9. The installed cost of the unit is $5000.

ANSWER

Power Factor Improvement

First cost:
From Table 4-1, the multiplier is 0.266. The required correction is 2000 kW X 0.266 = 532 Kvar. The capacitor cost is $40/Kvar X 532 Kvar = $21,280.

Annual savings:
With correction—
Billing Kva = 2000/0.8 = 2,500 Kva.

Without correction—
Billing Kva = 2000/0.9 = 2222 Kva

Savings = (2500 – 2222) X $9/Kva/month X 12 = $30,024

Since the payback period is less than one year, investment is justified.

Load Management

In load shedding, the peak demand is reduced (2500/0.9 – 2000/0.9) without considering energy usage savings; the net savings are:

$9(2777 – 2222) X 12 = $54,000

With a payback period of a fraction of the year, investment is justified.

5

Utility and Process System Optimization

The energy manager should analyze the total utility needs and the process for energy utilization opportunities. The overall heat and material balance and process flow diagram are very important tools. Each subprocess must also be analyzed in detail.

In this chapter, waste heat recovery, boiler operation, utility, and process systems will be discussed.

BASIS OF THERMODYNAMICS

Thermodynamics deals with the relationships between heat and work. It is based on two basic laws of nature; the first and second laws of thermodynamics. The principles are used in the design of equipment such as steam engines, turbines, pumps, and refrigerators, and in practically every process involving a flow of heat or a chemical equilibrium.

First Law: The first law states that energy can neither be created nor destroyed, thus, it is referred to as the law of conservation of energy. Equation 5-4 expresses the first law for the steady state condition.

$$E_2 - E_1 = Q - W \qquad (5\text{-}1)$$

where

$E_2 - E_1$ is the change in stored energy at the boundary states 1 and 2 of the system
Q is the heat added to the system.
W is the work done by the system.

Figure 5-1 illustrates a thermodynamic process where mass enters and leaves the system. The potential energy (Z) and the kinetic energy ($V^2/64.2$) plus the enthalpy represents the stored energy of the mass. Note, "Z" is the elevation above the reference point in feet, and "V" is the velocity of the mass in ft/sec. In the case of the steam turbine, the change in Z, V, and Q are small in comparison to the change in enthalpy. Thus, the energy equation reduces to

$$W/778 = h_1 - h_2 \tag{5-2}$$

where

W is the work done in ft · lb/lb
h_1 is the enthalpy of the entering steam Btu/lb
h_2 is the enthalpy of the exhaust steam, Btu/lb
And 1 Btu equals 778 ft · lb.

Second Law: The second law qualifies the first law by discussing the conversion between heat and work. All forms of energy, includ-

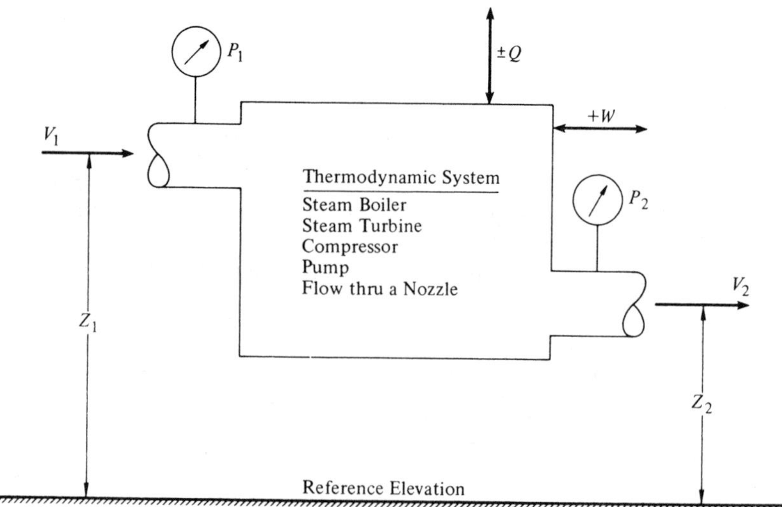

Fig. 5-1 System illustrating conservation of energy.

ing work, can be converted to heat, but the converse is not generally true. The Kelvin-Planck statement of the second law of thermodynamics says essentially the following: Only a portion of the heat from a heat work cycle, such as a steam power plant, can be converted to work. The remaining heat must be rejected as heat to a sink of lower temperature; to the atmosphere, for instance.

The Clausius statement, which also deals with the second law, states that heat, in the absence of some form of external assistance, can only flow from a hotter to a colder body.

THE CARNOT CYCLE

The Carnot cycle is of interest because it is used as a comparison of the efficiency of equipment performance. The Carnot cycle offers the maximum thermal efficiency attainable between any given temperatures of heat source and sink. A thermodynamic cycle is a series of processes forming a closed curve on any system of thermodynamic coordinates. The Carnot cycle is illustrated on a temperature-entropy diagram, Fig. 5-2A, and on the Mollier Diagram for super-heated steam, Fig. 5-2B.

The cycle consists of the following:

1. Heat addition at constant temperature, resulting in expansion work and changes in enthalpy.

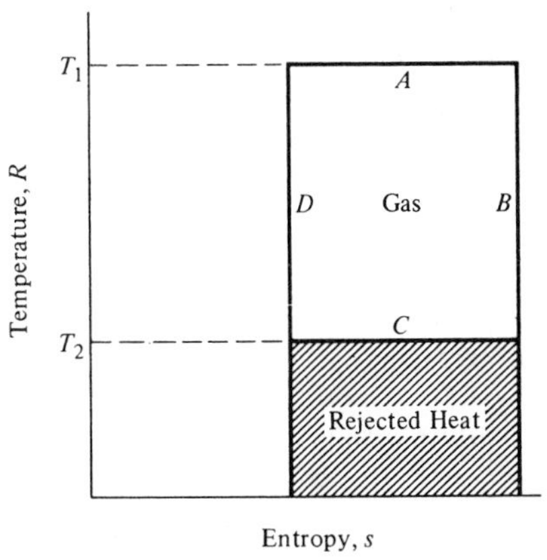

Fig. 5-2A Temperature–entropy diagram; gas.

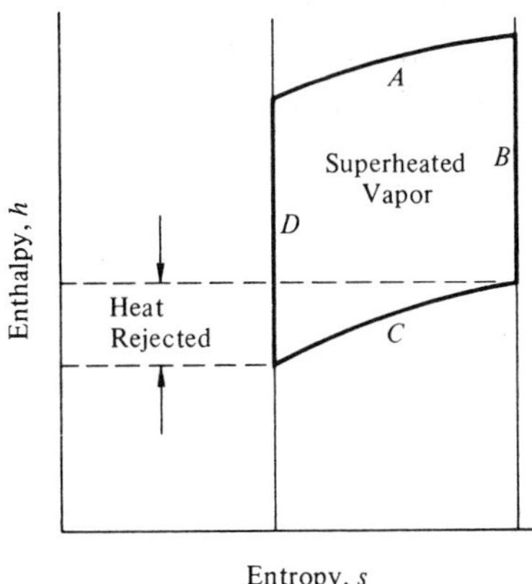

Fig. 5-2B Mollier diagram; superheated vapor.

Fig. 5-2 Carnot cycles.

2. Adiabatic isentropic expansion (change in entropy is zero) with expansion work and an equivalent decrease in enthalpy.
3. Constant temperature heat rejection to the surroundings, equal to the compression work and any changes in enthalpy.
4. Adiabatic isentropic compression returning to the starting temperature with compression work and an equivalent increase in enthalpy.

The Carnot cycle is an example of a reversible process and has no counterpart in practice. Nevertheless, this cycle illustrates the principles of thermodynamics. The thermal efficiency for the Carnot cycle is illustrated by Eq. 5-3.

$$\text{Thermal efficiency} = \frac{T_1 - T_2}{T_1} \tag{5-3}$$

where

T_1 = Absolute temperature of heat source, °R (Rankine)
T_2 = Absolute temperature of heat sink, °R

and absolute temperature is given by Eq. 5-4.

$$\text{Absolute temperature} = 460 + \text{temperature in Fahrenheit.} \tag{5-4}$$

T_2 is usually based on atmospheric temperature, which is taken as 500°R.

SIM 5-1

Compute the ideal thermal efficiency for a steam engine, based on the Carnot cycle and a saturated steam temperature of 540°F.

ANSWER

$$\text{Thermal efficiency} = \frac{(540 + 460) - 500}{(540 + 460)} = 0.5.$$

Properties of Steam Pressure and Temperature

Water boils at 212°F when it is in an open vessel under atmospheric pressure equal to 14.7 psia (pounds per square inch, absolute). Absolute pressure is the amount of pressure exerted by a system on its boundaries and is used to differentiate it from gage pressure. A pressure gage indicates the difference between the pressure of the system and atmospheric pressure.

$$\text{psia} = \text{psig} + \text{atmospheric pressure in psia} \tag{5-5}$$

Changing the pressure of water changes the boiling temperature. Thus, water can be vaporized at 170°F, at 300°F, or any other temperature, as long as the applied pressure corresponds to that boiling point.

Solid, Liquid, and Vapor States of a Liquid

Water, as well as other liquids, can exist in three states: solid, liquid, and vapor. In order to change the state from ice to water or from water to steam, heat must be added. The heat required to change a solid to liquid is called the *latent heat of fusion.* The heat required to change a liquid to a vapor is called the *latent heat of vaporization.*

In condensing steam, heat must be removed. The quantity is exactly equal to the latent heat that went into the water to change it to steam.

Heat supplied to a fluid, during the change of state to a vapor, will not cause the temperature to rise; thus, it is referred to as the latent heat of vaporization. Heat given off by a substance when it condenses from steam to a liquid is called *sensible* heat. Physical properties of water, such as the latent heat of vaporization, also change with variations in pressure.

Steam properties are given in the ASME steam tables.

USE OF THE SPECIFIC HEAT CONCEPT

Another physical property of a material is the *specific heat.* The specific heat is defined as the amount of heat in Btu required to raise one pound of a substance one degree F. For water, it can be seen from the previous examples that one Btu of heat is required to raise one lb water 1°F; thus, the specific heat of water $C_p = 1$. Specific heats for other materials are illustrated in Table 5-1. The following three equations are useful in heat recovery problems.

$$q = wC_p \Delta T \tag{5-6}$$

where

q = quantity of heat, Btu
w = weight of substance, lb
C_p = specific heat of substance, Btu/lb-°F
ΔT = temperature change of substance, °F.

$$q = MC_p \Delta T \tag{5-7}$$

TABLE 5-1 Specific Heat of Various Substances.

SUBSTANCE	SPECIFIC HEAT
SOLIDS	
ALUMINUM	0.230
ASBESTOS	0.195
BRASS	0.086
BRICK	0.220
BRONZE	0.086
CHALK	0.215
CONCRETE	0.270
COPPER	0.093
CORK	0.485
GLASS, CROWN	0.161
GLASS, FLINT	0.117
GLASS, THERMOMETER	0.199
GOLD	0.030
GRANITE	0.192
GYPSUM	0.259
ICE	0.480
IRON, CAST	0.130
IRON, WROUGHT	0.114
LEAD	0.031
LEATHER	0.360
LIMESTONE	0.216
MARBLE	0.210
MONEL METAL	0.128
PORCELAIN	0.255
RUBBER	0.481
SILVER	0.055
STEEL	0.118
TIN	0.045
WOOD	0.330
ZINC	0.092

SUBSTANCE	SPECIFIC HEAT
LIQUIDS	
ALCOHOL	0.600
AMMONIA	1.100
BRINE, CALCIUM (20% SOLUTION)	0.730
BRINE, SODIUM (20% SOLUTION)	0.810
CARBON TETRACHLORIDE	0.200
CHLOROFORM	0.230
ETHER	0.530
GASOLINE	0.700
GLYCERINE	0.576
KEROSENE	0.500
MACHINE OIL	0.400
MERCURY	0.033
PETROLEUM	0.500
SULPHURIC ACID	0.336
TURPENTINE	0.470
WATER	1.000
WATER, SEA	0.940
GASES	
AIR	0.240
AMMONIA	0.520
BROMINE	0.056
CARBON DIOXIDE	0.200
CARBON MONOXIDE	0.243
CHLOROFORM	0.144
ETHER	0.428
HYDROGEN	3.410
METHANE	0.593
NITROGEN	0.240
OXYGEN	0.220
SULPHUR DIOXIDE	0.154
STEAM (SUPERHEATED, 1 PSI)	0.450

Reprinted by permission of The Trane Company.

where

q = quantity of heat, Btu/hr (Btu/hr is sometimes abbreviated as Btuh)
M = Flow rate, lb/hr
C_p = Specific heat, Btu/lb-°F
ΔT = temperature change of substance °F.

$$q = M\Delta h \tag{5-8}$$

where

Δh = change in enthalpy of the fluid
q and M are defined above.

PRACTICAL APPLICATIONS FOR ENERGY CONSERVATION

The Steam Balance

The first step in evaluating energy conservation measures is to compile a flowsheet and steam balance. The steam balance indicates ways in which steam usage can be minimized. To start a balance, the following is evaluated:

1. Requirements of users.
2. Steam pressure levels to satisfy process needs.
3. Turbine drive operating pressures.
4. Pressure ratings for piping valves and fittings.

Fig. 5-3 illustrates a simple block heat balance diagram for a steam generating process. To conserve energy, losses must be minimized and furnace efficiency must be maximized.

In Chapter 6, heat losses will be treated in detail. The remaining portion of this chapter will indicate ways to reduce steam consumption and apply systems efficiently.

Using the Steam Turbine

Compared to other prime movers, the steam turbine remains a flexible component for most steam systems. Sizes vary from single stage units of 50 to 600 hp for driving pumps, fans, and compressors, to multi-stage units rated as high as 50,000 hp. The large units can drive huge process compressor trains. Turbines operate in the following way:[20] Steam enters the turbine through steam inlet valves and is expanded through the nozzles. It then impinges on the blades of the rotor and exits through exhaust connections. The process of ex-

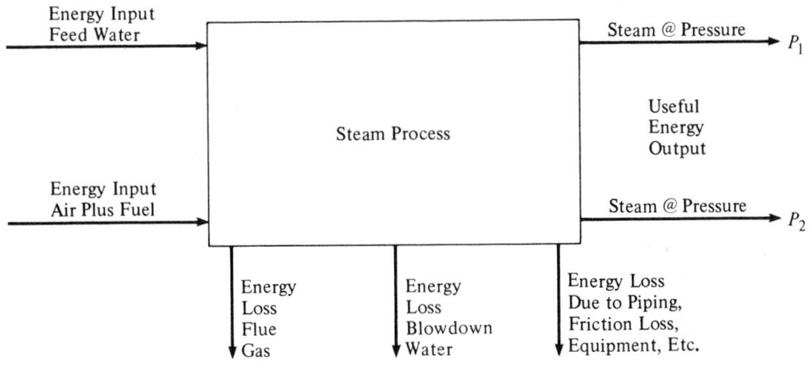

Fig. 5-3 Heat balance for steam process.

panding the steam through the nozzles changes the heat energy into mechanical energy.

Turbines can operate with inlet steam from 200 psig to over 900 psig. The throttling of the inlet steam occurs under an adiabatic (no heat loss) process. A throttling process reduces the pressure while the enthalpy remains constant. The mechanical energy can drive process equipment or a generator to produce electricity.

Some of the applications of a steam turbine are as follows:

1. Used as a pressure reducing valve, but instead of wasting energy, output is used to create electricity
2. Permits generation of high pressure and high temperature steam.
3. Used as a variable speed drive for fans and equipment.
4. Used for reliability applications where standby power is required.

Returning Condensate to the Boiler

When condensate is returned to the boiler plant, the amount of fuel used for steam generation is reduced by 10 to 30%[4]. Condensate returned to the boiler plant reduces water pollution and saves:

1. Energy and chemicals used for water treating.
2. Treated makeup boiler feed water.

Returning condensate is a common practice in new plant design, but many existing plants have not put into practice this valuable energy conservation technique.

SIM 5-2

An 80% efficient boiler uses No. 2 fuel oil to generate steam. What is the yearly (8000 h/yr) savings, excluding piping amortization, of returning 20,000 lb/h of 25 psig condensate to the boiler plant, instead of using 70°F make-up water? Assume the fuel costs \$6.00 per million Btu.

ANSWER

h_f (40 psia) = 236 Btu/lb
h_f (70°) = 38 Btu/lb
$q = 20{,}000\ (236\text{-}38) = 3.96 \times 10^6$ Btuh

$$q\ \text{(Fuel oil requirement)} = \frac{3.96 \times 10^6}{0.8} = 4.95 \times 10^6\ \text{Btuh}$$

Yearly energy savings before taxes = 4.95 × 8000 × \$6.00 = \$237,600.

Flashing Condensate to Lower Pressure

High pressure condensate can be flashed to lower pressure steam. This is often desirable when remote locations discourage returning condensate to the boiler plant and low pressure steam is required.

Fig. 5-4 should be used to calculate the amount of low pressure steam that is produced from higher pressure condensate. To use the figure, simply take the difference between the percent condensate flashed to 0 psig for the steam and condensate pressures in question. Multiply this percent by the pounds per hour of high pressure condensate available.

SIM 5-3

Compute the amount of 25 psig steam which is generated by flashing 10,000 lb/hr of 125 psig condensate. What is the yearly (8000 h/yr) savings, excluding piping amortization, for this process, as opposed to using 70°F water? The boiler is 80% efficient and uses No. 2 fuel oil. Assume fuel costs $6.00/million Btu.

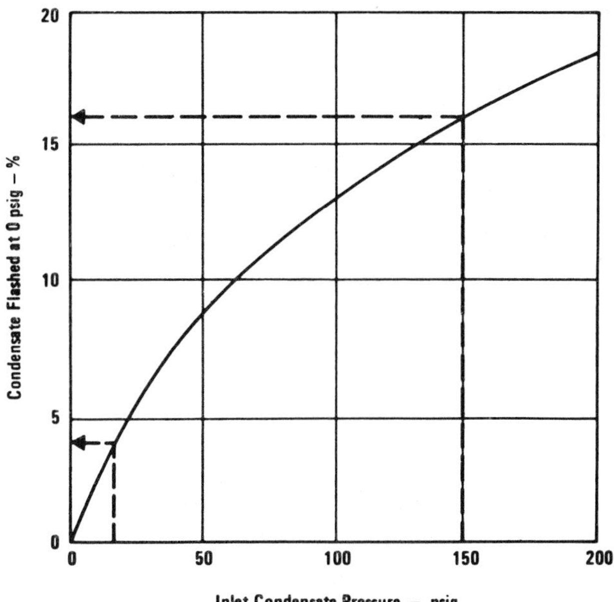

Fig. 5-4 Steam condensate flashing (calculated from steam tables). (Adapted from the *NBS Handbook 115.*)

ANSWER

From Fig. 5-4:

At 125 psig, condensate flashed to 0 psig	15%
At 25 psig, condensate flashed to 0 psig	5.5%
Percent of 25 psig steam available	= 9.5%

The potential amount of 25 psig steam available from flashing the 125 psig condensate is:

$$10{,}000 \text{ lb/hr} \times .095 = 950 \text{ lb/hr.}$$

The annual energy savings:

h_g (40 psia) = 1169.8 Btu/lb

h_f (70°) = 38 Btu/lb

$q = 950 \times (1169.8 - 38)$

$q = 1.075 \times 10^6$ Btuh

$$\text{Fuel oil requirement } q = \frac{1.075 \times 10^6}{0.8} = 1.34 \times 10^6 \text{ Btuh}$$

Yearly energy savings = 1.34 × 8000 × \$6.00
= \$64,320.

FURNACE EFFICIENCY

In order to produce steam, a boiler requires a source of heat at a sufficient temperature level. Fossil fuels, such as coal, oil, and gas are generally burned for this purpose, in the furnace of a boiler. The combustion of fossil fuels is defined as the rapid chemical combination of oxygen with the combustible elements of the fuel. The three combustible elements of fossil fuels are carbon, hydrogen, and sulfur. Sulfur is a major source of pollution and corrosion but is a minor source of heat.

Air is usually the source of oxygen for boiler furnaces.

The combustion reactions release about 61,000 Btu/lb of hydrogen burned and 14,100 Btu/lb of carbon burned.

Good combustion releases all of this heat while minimizing losses from excess air and combustion imperfections.

Unburned fuel, leaving carbon in the ash, or incompletely burned carbon represents a loss. A greater loss is the heat loss up the stack. To assure complete combustion, it is necessary to use more than theoretical air requirements. For an "ideal" union of gas and air,

TABLE 5-2 Furnace Efficiency Check List.

1. Excess air should be monitored to keep it as low as is practical.
2. Portable or permanent oxygen analyzers should be used on furnaces to help minimize oxygen in flue gas.
3. Fuel oil temperature should be monitored and checked against manufacturer's recommendations for the fuel specification used. Proper fuel oil temperature insures good atomization.
4. Atomizing steam should be dry.
5. If fuel gas is at dew point, knock out drum at the furnace with steam tracing to the burners should be used.
6. Atomizing steam rate should be at the minimum required for acceptable combustion.
7. Furnace systems should be evaluated for heat recovery, i.e., installing air preheaters or waste heat boilers.
8. Furnace systems should be evaluated for using preheated combustion air from gas-turbine exhausts.
9. Furnace systems should be evaluated for using high pressure gas for atomization instead of steam.
10. Furnace system should incorporate automated controls and recorders to maximize fuel usage and help the operator manage his system.
11. In particular, for existing furnace installations, the following should be checked:
    (a) Reduce all air leaks (infrared photography can be used to check for leaks).
    (b) Burners and soot blowers should be inspected and maintained on a scheduled basis.
    (c) Consider replacing burners with more efficient models.

excess air would not be required. In order to hold down stack losses, it is necessary to keep excess air to a minimum. Air not used in fuel combustion leaves the unit at stack temperature. When this air is heated from room temperature to stack temperature, the heat required serves no useful purpose and is therefore lost heat. A checklist of items to consider for maximum furnace efficiency[29] is illustrated in Table 5-2.

The Effect of Flue Gas and Combustion Air Temperature

Two common energy wastes in furnaces, boiler plants, and other heat processing equipment are high flue gas temperature and nonpreheated combustion air.

A common heat recovery system that saves energy is illustrated in Fig. 5-5. A portion of the flue gas heat is recovered through the use of a heat exchanger, recouper, regenerator, or similar equipment. The heat of the flue gas is used to preheat the combustion air. As an example, without combustion air preheat, a furnace operating at

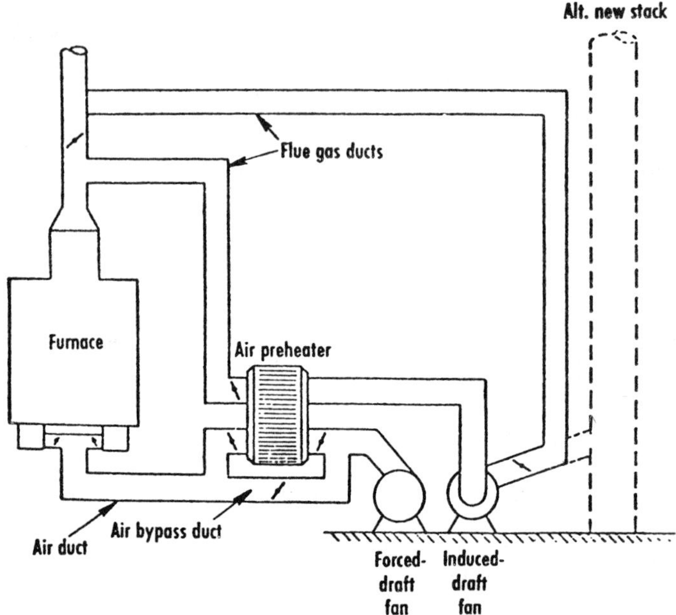

Fig. 5-5 Application of air preheater. (Reprinted by permission of *Oil and Gas Journal.*)

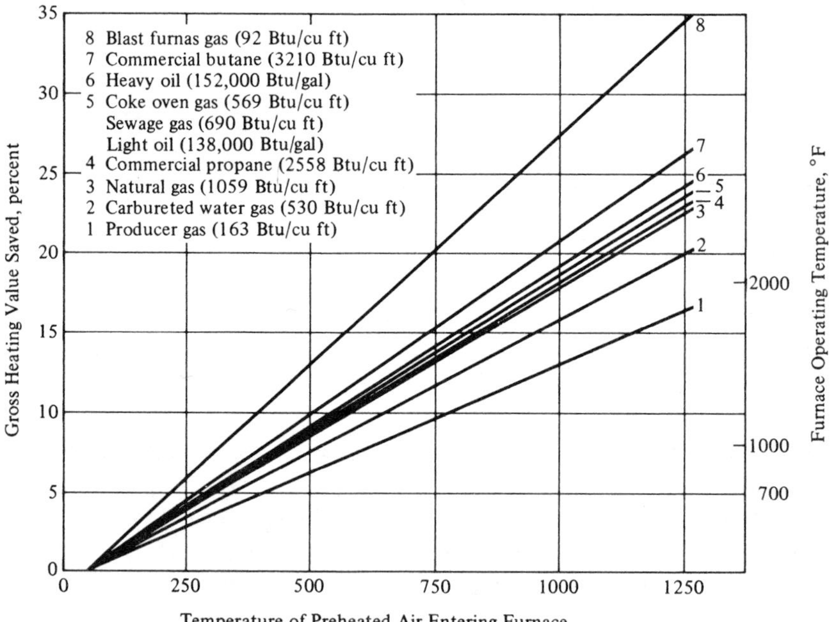

Fig. 5-6 Fuel savings resulting from use of preheated combustion air. (Figure adapted with permission from *Plant Engineering.*)

2000°F uses almost 40 percent[18] of its fuel just to heat the combustion air to the burning temperature.

Fig. 5-6 illustrates heat savings using preheated combustion air.

SIM 5-4

The plant engineer is studying the savings by adding a combustion air preheater to a furnace operating on heavy oil at 1500°F. Comment on the savings of preheating the combustion air to 750°F.

ANSWER

From Fig. 5-6, the net savings in fuel input is 13%.

Reducing Flue Gas Temperature

Even with a heat recovery system for preheating combustion air, flue gas temperatures should be maintained as low as possible. The amount of combustion air directly affects the oxygen available, which in turn increases the flue gas temperature and causes energy to be wasted. Fig. 5-7 illustrates the fuel savings which is achieved by re-

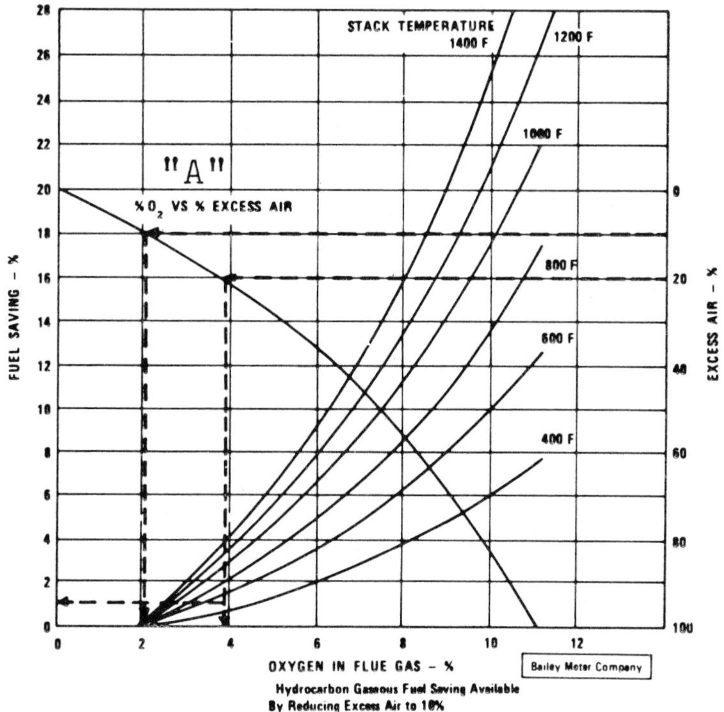

Fig. 5-7 The effect of reducing excess air for a hydrocarbon gaseous fuel. (Adapted from the *NBS Handbook 115.*)

ducing the excess air or oxygen content from the operating condition to 2% oxygen, corresponding to 10% excess air. New plant design should incorporate combustion control systems and oxygen analyzers to accomplish low excess air requirements.

To use Fig. 5-7, first find the oxygen (O_2) content in the flue gas. Note that oxygen content and percent excess air are related by curve "*A*." To find the savings that will result from reducing the O_2 content to 2%, simply find the intersection of the original O_2 level with the flue gas stack temperature. The fuel savings is read directly to the left. This curve is typical for hydrocarbon gaseous fuels such as natural gas.

Figure 5-8 is a similar curve for liquid petroleum fuels, where the

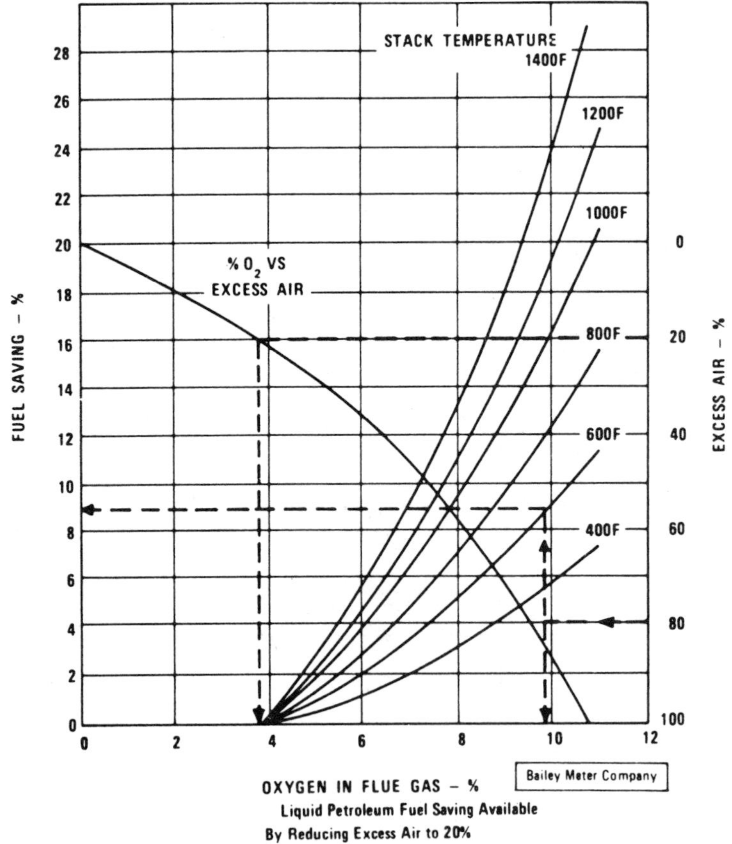

Fig. 5-8 The effect of reducing excess air for a liquid petroleum fuel. (Adapted from the *NBS Handbook 115.*)

oxygen content can be reduced to 4%, corresponding to 20% excess air.

STEAM TRACING

Steam tracing is commonly used to protect piping and equipment from freezeups and for process fluid requirements. It consists of a small tube in contact with the piping or equipment, through which steam is used as the heating medium. For systems requiring fifteen or more traces, an ethylene glycol solution can be used economically in place of steam. One such solution, SR-1, which is manufactured by Dow, freezes at minus thirty-four degrees Fahrenheit. The ethylene glycol solution is heated via a heat exchanger and pumped through the various tracers. The net savings in steam can be a thousand pounds per foot of tracer per year and achieve an overall energy savings[4] of 40%. A caution should be noted that too low an excess air value especially below 10% for gaseous fuels and 20% for liquid fuels can cause excessive corrosion of furnace tubes.

HEAT RECOVERY

Waste heat has been defined as heat that is rejected from a process at a temperature enough above the ambient temperature to permit the manager or engineer to extract additional value from it. Sources of waste energy can be divided according to temperature into three temperature ranges. The high temperature range refers to temperatures above 1200°F. The medium temperature range is between 450°F and 1200°F, and the low temperature range is below 450°F.

High- and medium-temperature waste heat can be used to produce process steam. If one has high-temperature waste heat, instead of producing steam directly, one should consider the possibility of using the high-temperature energy to do useful work before the waste heat is extracted. Both gas and steam turbines are useful and fully developed heat engines.

In the low temperature range, waste energy that would be otherwise useless can sometimes be made useful by application of mechanical work through a device called the heat pump. An interesting application of this is in petroleum distillation, where the working fluid of the heat pump can be the liquid being distilled. (This application was developed by the British Petroleum Co.).

Sources of Waste Heat

The combustion of hydrocarbon fuels produces product gases in the high temperature range. The maximum theoretical temperature possible in atmospheric combustors is somewhat under 3500°F, while measured flame temperatures in practical combustors are just under 3000°F. Secondary air or some other dilutant is often admitted to the combustor to lower the temperature of the products to the required process temperature, for example to protect equipment, thus lowering the practical waste heat temperature.

Table 5-3 below gives temperatures of waste gases from industrial process equipment in the high temperature range. All of these result from direct fuel fired processes.

Table 5-4 gives the temperatures of waste gases from process equipment in the medium temperature range. Most of the waste heat in this temperature range comes from the exhausts of directly fired process units. Medium temperature waste heat is still hot enough to allow consideration of the extraction of mechanical work from the waste heat, by a steam or gas turbine. Gas turbines can be economically utilized in some cases at inlet pressures in the range of 15 to 30 lb/in.^{2}g. Steam can be generated at almost any desired pressure and steam turbines used when economical.

Table 5-5 lists some heat sources in the low temperature range. In this range it is usually not practicable to extract work from the source, though steam production may not be completely excluded

TABLE 5-3

Type of Device	Temperature, °F
Nickel refining furnace	2500–3000
Aluminum refining furnace	1200–1400
Zinc refining furnace	1400–2000
Copper refining furnace	1400–1500
Steel heating furnaces	1700–1900
Copper reverberatory furnace	1650–2000
Open hearth furnace	1200–1300
Cement kiln (Dry process)	1150–1350
Glass melting furnace	1800–2800
Hydrogen plants	1200–1800
Solid waste incinerators	1200–1800
Fume incinerators	1200–2600

TABLE 5-4

Type of Device	Temperature, °F
Steam boiler exhausts	450–900
Gas turbine exhausts	700–1000
Reciprocating engine exhausts	600–1100
Reciprocating engine exhausts (turbocharged	450–700
Heat treating furnaces	800–1200
Drying and baking ovens	450–1100
Catalytic crackers	800–1200
Annealing furnace cooling systems	800–1200

if there is a need for low-pressure steam. Low-temperature waste heat may be useful in a supplementary way for preheating purposes. Taking a common example, it is possible to use economically the energy from an air conditioning condenser operating at around 90°F to heat the domestic water supply. Since the hot water must be heated to about 160°F, obviously the air conditioner waste heat is not hot enough. However, since the cold water enters the domestic

TABLE 5-5

Source	Temperature, °F
Process steam condensate	130–190
Cooling water from:	
Furnace doors	90–130
Bearings	90–190
Welding machines	90–190
Injection molding machines	90–190
Annealing furnaces	150–450
Forming dies	80–190
Air compressors	80–120
Pumps	80–190
Internal combustion engines	150–250
Air conditioning and refrigeration condensers	90–110
Liquid still condensers	90–190
Drying, baking and curing ovens	200–450
Hot processed liquids	90–450
Hot processed solids	200–450

water system at about 50°F, energy interchange can take place raising the water to something less than 90°F. Depending upon the relative air conditioning lead and hot water requirements, any excess condenser heat can be rejected and the additional energy required by the hot water provided by the usual electrical or fired heater.

How to Use Waste Heat

To use waste heat from sources such as those above, one often wishes to transfer the heat in one fluid stream to another (e.g., from flue gas to feedwater or combustion air). The device that accomplishes the transfer is called a heat exchanger. In the discussion immediately below is a listing of common uses for waste heat energy and in some cases, the name of the heat exchanger that would normally be applied in each particular case.

The equipment that is used to recover waste heat can range from something as simple as a pipe or duct to something as complex as a waste heat boiler. Here we categorize and describe some waste recovery systems that are available commercially suitable for retrofitting in existing plants, with lists of potential applications for each of the described devices. These are developed technologies which have been employed for years in some industries.

1. Medium to high temperature exhaust gases can be used to preheat the combustion air for:

Boilers using air-preheaters
Furnaces using recuperators
Ovens using recuperators
Gas turbines using regenerators.

2. Low to medium temperature exhaust gases can be used to preheat boiler feedwater or boiler makeup water using *economizers*, which are simply gas-to-liquid water heating devices.

3. Exhaust gases and cooling water from condensers can be used to preheat liquid and/or solid feedstocks in industrial processes. Finned tubes and tube-in-shell *heat exchangers* are used.

4. Exhaust gases can be used to generate steam in *waste heat boilers* to produce electrical power, mechanical power, process steam, and any combination of above.

5. Waste heat may be transferred to liquid or gaseous process units

directly through pipes and ducts or indirectly through a secondary fluid such as steam or oil.

6. Waste heat may be transferred to an intermediate fluid by heat exchangers or waste heat boilers, or it may be used by circulating the hot exit gas through pipes or ducts. Waste heat can be used to operate an absorption cooling unit for air conditioning or refrigeration.

Waste Heat Recovery Equipment

Industrial heat exchangers have many pseudonyms. They are sometimes called recuperators, regenerators, waste heat steam generators, condensers, heat wheels, temperature and moisture exchangers, etc. Whatever name they may have, they all perform one basic function: the transfer of heat.

Heat exchangers are characterized as single or multipass gas to gas, liquid to gas, liquid to liquid, evaporator, condenser, parallel flow, counter flow, or cross flow. The terms *single* or *multipass* refer to the heating or cooling media passing over the heat transfer surface once or a number of times. Multipass flow involves the use of internal baffles. The next three terms refer to the two fluids between which heat is transferred in the heat exchanger, and imply that no phase changes occur in those fluids. Here the term *fluid* is used in the most general sense. Thus, we can say that these terms apply to nonevaporator and noncondensing heat exchangers. The term *evaporator* applies to a heat exchanger in which heat is transferred to an evaporating (boiling) liquid, while a *condenser* is a heat exchanger in which heat is removed from a condensing vapor. A parallel flow heat exchanger is one in which both fluids flow in approximately the same direction whereas in counterflow the two fluids move in opposite directions. When the two fluids move at right angles to each other, the heat exchanger is considered to be of the crossflow type.

The principal methods of reclaiming waste heat in industrial plants make use of heat exchangers. The heat exchanger is a system which separates the stream containing waste heat and the medium which is to absorb it, but allows the flow of heat across the separation boundaries. The reasons for separating the two streams may be any of the following.

1. A pressure difference may exist between the two streams of

fluid. The rigid boundaries of the heat exchanger can be designed to withstand the pressure difference.

2. In many, if not most, cases the one stream would contaminate the other, if they were permitted to mix. The heat exchanger prevents mixing.

3. Heat exchangers permit the use of an intermediate fluid better suited than either of the principal exchange media for transporting waste heat through long distances. The secondary fluid is often steam, but another substance may be selected for special properties.

4. Certain types of heat exchangers, specifically the heat wheel, are capable of transferring liquids as well as heat. Vapors being cooled in the gases are condensed in the wheel and later re-evaporated into the gas being heated. This can result in improved humidity and/or process control, abatement of atmospheric air pollution, and conservation of valuable resources.

The various names or designations applied to heat exchangers are partly an attempt to describe their function and partly the result of tradition within certain industries. For example, a recuperator is a heat exchanger which recovers waste heat from the exhaust gases of a furnace to heat the incoming air for combustion. This is the name used in both the steel and the glass making industries. The heat exchanger performing the same function in the steam generator of an electric power plant is termed an air preheater, and in the case of a gas turbine plant, a regenerator.

However, in the glass and steel industries, the word regenerator refers to two chambers of brick checkerwork which alternately absorb heat from the exhaust gases and then give up part of that heat to the incoming air. The flows of flue gas and of air are periodically reversed by valves so that one chamber of the regenerator is being heated by the products of combustion while the other is being cooled by the incoming air. Regenerators are often more expensive to buy and more expensive to maintain than are recuperators, and their application is primarily in glass melt tanks and in open hearth steel furnaces.

It must be pointed out, however, that although their functions are similar, the three heat exchangers mentioned above may be structurally quite different as well as different in their principal modes of heat transfer.

The specification of an industrial heat exchanger must include the heat exchange capacity, the temperatures of the fluids, the allowable pressure drop in each fluid path, and the properties and volumetric flow of the fluids entering the exchanger. These specifications will determine construction parameters and thus the cost of the heat exchanger. The final design will be a compromise between pressure drop, heat exchanger effectiveness, and cost. Decisions leading to that final design will balance out the cost of maintenance and operation of the overall system against the fixed costs in such a way as to minimize the total. Advice on selection and design of heat exchangers is available from vendors.

The essential parameters that should be known in order to make an optimum choice of waste heat recovery devices are:

- Temperature of waste heat fluid
- Flow rate of waste heat fluid
- Chemical composition of waste heat fluid
- Minimum allowable temperature of waste heat fluid
- Temperature of heated fluid
- Chemical composition of heated fluid
- Maximum allowable temperature of heated fluid
- Control temperature, if control required.

Table 5-6 presents the collation of a number of significant attributes of the most common types of industrial heat exchangers in matrix form. This matrix allows rapid comparisons to be made in selecting competing types of heat exchangers. The characteristics given in the table for each type of heat exchanger are: allowable temperature range, ability to transfer moisture, ability to withstand large temperature differentials, availability as packaged units, suitability for retrofitting, and compactness and the allowable combinations of heat transfer fluids.

THE MOLLIER DIAGRAM

A visual tool for understanding and using the properties of steam is illustrated by the Mollier Diagram, Fig. 5-9. The Mollier Diagram enables one to find the relationship between temperature, pressure, enthalpy, and entropy, for steam. Constant temperature and pressure curves illustrate the effect of various processes on steam.

For a constant temperature process (isothermal), the change in

TABLE 5-6 Operation and Application Characteristics of Industrial Heat Exchangers.

Commercial Heat Transfer Equipment / Specifications for Waste Recovery Unit	Low Temperature, Sub-zero–250°F	Intermediate Temperature, 250°F–1200°F	High Temperature, 1200°F–2000°F	Recovers Moisture	Large Temperature Differentials Permitted	Packaged Units Available	Can Be Retrofit	No Cross-contamination	Compact Size	Gas-to-gas Heat Exchange	Gas-to-liquid Heat Exchanger	Liquid-to-liquid Heat Exchanger	Corrosive Gases Permitted with Special Construction
Radiation recuperator			•		•	1	•	•		•			•
Convection recuperator		•	•		•	•	•	•		•			•
Metallic heat wheel	•	•		2		•	•	3	•	•			•
Hygroscopic heat wheel	•			•		•	•	3	•	•			
Ceramic heat wheel		•	•		•	•	•		•	•			•

Passive regenerator	•	•			•	•	•	•		•			•
Finned-tube heat exchanger	•	•			•	•	•	•	•		•		4
Tube shell-and-tube exchanger	•	•			•	•	•	•	•		•	•	
Waste heat boilers	•	•				•	•	•			•		4
Heat pipes	•	•			5	•	•	•	•	•			•

[1] Off-the-shelf items available in small capacities only.
[2] Controversial subject. Some authorities claim moisture recovery. Do not advise depending on it.
[3] With a purge section added, cross-contamination can be limited to less than 1 percent by mass.
[4] Can be constructed of corrosion-resistant materials, but consider possible extensive damage to equipment caused by leaks or tube ruptures.
[5] Allowable temperatures and temperature differential limited by the phase equilibrium properties of the internal fluid.

MOLLIER CHART

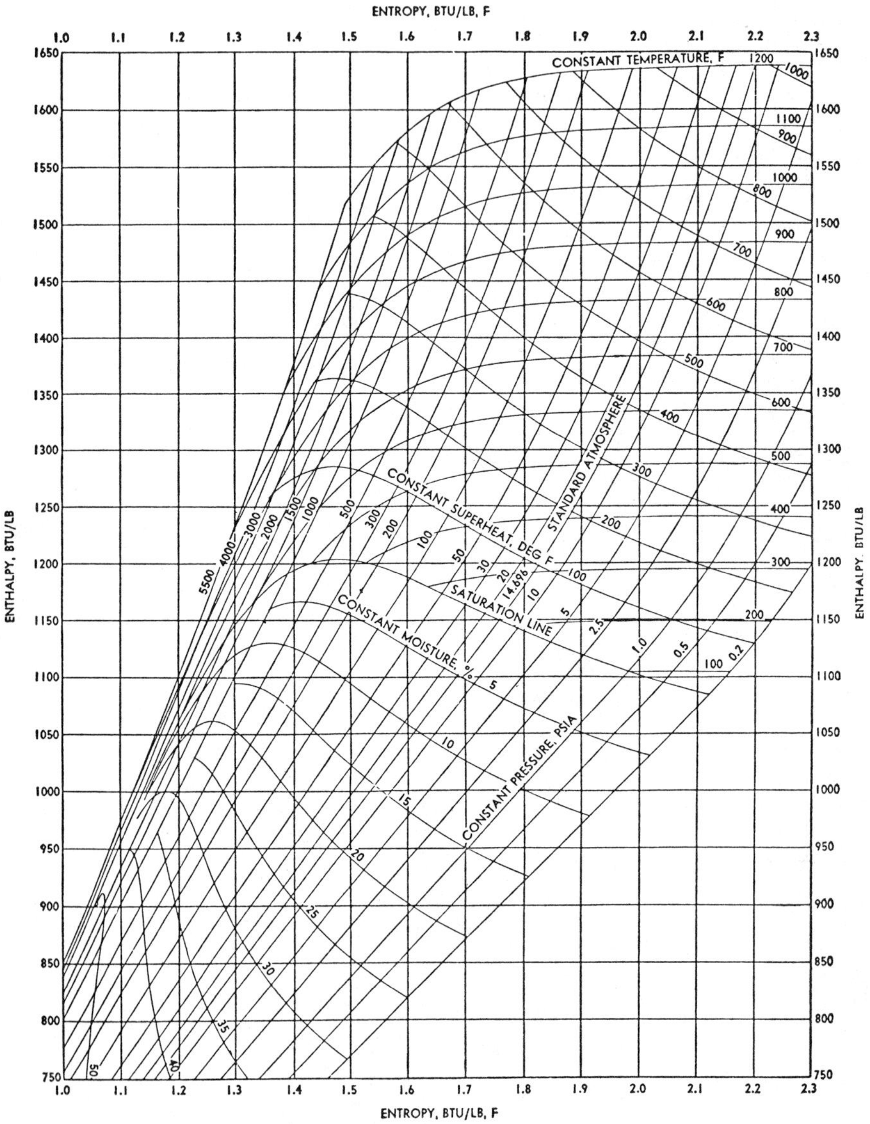

ENTROPY, BTU/LB, F

ENTHALPY, BTU/LB

ENTROPY, BTU/LB, F

Fig. 5-9 Mollier diagram. (*Courtesy of the Babcock & Wilcox Company and ASME.*)

entropy is equal to the heat added (or subtracted) divided by the temperature at which the process is carried out. This is a simple way of explaining the physical meaning of entropy. The *change* in entropy is of interest. Increases in entropy are a measure of the portion of heat in a process that is unavailable for conversion to work. Entropy has a close relation to the second law of thermodyamics, discussed later in this chapter.

Another item of interest from the Mollier Diagram is the saturation line. The saturation line indicates temperature and pressure relationships corresponding to saturated steam. Below this curve, steam contains a percent moisture, as indicated by the percent moisture curves. Steam at temperatures above the saturation curve is referred to as superheated. As an example, this chart indicates that at 212°F and 14.696 psia, the enthalpy h_g is 1150 Btu/lb.

If additional heat is added to raise the temperature of steam over the point at which it was evaporated, the steam is termed superheated. Thus, steam at the same temperature as boiling water is saturated steam. Steam at a temperature higher than boiling water, at the same pressure, is superheated steam.

Steam cannot be superheated in the presence of water because the heat supplied will only evaporate the water. Thus, the water will be evaporated prior to becoming superheated. Superheated steam is condensed by first cooling it down to the boiling point corresponding to its pressure. When the steam has been de-superheated, further removal of heat will result in its condensation.

In the generation of power, superheated steam has many uses.

STEAM GENERATION USING WASTE HEAT RECOVERY

A heat recovery system which uses the wasted exhaust from a gas turbine has proved to be very efficient. One of the largest power plants[7] (300,000 KW) using heat recovery from gas turbines was built in 1971. The heat recovery concept works as follows:

A heat recovery (waste heat) boiler is installed in the gas turbine hot air exhaust.

The unfired heat recovery boiler produces steam which drives a condensing steam turbine.

Both the gas turbine and steam turbine can drive electric generators. Thus, the wasted gas turbine exhaust is used to generate up to 50% more electric power. One such process[8] is illustrated in Fig. 5-10.

SIM 5-8

The plant engineer is evaluating two alternate power generation systems for the boiler house as illustrated in Fig. 5-11. An expansion to the Ajax plant requires 78,300 lb/hr of 150 psig process steam and 11,000 lb/hr of 30 psig process steam. The expansion also requires 3000 hp of additional drivers for pumps, compressors, fans, etc.

Alternate #1 requires the generation of 150 psig saturated steam with 11,000 lb/hr to be depressured to 30 psig. The 3000 driver hp is to be provided with electric motors.

Alternate #2 requires the generation of 600 psig 600°F superheated steam. This superheated steam is to be used to drive turbines to provide the 3000 driver hp and the turbine exhaust steam is to be used satisfy the process steam requirements.

The following assumptions are to be made.

1. Fuel oil costs $4.00/10^6 BTU.
2. Electricity costs $0.045/Kwh.

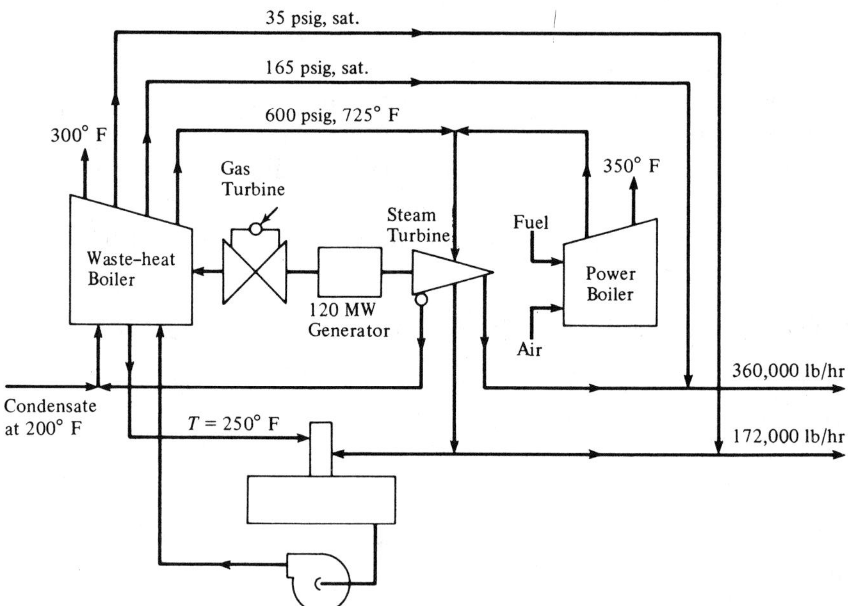

Fig. 5-10 Steam generation using waste heat recovery. [Reprinted by special permission from *Chemical Engineering* (January 21, 1974) Copyright (c) (1974) by McGraw-Hill, New York, N.Y.]

ALTERNATE #1: Steam Generation without Turbines.

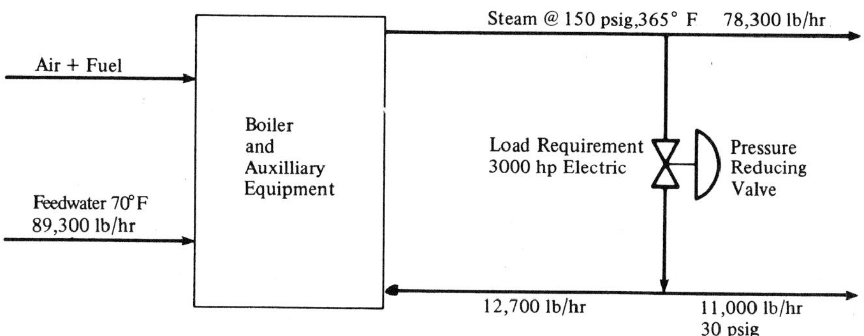

ALTERNATE #2: Steam Generation with Turbines.

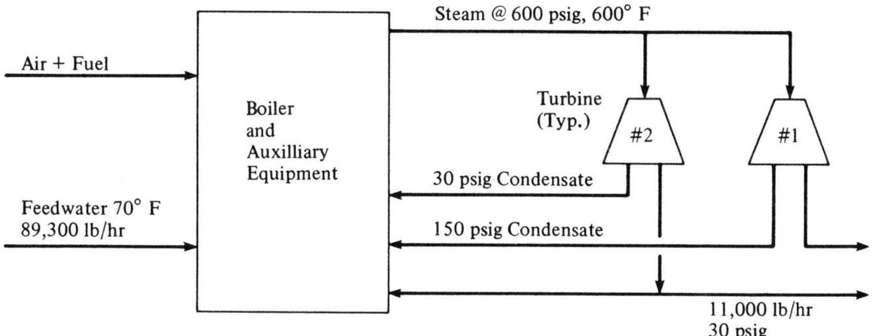

Fig. 5-11 Steam generation alternatives.

3. Temperature of boiler make-up 70°F.
4. Turbine efficiency is 0.5.
5. Neglect effect of heating combustion air.
6. Boiler efficiency is 0.8.
7. 4000 hours per year of operation (16 hours/day, 250 days/year).
8. Neglect costs associated with pumping boiler feed water.
9. In alternate #1, 12700 lb/hr of 30 psig steam is returned for de-aeration of boiler make-up water.
10. In alternate #2, 30 psig condensate and 150 psig condensate from turbines, plus 30 psig steam is returned to the boiler for heat recovery and de-aeration of boiler make-up water.
11. Electrical efficiency is 0.8.

The following are required.

1. Compute a material and energy balance for each alternate system.
2. Determine the energy requirements in dollars per year for each system.

ANSWER

A material balance requires that lb/hr "in" equals the lb/hr "out." An energy balance requires the Btu/hr "in" equals the Btu/hr "out."

The steps to be taken are:

1. First find all missing material quantities.
2. Second, find the corresponding enthalpies from the Mollier Diagram (Fig. 5-9).
3. Third, compute the energy balance.

For Alternate #1

Step 1: The quantity of 150 psig is 102,000 lb/hr.

Step 2: The enthalpy of 150 psig (165 psia) steam is 1195.6 Btu/lb. The enthalpy of 30 psig (45 psia) steam is 1172 Btu/lb. The enthaply of water at 70° F is 38.04 Btu/lb.

Step 3: The energy balance for the system is:

$$(102{,}000 \text{ lb/hr})(1195.6 \text{ Btu/lb}) = (89{,}300 \text{ lb/hr} (38.04 \text{ Btu/lb}) + (12{,}700 \text{ lb/hr})(1172.0 \text{ Btu/lb}) + \text{input energy.}$$

Input energy = 121,951,200 Btu/hr - 3,396,972 Btu/hr - 14,884,400 = 103,669,828 Btu/hr. The fuel required is 103,669,828 ÷ 0.8 = 129,587,285 Btu/hr. The fuel cost is 129,587,285 X \$4.00 ÷ 10^6 = \$518.33/hour or \$2,073,353/year.

The electrical cost is:

$$\frac{(3000 \text{ hp} \times 4000 \text{ hr/yr})(0.746 \text{ Kw/hp})(\$0.045 \text{ Kwh})}{0.8} = \$503,550/\text{year.}$$

The total energy cost per year for alternate system #1 is \$2,073,353 + \$503,550 = \$2,576,903/year.

For Alternate #2

Steps 1 and 2: To analyze turbine performance, use the Mollier Diagram of Fig. 5-12. The first point "*A*" at 600 psig, 600°F, is located on the diagram (h = 1290 Btu/lb). The second point "*B*" is drawn vertically from "*A*" until it intersects the 165 psia constant pressure line (h = 1170 Btu/lb). The difference between inlet and exhaust enthalpy is the amount of energy available to the turbine to generate horsepower. The amount of condensate formed is indicated by the intersection with the constant moisture line. At point "*B*" the constant moisture is approximately 3 percent.

The material balance is then drawn for Turbine #1. Refer to Fig. 5-13A.

$$X = 78{,}300 + 0.03X$$

$$0.97X = 78{,}300$$

$$X = 80{,}721 \text{ lb/hr to turbine}$$

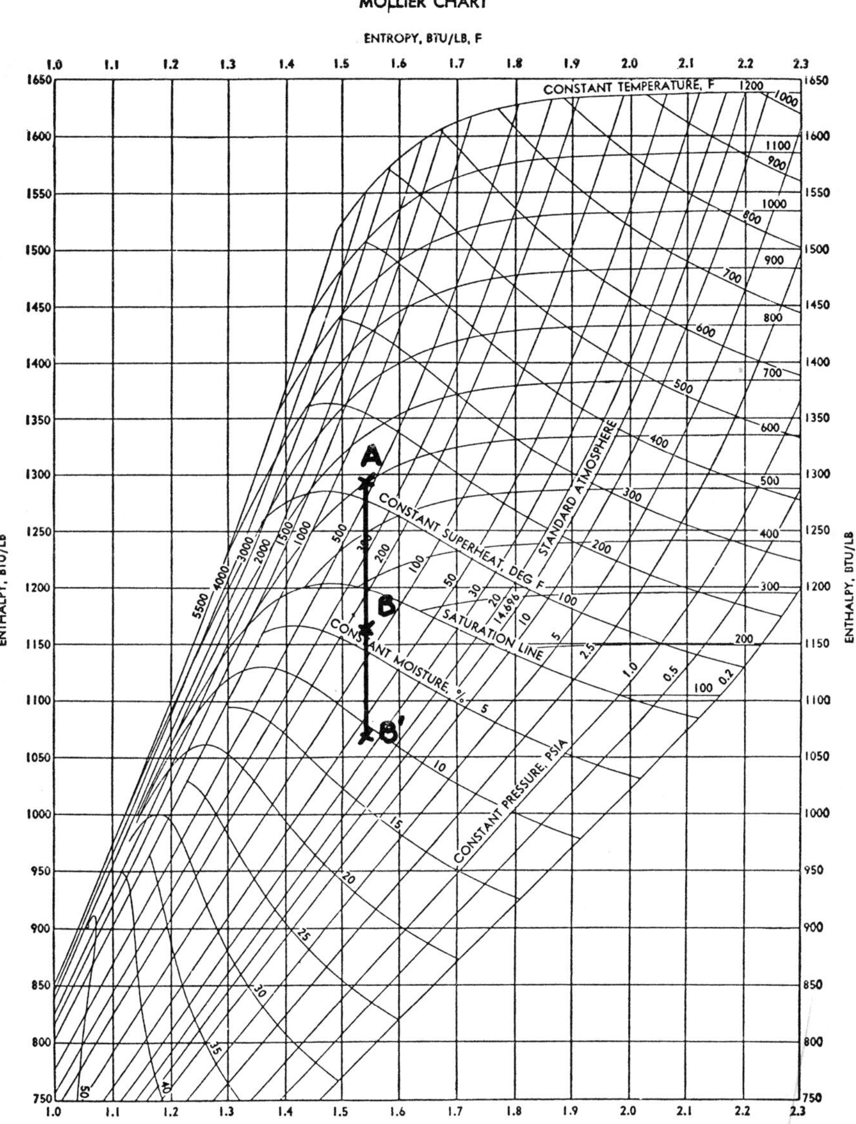

Fig. 5-12 Turbine performance by use of Mollier Diagram.

150 psig condensate = 2,421 lb/hr

$$\text{hp} = \frac{\text{Steam (lb/hr)} \times \Delta h \times n}{2545} \tag{5-8}$$

Equation 5-8 based on energy Equation 5-2. Turbine efficiency (n) and conversion factor (2545 Btu/hp) are included.

For Turbine No. 1

$$\text{hp} = \frac{80721 \times (1290 - 1170) \times 0.5}{2545}$$

$$\text{hp} = 1903$$

Material Balance for Turbine No. 2

Steps 1 & 2: Since the total horsepower requirements is 3000, Turbine No. 2 must supply 3000 - 1903 = 1097. This determines the steam rate to the turbine. To find the % moisture draw a vertical line from point "A" of Fig. 5-12 until it intersects the 45 psia constant pressure line. (h = 1070) and the constant moisture = 11%. Thus,

$$\text{hp} = \frac{X\,(1290 - 1070) \times 0.5}{2545} = 1097$$

$$X = 25{,}380 \text{ lb/hr}$$

A material balance for Turbine No. 2 is drawn as illustrated in Fig. 5-13B.

30 psig condensate = 0.11 (25,380) = 2791 lb/hr

30 psig steam = 25,380 - 11,000 - 2791 = 11, 589 lb/hr

The total balance is illustrated in Fig. 14.

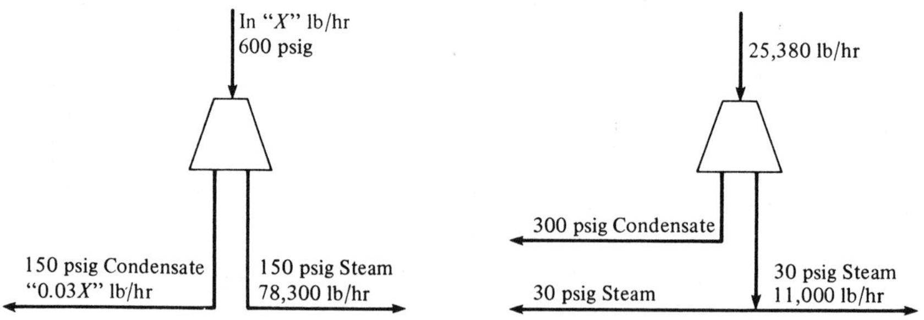

Fig. 5-13A Material balance for turbine #1. **Fig. 5-13B** Material balance for turbine #2.

Fig. 5-13 Material balances for turbines.

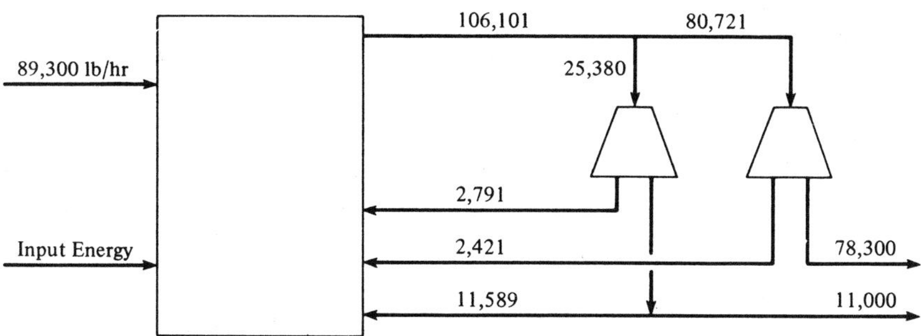

Fig. 5-14 Total material balance for steam generation.

Step 3

The energy balance of the system becomes:

$$\begin{aligned} 106{,}101\,(1290) &= 89{,}300\,(38) \\ &\quad + 2791\,(243) + 11589\,(1070) \\ &\quad + 2421\,(338) \\ &\quad + \text{input energy} \\ \text{Input energy} &= 136.8 \times 10^6 - 3.39 \times 10^6 \\ &\quad - .67 \times 10^6 - 12.4 \times 10^6 \\ &\quad - .81 \times 10^6 \\ &= 119.53 \times 10^6 \text{ Btu/hr} \end{aligned}$$

Fuel required is $119.53 \times 10^6 \div 0.8 = 149.4 \times 10^6$ Btu/hr.
The fuel cost = 149.4 × \$4.0 × 4000 = \$2,390,400/year.

Savings for Alternate No. 2 without equipment cost amortization = \$2,576,903 - \$2,390,400 = \$186,503.

PUMPS AND PIPING SYSTEMS

Rules of thumb have governed the selection of pipe sizing for pump applications. With the cost of energy increasing, many of these rules of thumb are not valid. The basic energy balance (Equation 5-1) applies as well to fluid flow. Basically, the size of discharge line piping from the pump determines the friction loss through the pipe that the pump must overcome. The greater the line loss, the more pump horsepower required. If the line is short or has a small flow, this loss may not be significant in terms of the total system head requirements. On the other hand, if the line is long and has a large flow rate, the line loss will be significant. Therefore, an economic analysis should be made on each pumping system. Table 5-7 illus-

TABLE 5-7 Flow of Water Through Schedule 40 Steel Pipe.

Discharge		Pressure Drop per 100 feet and Velocity in Schedule 40 Pipe for Water at 60 F.															
		Veloc-ity	Press. Drop	Veloc-ity	Press. Drop	Veloc-ity	Press. Drop	Veloc-ity	Press. Drop	Veloc-ity	Press. Drop	Veloc-ity	Press. Drop	Veloc-ity	Press. Drop	Veloc-ity	Press. Drop
Gallons per Minute	Cubic Ft. per Second	Feet per Second	Lbs. per Sq. In.	Feet per Second	Lbs. per Sq. In.	Feet per Second	Lbs. per Sq. In.	Feet per Second	Lbs. per Sq. In.	Feet per Second	Lbs. per Sq. In.	Feet per Second	Lbs. per Sq. In.	Feet per Second	Lbs. per Sq. In.	Feet per Second	Lbs. per Sq. In.
		1/8"		1/4"		3/8"		1/2"									
.2	0.000446	1.13	1.86	0.616	0.359												
.3	0.000668	1.69	4.22	0.924	0.903	0.504	0.159	0.317	0.061								
										3/4"							
.4	0.000891	2.26	6.98	1.23	1.61	0.672	0.345	0.422	0.086								
.5	0.00111	2.82	10.5	1.54	2.39	0.840	0.539	0.528	0.167	0.301	0.033						
.6	0.00134	3.39	14.7	1.85	3.29	1.01	0.751	0.633	0.240	0.361	0.041						
.8	0.00178	4.52	25.0	2.46	5.44	1.34	1.25	0.844	0.408	0.481	0.102						
												1"		1¼"		1½"	
1	0.00223	5.65	37.2	3.08	8.28	1.68	1.85	1.06	0.600	0.602	0.155	0.371	0.048				
2	0.00446	11.29	134.4	6.16	30.1	3.36	6.58	2.11	2.10	1.20	0.526	0.743	0.164	0.429	0.044		
3	0.00668			9.25	64.1	5.04	13.9	3.17	4.33	1.81	1.09	1.114	0.336	0.644	0.090	0.473	0.043
4	0.00891			12.33	111.2	6.72	23.9	4.22	7.42	2.41	1.83	1.49	0.565	0.858	0.150	0.630	0.071
		2"															
5	0.01114					8.40	36.7	5.28	11.2	3.01	2.75	1.86	0.835	1.073	0.223	0.788	0.104
				2½"													
6	0.01337	0.574	0.044			10.08	51.9	6.33	15.8	3.61	3.84	2.23	1.17	1.29	0.309	0.946	0.145
8	0.01782	0.765	0.073			13.44	91.1	8.45	27.7	4.81	6.60	2.97	1.99	1.72	0.518	1.26	0.241
10	0.02228	0.956	0.108	0.670	0.046			10.56	42.4	6.02	9.99	3.71	2.99	2.15	0.774	1.58	0.361
						3"											
15	0.03342	1.43	0.224	1.01	0.094					9.03	21.6	5.57	6.36	3.22	1.63	2.37	0.755
								3½"									
20	0.04456	1.91	0.375	1.34	0.158	0.868	0.056			12.03	37.8	7.43	10.9	4.29	2.78	3.16	1.28
										4"							
25	0.05570	2.39	0.561	1.68	0.234	1.09	0.083	0.812	0.041			9.28	16.7	5.37	4.22	3.94	1.93
30	0.06684	2.87	0.786	2.01	0.327	1.30	0.114	0.974	0.056			11.14	23.8	6.44	5.92	4.73	2.72
35	0.07798	3.35	1.05	2.35	0.436	1.52	0.151	1.14	0.704	0.882	0.041	12.99	32.2	7.51	7.90	5.52	3.64
40	0.08912	3.83	1.35	2.68	0.556	1.74	0.192	1.30	0.095	1.01	0.052	14.85	41.5	8.59	10.24	6.30	4.65
45	0.1003	4.30	1.67	3.02	0.668	1.95	0.239	1.46	0.117	1.13	0.064			9.67	12.80	7.09	5.85
												5"					
50	0.1114	4.78	2.03	3.35	0.839	2.17	0.288	1.62	0.142	1.26	0.076			10.74	15.66	7.88	7.15
60	0.1337	5.74	2.87	4.02	1.18	2.60	0.406	1.95	0.204	1.51	0.107			12.89	22.2	9.47	10.21
70	0.1560	6.70	3.84	4.69	1.59	3.04	0.540	2.27	0.261	1.76	0.143	1.12	0.047			11.05	13.71
80	0.1782	7.65	4.97	5.36	2.03	3.47	0.687	2.60	0.334	2.02	0.180	1.28	0.060			12.62	17.59
														6"			
90	0.2005	8.60	6.20	6.03	2.53	3.91	0.861	2.92	0.416	2.27	0.224	1.44	0.074			14.20	22.0
100	0.2228	9.56	7.59	6.70	3.09	4.34	1.05	3.25	0.509	2.52	0.272	1.60	0.090	1.11	0.036	15.78	26.9
125	0.2785	11.97	11.76	8.38	4.71	5.43	1.61	4.06	0.769	3.15	0.415	2.01	0.135	1.39	0.055	19.72	41.4
150	0.3342	14.36	16.70	10.05	6.69	6.51	2.24	4.87	1.08	3.78	0.580	2.41	0.190	1.67	0.077		
175	0.3899	16.75	22.3	11.73	8.97	7.60	3.00	5.68	1.44	4.41	0.774	2.81	0.253	1.94	0.102		
200	0.4456	19.14	28.8	13.42	11.68	8.68	3.87	6.49	1.85	5.04	0.985	3.21	0.323	2.22	0.130		
																8"	
225	0.5013	...	...	15.09	14.63	9.77	4.83	7.30	2.32	5.67	1.23	3.61	0.401	2.50	0.162	1.44	0.043
250	0.557	...	...	...	...	10.85	5.93	8.12	2.84	6.30	1.46	4.01	0.495	2.78	0.195	1.60	0.051
275	0.6127	...	...	...	...	11.94	7.14	8.93	3.40	6.93	1.79	4.41	0.583	3.05	0.234	1.76	0.061
300	0.6684	...	...	...	...	13.00	8.36	9.74	4.02	7.56	2.11	4.81	0.683	3.33	0.275	1.92	0.072
325	0.7241	...	...	...	...	14.12	9.89	10.53	4.09	8.19	2.47	5.21	0.797	3.61	0.320	2.08	0.083

350	0.7798			...	...	...	...	11.36	**5.41**	8.82	**2.84**	5.62	**0.919**	3.89	**0.367**	2.24	**0.095**
375	0.8355			...	...	...	...	12.17	**6.18**	9.45	**3.25**	6.02	**1.05**	4.16	**0.416**	2.40	**0.108**
400	0.8912			...	...	...	...	12.98	**7.03**	10.08	**3.68**	6.42	**1.19**	4.44	**0.471**	2.56	**0.121**
425	0.9469			...	...	...	...	13.80	**7.89**	10.71	**4.12**	6.82	**1.33**	4.72	**0.529**	2.73	**0.136**
450	1.003	10″		...	...	...	...	14.61	**8.80**	11.34	**4.60**	7.22	**1.48**	5.00	**0.590**	2.89	**0.151**
475	1.059	1.93	**0.054**			...	...	...	...	11.97	**5.12**	7.62	**1.64**	5.27	**0.653**	3.04	**0.166**
500	1.114	2.03	**0.059**			...	...	...	...	12.60	**5.65**	8.02	**1.81**	5.55	**0.720**	3.21	**0.182**
550	1.225	2.24	**0.071**			...	...	...	...	13.85	**6.79**	8.82	**2.17**	6.11	**0.861**	3.53	**0.219**
600	1.337	2.44	**0.083**			...	...	...	...	15.12	**8.04**	9.63	**2.55**	6.66	**1.02**	3.85	**0.258**
650	1.448	2.64	**0.097**	12″		...	...	...	...	...	...	10.43	**2.98**	7.22	**1.18**	4.17	**0.301**
700	1.560	2.85	**0.112**	2.01	**0.047**			...	...	...	...	11.23	**3.43**	7.78	**1.35**	4.49	**0.343**
750	1.671	3.05	**0.127**	2.15	**0.054**	14″		...	...	...	...	12.03	**3.92**	8.33	**1.55**	4.81	**0.392**
800	1.782	3.25	**0.143**	2.29	**0.061**			...	...	...	...	12.83	**4.43**	8.88	**1.75**	5.13	**0.443**
850	1.894	3.46	**0.160**	2.44	**0.068**	2.02	**0.042**	...	...	...	...	13.64	**5.00**	9.44	**1.96**	5.45	**0.497**
900	2.005	3.66	**0.179**	2.58	**0.075**	2.13	**0.047**	...	...	...	...	14.44	**5.58**	9.99	**2.18**	5.77	**0.554**
950	2.117	3.86	**0.198**	2.72	**0.083**	2.25	**0.052**			...	...	15.24	**6.21**	10.55	**2.42**	6.09	**0.613**
1 000	2.228	4.07	**0.218**	2.87	**0.091**	2.37	**0.057**	16″		...	...	16.04	**6.84**	11.10	**2.68**	6.41	**0.675**
1 100	2.451	4.48	**0.260**	3.15	**0.110**	2.61	**0.068**			...	...	17.65	**8.23**	12.22	**3.22**	7.05	**0.807**
1 200	2.674	4.88	**0.306**	3.44	**0.128**	2.85	**0.080**	2.18	**0.042**	...	...	...	...	13.33	**3.81**	7.70	**0.948**
1 300	2.896	5.29	**0.355**	3.73	**0.150**	3.08	**0.093**	2.36	**0.048**	...	...	...	...	14.43	**4.45**	8.33	**1.11**
1 400	3.119	5.70	**0.409**	4.01	**0.171**	3.32	**0.107**	2.54	**0.055**					15.55	**5.13**	8.98	**1.28**
1 500	3.342	6.10	**0.466**	4.30	**0.195**	3.56	**0.122**	2.72	**0.063**	18″				16.66	**5.85**	9.62	**1.46**
1 600	3.565	6.51	**0.527**	4.59	**0.219**	3.79	**0.138**	2.90	**0.071**					17.77	**6.61**	10.26	**1.65**
1 800	4.010	7.32	**0.663**	5.16	**0.276**	4.27	**0.172**	3.27	**0.088**	2.58	**0.050**			19.99	**8.37**	11.54	**2.08**
2 000	4.456	8.14	**0.808**	5.73	**0.339**	4.74	**0.209**	3.63	**0.107**	2.87	**0.060**	20″		22.21	**10.3**	12.82	**2.55**
2 500	5.570	10.17	**1.24**	7.17	**0.515**	5.93	**0.321**	4.54	**0.163**	3.59	**0.091**					16.03	**3.94**
3 000	6.684	12.20	**1.76**	8.60	**0.731**	7.11	**0.451**	5.45	**0.232**	4.30	**0.129**	3.46	**0.075**	24″		19.24	**5.59**
3 500	7.798	14.24	**2.38**	10.03	**0.982**	8.30	**0.607**	6.35	**0.312**	5.02	**0.173**	4.04	**0.101**			22.44	**7.56**
4 000	8.912	16.27	**3.08**	11.47	**1.27**	9.48	**0.787**	7.26	**0.401**	5.74	**0.222**	4.62	**0.129**	3.19	**0.052**	25.65	**9.80**
4 500	10.03	18.31	**3.87**	12.90	**1.60**	10.67	**0.990**	8.17	**0.503**	6.46	**0.280**	5.20	**0.162**	3.59	**0.065**	28.87	**12.2**
5 000	11.14	20.35	**4.71**	14.33	**1.95**	11.85	**1.21**	9.08	**0.617**	7.17	**0.340**	5.77	**0.199**	3.99	**0.079**	...	...
6 000	13.37	24.41	**6.74**	17.20	**2.77**	14.23	**1.71**	10.89	**0.877**	8.61	**0.483**	6.93	**0.280**	4.79	**0.111**	...	...
7 000	15.60	28.49	**9.11**	20.07	**3.74**	16.60	**2.31**	12.71	**1.18**	10.04	**0.652**	8.08	**0.376**	5.59	**0.150**	...	...
8 000	17.82	...	...	22.93	**4.84**	18.96	**2.99**	14.52	**1.51**	11.47	**0.839**	9.23	**0.488**	6.38	**0.192**	...	...
9 000	20.05	...	...	25.79	**6.09**	21.34	**3.76**	16.34	**1.90**	12.91	**1.05**	10.39	**0.608**	7.18	**0.242**	...	...
10 000	22.28	...	...	28.66	**7.46**	23.71	**4.61**	18.15	**2.34**	14.34	**1.28**	11.54	**0.739**	7.98	**0.294**	...	...
12 000	26.74	...	...	34.40	**10.7**	28.45	**6.59**	21.79	**3.33**	17.21	**1.83**	13.85	**1.06**	9.58	**0.416**	...	...
14 000	31.19	...	...	...	...	33.19	**8.89**	25.42	**4.49**	20.08	**2.45**	16.16	**1.43**	11.17	**0.562**	...	...
16 000	35.65	...	...	...	...	...	...	29.05	**5.83**	22.95	**3.18**	18.47	**1.85**	12.77	**0.723**	...	...
18 000	40.10	...	...	...	...	...	...	32.68	**7.31**	25.82	**4.03**	20.77	**2.32**	14.36	**0.907**	...	...
20 000	44.56	...	...	...	...	...	...	36.31	**9.03**	28.69	**4.93**	23.08	**2.86**	15.96	**1.12**	...	...

For pipe lengths other than 100 feet, the pressure drop is proportional to the length. Thus, for 50 feet of pipe, the pressure drop is approximately one-half the value given in the table . . . for 300 feet, three times the given value, etc.

Velocity is a function of the cross sectional flow area; thus, it is constant for a given flow rate and is independent of pipe length.

Reproduced from Technical Paper No. 410, Courtesy Crane Co.

trates typical pressure drops for various flows and pipe diameters. The table is based on Crane Technical[44] paper 410. To compute the brake horsepower of a pump:

$$\text{Brake horsepower} = \frac{\text{GPM} \times \Delta P}{1715 \times \eta} \tag{5-9}$$

The brake horsepower represents the actual horsepower required. where

ΔP is the differential pressure across the pump in psi.
η is the pump efficiency

GPM is the required flow in gallons per minute. Figure 5-15 illustrates a typical pumping system and the associated pressure drops. In order to compute the differential pressure across a pump, the pressure drops need to be computed and the discharge pressure, suction pressure, and the relative pumping elevations need to be known.

Figure 5-16 illustrates a pump and piping worksheet for computing brake horsepower. The Net Positive Suction Head represents the minimum head available at the pump suction that will permit it to operate. Note: The pump efficiency and the Net Positive Suction

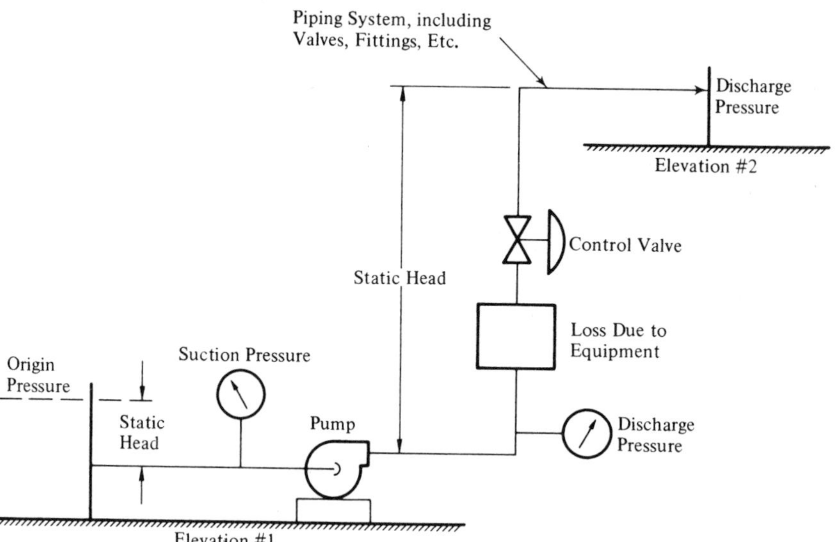

Fig. 5-15 Pressure drops associated with pipe and pumping systems.

Head of the pump should be checked against the actual pump selected.

Another equation which relates psi to head in feet is:

$$\text{Head in feet} = \frac{\text{psi} \times 2.31}{\text{specific gravity of fluid}} \tag{5-10}$$

SIM 5-11

Due to corrosion, the pump discharge piping required for six identical processes will need to be replaced. Based on the flow rate, either 8 in., 10 in., or 12 in. piping can be used. Comment on the most economical piping system, given the following:

Process and Design Conditions	
Delivery pressure	80 psia
Flow	2000 GMP
Static head discharge	13.4 psi
Suction pressure	15 psi
Suction static head	—
Suction line loss	1 psi
Pump efficiency	0.7
	Assume 150 hp motor will be applicable for the three schemes with an efficiency and P.F. of 0.9

Piping data
6–90° elbows
300 feet of pipe
2 gate valves

The additional cost to install the three systems are:

10″ - \$3000 $i = 10\%$
12″ - \$6900 $n = 10$

The cost for electricity is \$0.045/KWH.
Refer to Fig. 5-15 for pump system and exclude control valve.

ANSWER

	8″	10″	12″
1. Discharge pressure (psia)	80	80	80
2. Static head (psi)	13.4	13.4	13.4
Line Loss			
Straight pipe (ft)	300	300	300
6–90° EL - (ft)	87	111	132
2-Gate valves (ft)	17	22	26
Equivalent length (ft)	404	433	458

No. of Fittings or Total Length		Le	Equivalent Length Total L1 Suction	Equivalent Length Total L1 Discharge
______	ft. of straight pipe		______	______
______	90° Ells X	______	______	______
______	45° Ells X	______	______	______
______	Tee Run X	______	______	______
______	Tee Branch X	______	______	______
______	Gate Valves X	______	______	______
______	Globe Valves X	______	______	______
______	Check Valves X	______	______	______
	Total		______	______

Delivery Pressure	______ psia
Static Head	______ psi
Line Loss = $L_1 \times P_1$ (Table 5-3)	______ psi
ΔP Control Valves (Allow 10 psi minimum per valve)	______ psi
ΔP Other Losses	______ psi
1. Total Discharge Pressure	______ psia
Suction Pressure	
Origin Pressure	______ psia
Static Head	______ psi
−Line Loss $L_1 \times P_1$	______ psi
2. Total Suction	______ psia
Differential Pressure	
3. Discharge-Suction Pressure ΔP	______ psi
Brake Horsepower $\frac{GPM \times \Delta P}{1715 \times \eta}$	______ psi

Fig. 5-16 Pump and piping worksheet.

Valve or Fitting Type	Pipe Size (Inches) Le—Values													
	1	1½	2	3	4	6	8	10	12	14	16	18	20	24
*90° El	1.5	3	3.5	6	7.5	11	14.5	18.5	22	22.5	29.5	33	37	44
*45° El	1	1.5	1.5	2.5	3.5	5	6.5	8.5	10	11.5	13.5	15	16.5	20
*Tee Run	1	1	1.5	2	3	4	5	7	8	9	11	12	13	16
*Tee Branch	5	7	10	14.5	19	29	39	48	58	68	78	87	97	116
Gate Valve (Full Open)	1	1.5	2	3	4.5	6.5	8.5	11	13	15	17	19.5	22	26
Globe Valve	28	42.5	57	85	113	170	227	283	340	397	453	510	567	686
Check Valve	11	17	22.5	34	45	67.5	90	112.5	135	157.5	180	202.5	225	270

*Standard Radius Fittings—Butt Welded

Note: For energy comparisons only exit and entrance losses, reducers and expanders have been neglected.

Fig. 5-16 (continued)

P_1 (psi/100 ft) from Table 5-3	2.55	0.808	0.339
3. Line loss total (psi)	10.3	3.49	1.55
Total discharge pressure (psi) (1 + 2 + 3)	103.7	96.89	94.95
Origin pressure (psi)	15	15	15
Static loss			
Line loss (suction)	1	1	1
4. Total suction pressure (psi)	14	14	14
5. Differential pressure (psi)	89.7	82.89	80.95
6. Brake horsepower	149.4	138	134.8
7. Yearly energy cost $\left(0.045 \times 8000 \times \frac{hp \times .746}{n \times PF}\right)$	\$49,533	\$45,753	\$44,691
8. Additional annual owning cost $P \times CR = P \times .162$		486	1,116
9. Total yearly cost	\$49,533	\$46,239	\$45,807
		Choice	———

For six identical systems, a yearly savings before taxes of \$22,356 is possible by using the 12″ piping system.

DISTILLATION COLUMNS

In chemical plants, a large percentage of the steam generated is used for distillation processes. A distillation column is, essentially, a column filled with trays. The liquid to be separated is heated in an exchanger called a reboiler, as illustrated in Fig. 5-17. The reboiler is usually heated by steam or another heated fluid. The overhead vapors are condensed prior to discharge to the remainder of the process. A portion of the overhead is refluxed back to the column to improve separation efficiency. The traditional design uses a minimum number of trays inside the column in order to minimize first cost. The reflux back to the column is usually a set flow rate regardless of column feed rate.

Several ways to minimize steam usage are as follows:

1. The reflux to the column should be based on a fixed ratio overhead rate with overhead and bottom concentrations at minimum quality requirements. It should be noted that steam usage is directly dependent on the quantity of overhead vapors and bottoms liquids. A small increase in overhead and bottoms can

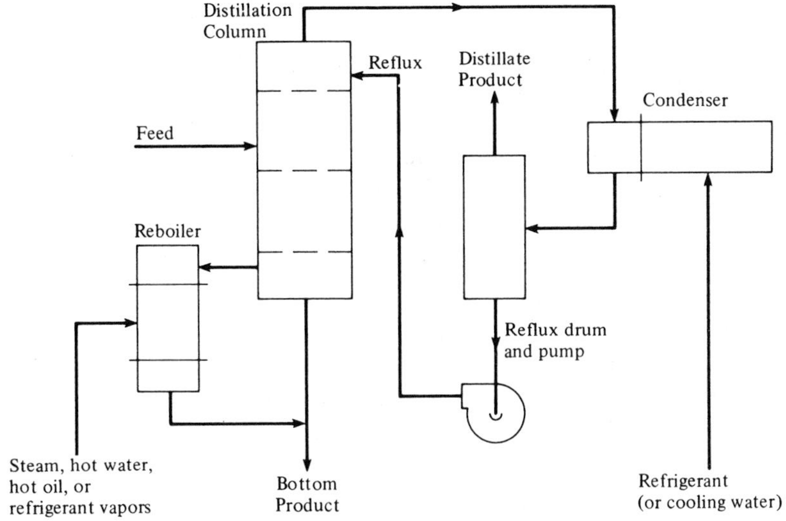

Fig. 5-17 Typical flowsheet of distillation process.

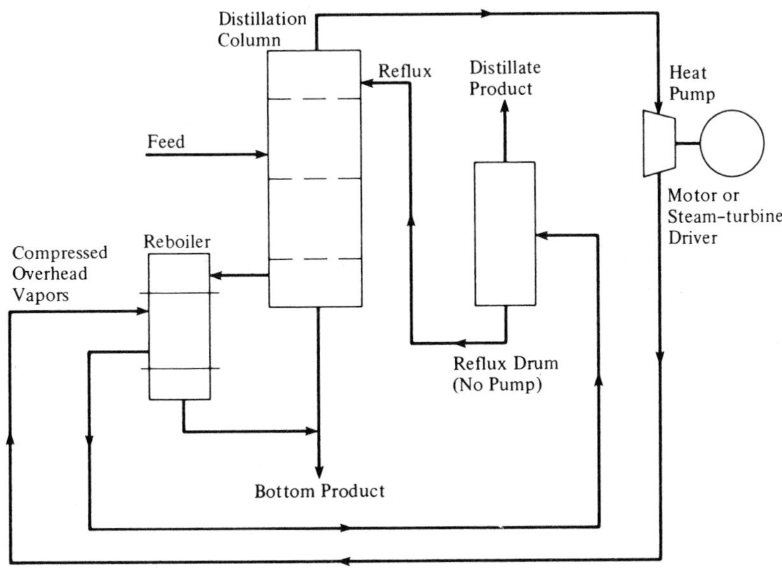

Fig. 5-18 Typical flowsheet of vapor recompression distillation.

increase steam usage significantly. Thus, by minimizing reflux, the energy usage will be decreased.

2. Distillation columns designers should lean to using more trays in the columns. More trays will reduce required reflux rate and, therefore, reduce steam usage.
3. Distillation columns should be operated near flooding conditions. At reduced feed rates, separation efficiency decreases. To correct this condition, the normal top operating pressure should be reduced.
4. The feed location to the distillation column should be designed for the temperature and composition of the incoming liquid. If the process changes, the location should be changed.
5. For processes such as the separation of ethane from ethylene, or propane from propylene, or distillation processes run at lower pressures, vapor re-compression distillation should be considered.[8] Figure 5-17 shows a typical flow sheet for vapor recompression. The vapors from the column are compressed and then routed to the re-boiler. The overhead vapor temperature is raised by the heat of compression so that it has the required ΔT to drive the re-boiler. The condensed overhead vapors are then sent to the reflux drum and pressure controlled to provide reflux to the column without a reflux pump. The only additional energy requirements are for the compressor driver, which can be a steam turbine or heat pump.

INCORPORATION OF ENERGY UTILIZATION IN PROCUREMENT SPECIFICATIONS

The distillation column's first cost is greatly influenced by the number of trays inside the vessel. The more trays, the higher the first cost, but the lower the energy cost.

To design an energy-efficient pump-piping system requires more man hours. Optimum discharge pipe design as illustrated saves on lower operating costs.

These two examples emphasize that procurement specifications that encourage choices based on lowest lump sum bids greatly penalize an energy utilization program. The best time to save energy is during initial design. Procurement specifications should reflect "cost plus" bids where the client can ensure that energy utilization measures are considered during the design phase.

6
Heat Transfer

THE IMPORTANCE OF UNDERSTANDING THE PRINCIPLES OF HEAT TRANSFER

The principles of heat transfer are the fundamental building blocks needed to understand how heat losses occur and how they can be minimized. Heat transfer finds application in equipment sizing as well. For instance, a heat exchanger is used to transfer heat from one fluid to another. Thus, heat transfer applications are involved with energy transfer in equipment, piping systems, and building. In Chapter 7, heat transfer applications to building design will be presented. In this chapter, you will learn three modes of heat transfer, see how much energy is lost from uninsulated tanks and pipes, and discover how to use economic insulation thickness tables to reduce heat losses.

THREE WAYS HEAT IS TRANSFERRED

Heat transfer is determined by the effects of conduction, radiation and convection. The three modes can be thought of simply as follows:

Conduction—Heat transfer is based on one space surrendering heat while another one gains it by the ability of the dividing surface to conduct heat.

Radiation—Heat transfer is based on the properties of light, where no surface or fluid is needed to carry heat from one object to another.

Convection—Heat transfer is based on the exchange of heat between a fluid, gas, or liquid as it transverses a conducting surface.

Heat Transfer By Conduction

Various metals conduct heat differently. Metals are the best conductors of heat, while wood, asbestos, and felt are the poorer ones. The physical property which relates the ability of a material to transmit heat is referred to as the *conductivity* K of a material. The units of K is expressed as Btu · in/hr · ft^2 · °F or Btu · in per hr · ft^2 · degree F. Thus, K represents the amount of heat in Btu flowing through a one inch thick, one square foot area of homogeneous material in one hour with a temperature differential of one degree. Table 6-1 indicates typical conductivities of various metals at room temperature. Thermal conductivities of most pure metals decrease with an increase in temperature.

TABLE 6-1 Thermal Conductivity at Room Temperature for Various Metals. Btu in/ft^2 · hr · °F.

Description	K
Aluminum (alloy 1100)	1536
Aluminum bronze (76% Cu, 22% Zn, 2% Al)	696
Brass:	
red (85% Cu, 15% Zn)	1044
yellow (65% Cu, 35% Zn)	828
Bronze	204
Copper (electrolytic)	2724
Gold	2064
Iron:	
Cast	331
Wrought	418.8
Nickel	412.8
Platinum	478.8
Silver	2940
Steel (mild)	314.4
Zinc:	
Cast	780
Hot rolled	744

For heat transfer calculations, it is satisfactory to assume a constant conductivity at the average temperature of the material. As an example, 0.23 carbon has a K of 350 at 100°F and varies linearily to 300 at 700°F.

Conduction Through A Flat Surface

When a flat plate is heated on one side and cooled on the other, as indicated in Fig. 6-1, a flow of heat from the hot side to the cold side will occur. The flow of heat is defined as:

$$q = \frac{K}{d} A \,(t_2 - t_1) = \frac{KA}{d}\,\Delta T \tag{6-1}$$

where

q is the rate of heat flow Btuh
d is the thickness of the material in inches
A is the area of the plate in ft^2
$t_2 - t_1$ is the temperature difference ΔT, causing heat flow, °F.

Figure 6-2 illustrates a composite of several materials or a nonhomogeneous material.

The flow of heat for this case is defined as:

$$q = UA\,(t_0 - t_5) = UA\,\Delta T \text{ Btuh} \tag{6-2}$$

where U is referred to as the conductance or coefficient of transmission of the material. Btu per hr · ft^2 · °F. In Fig. 6-2, a surface film

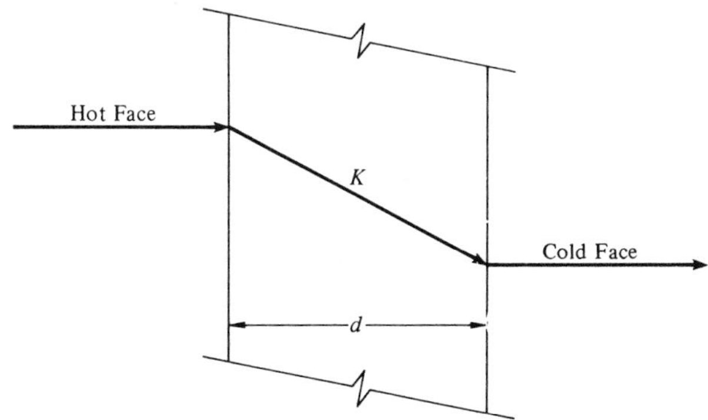

Fig. 6-1 Temperature distribution for the single wall with conductivity K.

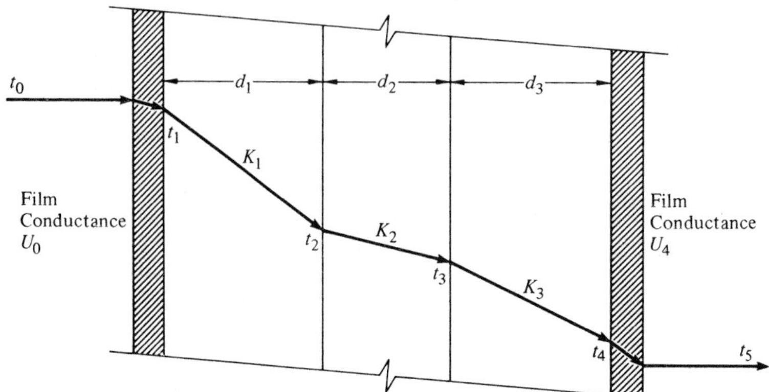

Fig. 6-2 Temperature distribution for the composite wall.

conductance is introduced. The surface or film conductance is the amount of heat transferred in Btu per hour from a surface to air or from air to a surface per square foot for one degree difference in temperature. The flow of heat for the composite material can also be specified in terms of the conductivity of the material and the conductance of the air film:

$$q = \frac{A(t_0 - t_5)}{1/U_0 + d_1/K_1 + d_2/K_2 + d_3/K_3 + 1/U_4}. \tag{6-3}$$

The conductance of a homogeneous material is expressed as

$$U = \frac{K}{d}. \tag{6-4}$$

The reciprocal of conductance is referred to as the resistance (R). For a unit length and through a unit area, R becomes:

$$R = \frac{1}{U} = \frac{d}{K} \tag{6-5}$$

The resistance of a material is directly analogous to that of electrical circuits. Heat transfer problems are solved using electrical analogies. Referring to Fig. 6-2, each material is thought of as having a heat flow resistance of d/K or $1/U$.

In series circuits, resistances are added; thus

$$R = R_1 + R_2 + R_3 + R_4 + R_5 \tag{6-6}$$

$$R = \frac{1}{U_0} + \frac{d_1}{K_1} + \frac{d_2}{K_2} + \frac{d_3}{K_3} + \frac{1}{U_4} \tag{6-7}$$

SIM 6-1

What is the heat loss through six inches of mild steel at room temperature, given the following:

$$\Delta T = 15^\circ\text{F}$$

$$\text{Area} = 100 \text{ ft}^2$$

Exclude the effect of surface film conductance.

ANSWER

From Table 6-1, $K = 314.4$

$$q = \frac{314.4}{6} \times 100 \times 15 = 78{,}600 \text{ Btuh.}$$

SIM 6-2

The surface film resistance of galvanized steel for still air is 1.85. Calculate the heat flow when this factor is included in SIM 6-1.

ANSWER

$$q = \frac{A\Delta T}{R}$$

$$R = 2\,(1.85) + \frac{6}{314.4} = 3.70 + 0.019 = 3.719$$

$$q = \frac{100 \times 15}{3.719} = 403.3 \text{ Btuh.}$$

SIM 6-3

A 1 in. layer of felt fiberglass insulation having a conductivity of 0.3 is added to the plate of SIM 6-1. The outside air film resistance then becomes 0.68 while the inside air film resistance remains at 1.85. Comment on the heat loss of the plate with insulation.

ANSWER

$$R = 1.85 + 0.68 + \frac{1}{0.3} + \frac{6}{314} = 5.88$$

$$q = \frac{100 \times 15}{5.88} = 255 \text{ Btuh.}$$

Conduction Through A Cylindrical Surface

A common problem facing the engineer is the computation of heat flow through a cylindrical surface. The surface area of a cylinder is

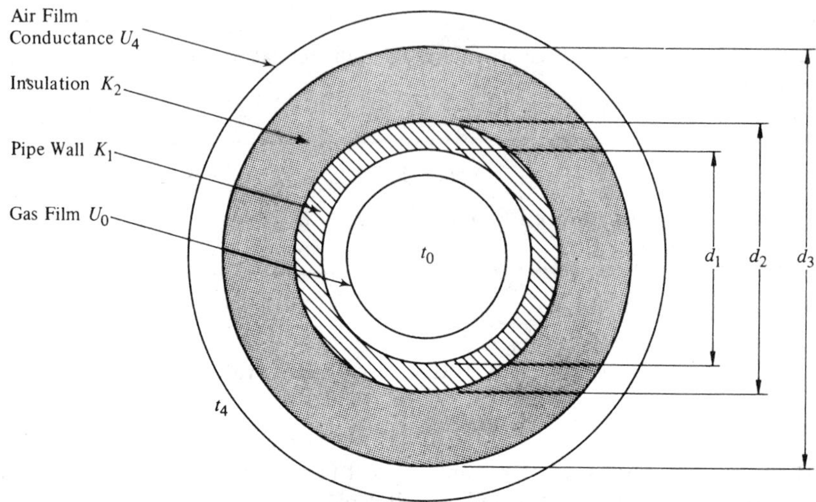

Fig. 6-3 Heat flow through the composite cylindrical wall.

not constant as the distance from the center increases; thus, the basic heat flow equation must be modified.

Figure 6-3 indicates the heat flow through a composite cylinder. In this case, the heat flow is: (temperatures, conductivities, and film conductances are defined in the figure)

$$q = \frac{A(t_0 - t_4)}{(d_3/d_1)(1/U_0) + (d_3/2K_1)\log_e d_2/d_1 + (d_3/2K_2)\log_e d_3/d_2 + 1/U_4} \tag{6-8}$$

where

$$A = \pi L d_3$$

$$L = \text{length of pipe.}$$

Heat Transfer By Radiation

Radiation is the transfer of radiant energy from a source to a receiver. Radiation from a source is partially absorbed by the receiver and partially reflected. The radiation emitted depends upon its surface emissivity, area, and temperature, as illustrated by the following equation:

$$q = \epsilon \sigma A T^4 \tag{6-9}$$

where

q = rate of heat, flow by radiation Btu/hr
ϵ = Emissivity of a body, which is defined as the ratio of energy radiated by the actual body to that of a black body. $\epsilon = 1$, for a black body
σ = Stefan Boltzmann Constant, 1.71×10^{-9} Btu/ft^2 · hr · T^4
A = surface area of body in square feet.

Heat Transfer By Convection

Convection is the transfer of heat between a fluid, gas, or liquid. Equation (6-2) is indicative of the basic form of convective heat transfer. U, in this case, represents the convection film conductance, Btu/ft^2 · hr · °F.

Heat transferred for heat exchanger applications is predominently a combination of conduction and convection expressed as:

$$q = U_0 A \Delta T_m \tag{6-10}$$

where

q = rate of heat flow by convection, Btu/hr
U_0 = is the overall heat transfer coefficient Btu/ft^2 · hr · °F
A = is the area of the tubes in square feet
ΔT_m = is the logarithmic means temperature difference and represents the situation where the temperature of two fluids change as they transverse the surface.

$$\Delta T_m = \frac{\Delta T_1 - \Delta T_2}{\text{Log}_e\,[\Delta T_1/\Delta T_2]} \tag{6-11}$$

To understand the different logarithmic mean temperature relationships, Fig. 6-4 should be used. Referring to Fig. 6-4, the ΔT_m for the counterflow heat exchanger is:

$$\Delta T_m = \frac{(t_1 - t_2') - (t_2 - t_1')}{\text{Log}_e\,[t_1 - t_2'/t_2 - t_1']}. \tag{6-12a}$$

The ΔT_m for the parallel flow heat exchanger is:

$$\Delta T_m = \frac{(t_1 - t_1') - (t_2 - t_2')}{\text{Log}_e\,[t_1 - t_1'/t_2 - t_2']}. \tag{6-12b}$$

A. Counterflow

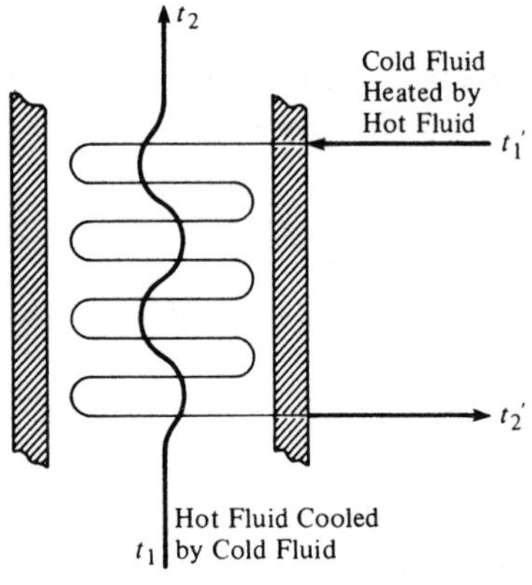

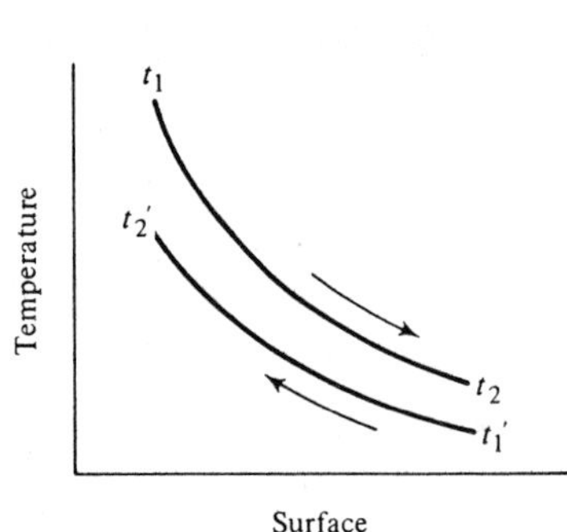

B. Parallel Flow

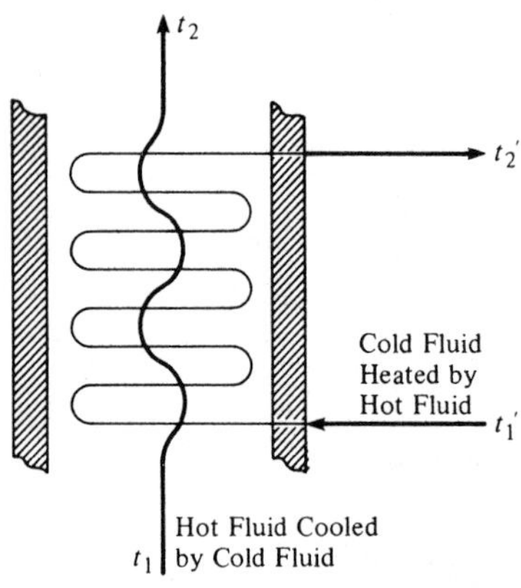

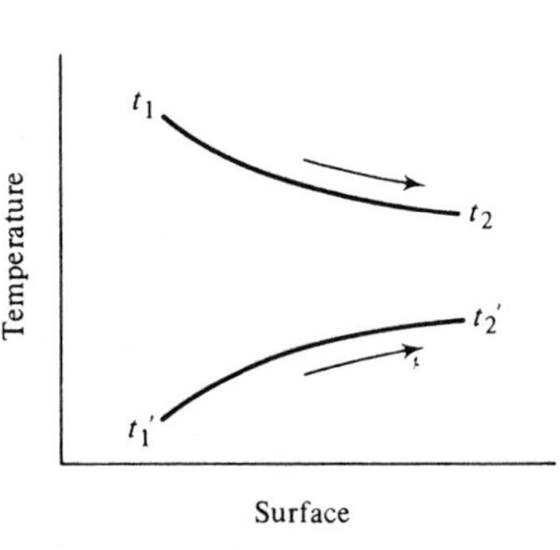

Fig. 6-4 Temperature relationships for heat exchangers.

Finned-Tube Heat Exchangers

When waste heat in exhaust gases is recovered for heating liquids for purposes such as providing domestic hot water, for heating the feed-water for steam boilers, or for hot water space heating, the finned-tube heat exchanger is generally used. Round tubes are connected together in bundles to contain the heated liquid and fins are welded or otherwise attached to the outside of the tubes to provide addi-

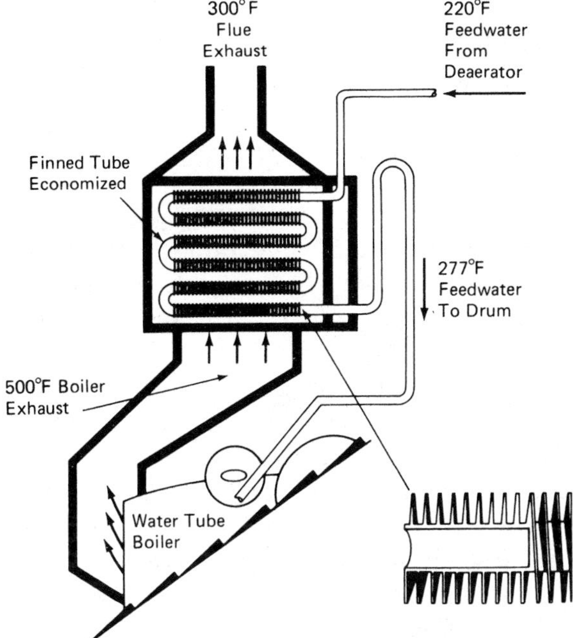

Fig. 6-5 Finned tube gas to liquid regenerator (economizer).

tional surface area for removing the waste heat in the gases. Figure 6-5 shows the usual arrangement for the finned-tube exchanger positioned in a duct and details of a typical finned-tube construction. This particular type of application is more commonly known as an economizer. The tubes are often connected all in series but can also be arranged in series-parallel bundles to control the liquid side pressure drop. The air side pressure drop is controlled by the spacing of the tubes and the number of rows of tubes within the duct. Finned-tube exchangers are available prepackaged in modular sizes or can be made up to custom specifications very rapidly from standard components. Temperature control of the heated liquid is usually provided by a bypass duct arrangement which varies the flow rate of hot gases over the heat exchanger. Materials for the tubes and the fins can be selected to withstand corrosive liquids and/or corrosive exhaust gases.

Finned-tube heat exchangers are used to recover waste heat in the low to medium temperature range from exhaust gases for heating liquids. Typical applications are domestic hot water heating, heating boiler feedwater, hot water space heating, absorption-type refrigeration or air conditioning, and heating process liquids.

Shell and Tube Heat Exchanger

When the medium containing waste heat is a liquid or a vapor which heats another liquid, then the shell and tube heat exchanger must be used since both paths must be sealed to contain the pressures of their respective fluids. The shell contains the tube bundle, and usually internal baffles, to direct the fluid in the shell over the tubes in multiple passes. The shell is inherently weaker than the tubes so that the higher pressure fluid is circulated in the tubes while the lower pressure fluid flows through the shell. When a vapor contains the waste heat, it usually condenses, giving up its latent heat to the liquid being heated. In this application, the vapor is almost invariably contained within the shell. If the reverse is attempted, the condensation of vapors within small diameter parallel tubes causes flow instabilities. Tube and shell heat exchangers are available in a wide range of standard sizes with many combinations of materials for the tubes and shells.

Typical applications of shell and tube heat exchangers include heating liquids with the heat contained by condensates from refrigeration and air conditioning systems; condensate from process steam; coolants from furnace doors, grates, and pipe supports; coolants from engines, air compressors, bearings, and lubricants; and the condensates from distillation processes.

HOW TO ESTIMATE THE HEAT LOSS OF A VESSEL OR TANK

Heat loss calculations from a vessel or tank are complex, since conduction, convection, and radiation flows occur simultaneously. Figure 6-6 indicates the heat loss associated with a tank. As indicated, the total heat loss is the sum of the following:

$$q = q_1 + q_2 + q_3 + q_4 \tag{6-13}$$

where

q_1 is the heat loss from liquid in the tank through the tank sidewalls to atmosphere (Btuh)

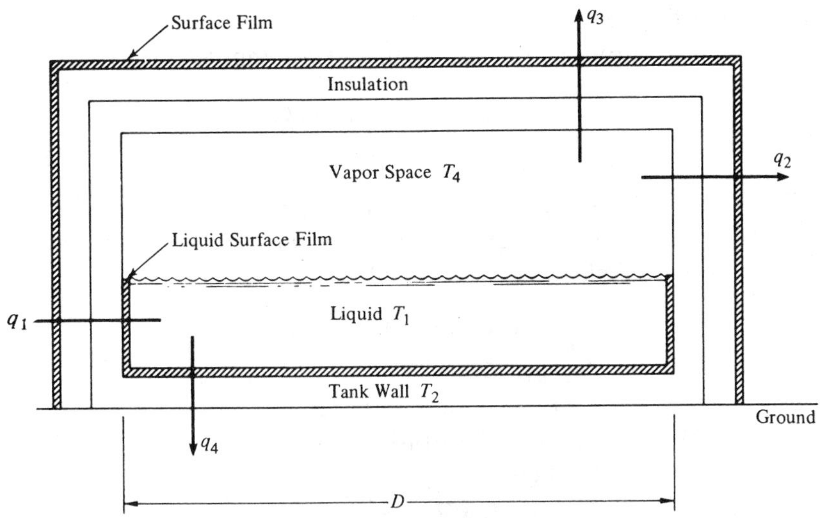

Fig. 6-6 Losses from tank or vessel.

q_2 is the heat loss from vapor in the tank through the sidewalls to atmosphere (Btuh)

q_3 is the heat loss from the vapors in the tank through the roof to atmosphere

q_4 is the heat loss from the bottom of the tank to the ground (Btuh).

To simplify calculations, the following is assumed:

1. Conductances of tank walls are neglected. (For metal tanks, as indicated in SIM 6-2, this assumption is valid. For other materials, the tank resistance should be added.)
2. Liquid and vapor surface conductances are neglected for uninsulated tanks. Heat loss is based on air surface conductance.
3. Liquid and vapor conductances are neglected for insulated tanks. Heat loss is based on insulation conductance and surface air conductance.

A simplified procedure[15] to determine approximate heat losses will be illustrated. This procedure simplifies computations of the various film coefficients. As an approximation, Fig. 6-6 is used to compute the heat loss for uninsulated tanks based on the surface film conductances. This figure is used to compute q_1, q_2 and q_3 by equating q_t, to the q in question. The total heat loss is computed by multiplying the heat loss from Fig. 6-6 by the surface area. Another term, ΔT_W, is introduced in this figure. To compute ΔT_W:

$$\Delta T_W = (T_W - T_A)W \tag{6-14}$$

where

T_W is the fluid temperature, °F, T_A is the ambient temperature 70°F, and W is the correction factor which takes into account the process fluid.

W is defined as

Process Fluid	*W*
Agitated Liquid	1
Aqueous Solution	0.9
Fuel Oil	0.7
Condensing Vapor	1.0
Non-condensing Vapor	0.2
Fouling Liquids	0.4

Figure 6-7 is based on an ambient air temperature of 70°F, a surface emissivity of 0.9, and a wind velocity of zero. To use Fig. 6-6 for other values, correct each term separately.

$$q_a \text{ (radiation)} = \frac{\text{New Emissivity}}{0.90}$$

$$\times \text{ value from graph (use for other than 0.9)} \tag{6-15}$$

$$q_c \text{ (convection)} = (1.28 \times \text{V} + 1)^{1/2}$$

$$\times \text{ value from graph (use for other than 0)} \tag{6-16}$$

$$q_t = q_a + q_c \tag{6-17}$$

Heat Loss To Ground

The last portion of Equation 6-13 deals with heat losses from the bottom of the tank when it rests on the ground. It is defined as:

$$q_4 = 2DK_4 \, (T_W - T_3) \tag{6-18}$$

where

D is the diameter of the tank in ft
K_4 is the conductance to ground

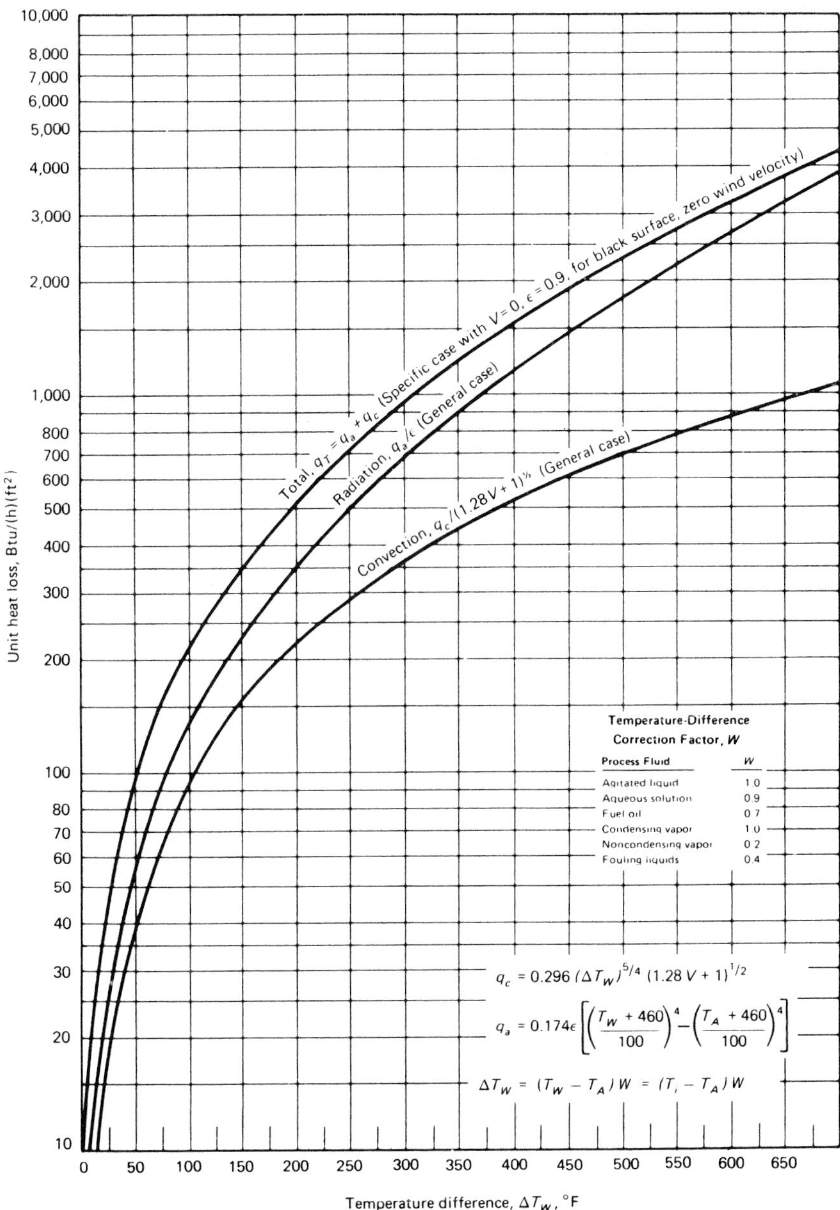

Fig. 6-7 Heat loss for uninsulated tanks and vessels. [Reprinted by special permission from *Chemical Engineering* (May 27, 1974) Copyright © 1974 by McGraw-Hill, Inc., New York, N.Y.]

Use K_4 = 0.8 unless otherwise known
T_W is the liquid temperature, °F
T_3 is the ground temperature, °F
Assume ambient temperatures unless the ground temperature is specifically known.

Reducing Heat Loss of Vessels or Tanks

Insulation is the common method used to reduce the heat loss from a vessel or tank. Figure 6-8 shows the effect of various insulation

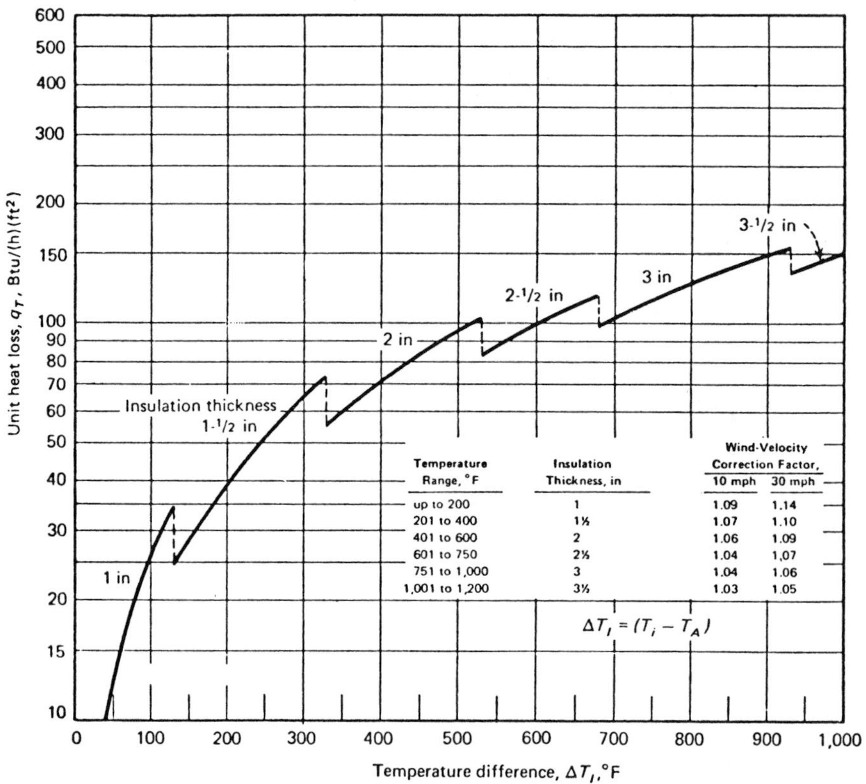

Temperature Range, °F	Insulation Thickness, in	Wind-Velocity Correction Factor, 10 mph	30 mph
up to 200	1	1.09	1.14
201 to 400	1½	1.07	1.10
401 to 600	2	1.06	1.09
601 to 750	2½	1.04	1,07
751 to 1,000	3	1.04	1.06
1,001 to 1,200	3½	1.03	1.05

INSULATED tanks, covered with calcium silicate, have heat losses based on negligible resistance to heat flow on process side. Values in chart are for wind velocity of zero, emissivity of 0.8, ambient air temperature of 70°F-

Fig. 6-8 Heat loss for calcium silicate insulated tanks. [Reprinted by special permission from *Chemical Engineering* (May 27, 1974) Copyright © 1974 by McGraw-Hill, Inc., New York, N.Y.]

thicknesses. This figure can be used as a guide to determine suggested economic thickness for various temperature differentials.

To use Fig. 6-8, the temperature difference factor ΔT_1 should be corrected for wind velocities other than zero. To correct ΔT_1, multiply by the correction factors for various wind velocities given below:

Wind Velocity Correction Factors

Insulation Thickness	10 mph	30 mph
1	1.09	1.14
$1\frac{1}{2}$	1.07	1.10
2	1.06	1.09
$2\frac{1}{2}$	1.04	1.07
3	1.04	1.06
$3\frac{1}{2}$	1.03	1.05

SIM 6-4

A 15 ft diameter by 15 ft high carbon tank, supported at grade, is used to contain 250°F non-agitated process fluid. The fluid level is 10 ft high. Calculate the annual savings by adding $1\frac{1}{2}$ in. insulation, given the following:

Plant operation–8760 hr/yr
Wind velocity–10 mph
Ambient temperature–70°
Emissivity of tank–0.20
Vapor space has non-condensable vapors
Heat content of steam–926.1 Btu/lb
Steam cost–$6/1000 lb

ANSWER

Uninsulant Tank

Wetted Area of Tank

$A_1 = \pi DH = \pi \times 15 \times 10 = 471 \text{ ft}^2$

$\Delta T = (250 - 70) = 180°\text{F}$

$\Delta T_W = 0.9 \times 180 = 162$

From Fig. 6-7

$q_c' = q_c \times (1.28 \times V + 1)^{1/2} = 180 \times 3.71 = 667.8 \text{ Btu/h} \cdot \text{ft}^2$

$q_a' = q_a \times \dfrac{0.2}{0.9} = 250 \times 0.222 = 55.5 \text{ Btu/h} \cdot \text{ft}^2$

$q_T = 667.8 + 55.5 = 723.3 \text{ Btu/h} \cdot \text{ft}^2$

$q_1 = 471 \times 723.3 = 340{,}674 \text{ Btuh.}$

Heat Loss For Vapor Space

$A_2 = \pi \times 15\,(15 - 10) = 235.6\ \text{ft}^2$

$\Delta T_W = \Delta T \times 0.20 = 36$

$q'_c = 25 \times 3.71 = 92.75\ \text{Btu/h} \cdot \text{ft}^2$

$q'_a = 35 \times 0.222 = 7.77\ \text{Btu/h} \cdot \text{ft}^2$

$q_T = 100.52\ \text{Btuh}$

$q_2 = 235.6 \times 100.52 = 23{,}682\ \text{Btuh.}$

Heat Loss For Roof

$$A = \frac{\pi D^2}{4} = 176.7\ \text{ft}^2$$

$q_3 = 176.6 \times 100.52 = 17{,}761\ \text{Btuh.}$

Heat Loss For Tank Bottom

$q_4 = 2 \times 15 \times 0.8\,(250 - 70) = 4{,}320\ \text{Btuh.}$

Total Heat Loss = 386,437 Btuh

Insulated Tank

$\Delta T = (250 - 70) = 180°\text{F}$

From Fig. 6-8,

$q_T = 35\ \text{Btu/h} \cdot \text{ft}^2$

$q'_T = 35 \times 1.07 = 37.45\ \text{Btu/h} \cdot \text{ft}^2$

Since the total tank is insulated,

$q = 37.45\,(A_1 + A_2 + A_3) = 37.45\,(471 + 235.6 + 176.7) = 33{,}079\ \text{Btuh}$

Total heat loss = 33,079 + 4,320 = 37,339 Btuh

Energy savings with insulation = 349,038 Btuh

$$\text{Yearly energy savings by reducing steam usage} = \frac{349{,}038 \times 8760}{926.1} \times \frac{\$6.00}{1000}$$

= \$19,809/year

HOW TO ESTIMATE THE HEAT LOSS OF PIPING AND FLAT SURFACES

A simple method used to estimate heat losses for horizontal bare steel pipes and flat surfaces is to use Table 6-2. By knowing the temperature difference between the pipe surface and the ambient air, the heat loss is determined.

SIM 6-5

An existing process indicated in Scheme 1 of Fig. 6-9, uses a heat exchanger to cool the discharge of the column before it enters the sewer. The traditional process supplies fluid directly to Column #1 at 104°F. The column is used as a stripper with 30 psig steam.

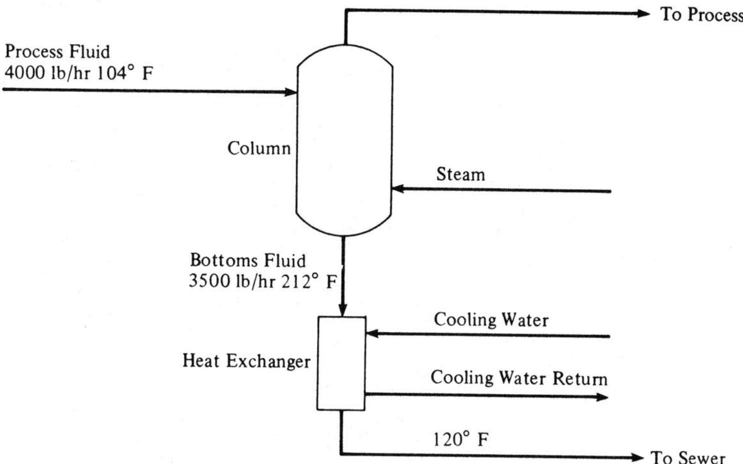

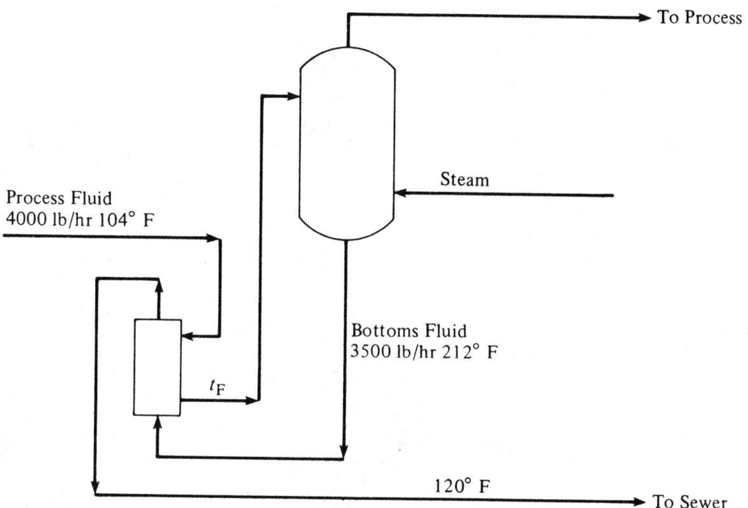

Fig. 6-9 Process schemes for Ajax plant.

TABLE 6-2 Heat Losses for Horizontal Bare Steel Pipe and Flat Surfaces.

In Btu per (Sq Ft of Pipe Surface) (Hour) (F Deg Temperature Difference Between Pipe and Air)

Pipe size inches	Linear Foot Factor	Temperature Difference F Deg Between Pipe Surface and Surrounding Air at 80 F																			
		50	100	150	200	250	300	350	400	450	500	550	600	650	700	750	800	850	900	950	1000
1/2	0.220	2.12	2.48	2.80	3.10	3.42	3.74	4.07	4.47	4.86	5.28	5.72	6.19	6.69	7.22	7.79	8.39	9.03	9.70	10.42	11.18
3/4	0.275	2.08	2.43	2.74	3.04	3.35	3.67	4.00	4.40	4.79	5.21	5.65	6.12	6.61	7.15	7.71	8.31	8.95	9.62	10.34	11.09
1	0.344	2.04	2.38	2.69	2.99	3.30	3.61	3.94	4.33	4.72	5.14	5.58	6.05	6.54	7.07	7.64	8.23	8.87	9.55	10.26	11.02
1 1/4	0.435	2.00	2.34	2.64	2.93	3.24	3.55	3.88	4.27	4.66	5.07	5.51	5.97	6.47	7.00	7.56	8.16	8.79	9.47	10.18	10.94
1 1/2	0.497	1.98	2.31	2.61	2.90	3.20	3.52	3.84	4.23	4.62	5.03	5.47	5.93	6.43	6.96	7.52	8.12	8.75	9.43	10.14	10.89
2	0.622	1.95	2.27	2.56	2.85	3.15	3.46	3.78	4.17	4.56	4.97	5.41	5.87	6.37	6.89	7.45	8.05	8.68	9.36	10.07	10.82
2 1/2	0.753	1.92	2.23	2.52	2.81	3.11	3.42	3.74	4.12	4.51	4.92	5.36	5.82	6.31	6.84	7.40	7.99	8.63	9.30	10.01	10.77
3	0.916	1.89	2.20	2.49	2.77	3.07	3.37	3.69	4.08	4.46	4.87	5.31	5.77	6.26	6.79	7.35	7.94	8.57	9.25	9.96	10.71
3 1/2	1.047	1.87	2.18	2.46	2.74	3.04	3.34	3.66	4.05	4.43	4.84	5.27	5.73	6.23	6.75	7.31	7.91	8.54	9.21	9.92	10.67
4	1.178	1.85	2.16	2.44	2.72	3.01	3.32	3.64	4.02	4.40	4.81	5.25	5.71	6.20	6.72	7.28	7.87	8.51	9.18	9.89	10.64
4 1/2	1.309	1.84	2.14	2.42	2.70	2.99	3.30	3.61	4.00	4.38	4.79	5.22	5.68	6.17	6.69	7.25	7.85	8.48	9.15	9.86	10.61
5	1.456	1.83	2.13	2.40	2.68	2.97	3.28	3.59	3.97	4.35	4.76	5.20	5.65	6.15	6.68	7.23	7.82	8.45	9.12	9.83	10.58
6	1.734	1.80	2.10	2.37	2.65	2.94	3.24	3.55	3.94	4.32	4.72	5.16	5.61	6.10	6.63	7.19	7.78	8.41	9.08	9.79	10.54
7	1.996	1.79	2.08	2.35	2.63	2.91	3.21	3.53	3.91	4.29	4.69	5.13	5.58	6.07	6.60	7.15	7.75	8.38	9.05	9.76	10.51
8	2.258	1.77	2.06	2.33	2.60	2.89	3.19	3.50	3.88	4.26	4.67	5.10	5.56	6.05	6.57	7.12	7.72	8.35	9.02	9.73	10.48
9	2.520	1.76	2.05	2.31	2.59	2.87	3.17	3.48	3.86	4.24	4.65	5.08	5.53	6.02	6.54	7.10	7.69	8.32	8.99	9.70	10.45
10	2.814	1.75	2.03	2.30	2.57	2.85	3.15	3.46	3.84	4.22	4.62	5.05	5.51	6.00	6.52	7.08	7.67	8.30	8.97	9.68	10.43
12	3.338	1.73	2.01	2.27	2.54	2.83	3.12	3.43	3.81	4.19	4.59	5.02	5.48	5.96	6.48	7.04	7.63	8.26	8.93	9.64	10.39
14	3.655	1.72	2.00	2.26	2.53	2.81	3.11	3.41	3.79	4.17	4.57	5.00	5.46	5.94	6.47	7.02	7.61	8.24	8.91	9.62	10.37
16	4.189	1.70	1.98	2.24	2.51	2.79	3.08	3.39	3.77	4.14	4.55	4.98	5.43	5.92	6.44	6.99	7.59	8.21	8.88	9.59	10.34
18	4.717	1.69	1.96	2.22	2.49	2.77	3.07	3.37	3.75	4.12	4.53	4.96	5.41	5.90	6.42	6.97	7.56	8.19	8.86	9.57	10.32
20	5.236	1.68	1.95	2.21	2.47	2.75	3.05	3.36	3.73	4.11	4.51	4.94	5.39	5.88	6.40	6.95	7.54	8.17	8.84	9.55	10.29
24	6.283	1.66	1.93	2.19	2.45	2.73	3.02	3.33	3.70	4.07	4.48	4.90	5.36	5.84	6.36	6.92	7.51	8.14	8.80	9.51	10.26

Vertical Surface	1.84	2.14	2.42	2.70	3.00	3.30	3.62	4.00	4.38	4.79	5.22	5.68	6.17	6.70	7.26	7.85	8.48	9.15	9.86	10.62
Horizontal Surface Facing Upward	2.03	2.37	2.67	2.97	3.28	3.59	3.92	4.31	4.70	5.12	5.56	6.02	6.52	7.05	7.61	8.21	8.85	9.52	10.24	10.99
Horizontal Surface Facing Downward	1.61	1.86	2.11	2.36	2.64	2.93	3.23	3.60	3.97	4.37	4.80	5.25	5.73	6.25	6.80	7.39	8.02	8.69	9.39	10.14

Notes:

1. To find losses per linear foot, multiply square foot losses by factors in column 2.
2. Area of flat surfaces are four square feet or more.
3. For pipe sizes larger than 24 inches, use the losses for 24 inch pipe.

Reprinted by permission from *ASHRAE Handbook of Fundamentals* 1972.

Scheme 2 illustrates a proposed modification. The proposed scheme replaces the bottom fluid exchanger with a larger exchanger and uses the bottom fluid to heat the process liquid; the bottom fluid in turn will be cooled. What recommendations should be made, given the following data?

The installed cost of heat exchanger #1 is $9000 with a 50 ft^2 area. The conductance U of the heat exchanger = 100

30 psig steam is used
Plant operates 8760 hours/yr
Steam cost $6.00/1000#

Note: To compute the cost for an exchanger of increased area, use the following approximation:

$$\text{New cost} = \left[\frac{\text{new area}}{\text{original area}}\right]^{0.6} \times \text{original price.} \tag{6-19}$$

Analysis

The first analysis indicates that the hot bottom fluid can be used to heat the process fluid.

The amount of heat supplied to the process fluid is

$$q = MCp\ \Delta T = 3500\,\#/\text{hr} \times 1(212 - 120) = 322{,}000 \text{ Btuh}$$

30 psig = 30 + 14.7 = 44.7 psia steam
From Steam Table 12-20, h_{fg} = 928.6 Btu/#

Energy Savings of Stripper

$$\#/\text{hr of 30 psig steam saved} = \frac{322{,}000 \text{ Btuh}}{928.6 \text{ Btu}/\#} = 347\ \#/\text{hr}$$

$$\text{Steam savings} = \frac{\$6.00}{1000} \times 347 \times 8760 = \$18{,}237$$

The temperature of the process fluid will rise to:

$$MCp\ \Delta T \text{ (process)} = 322{,}000 \text{ Btuh}$$

$$\Delta t = \frac{322{,}000}{4000 \times 1} = 80^\circ\text{F}$$

$$t_f = 104 + 80 = 184^\circ\text{F.}$$

The required area of heat exchanger will increase from Scheme 1

$$q = U_0 A\ \Delta T_m = 100\, A \Delta T_m = 322{,}000 \tag{6-20}$$

$$\Delta T_m = \frac{(212 - 184) - (120 - 104)}{\log_e \left(\dfrac{212 - 184}{120 - 104}\right)}$$

$$\Delta T_m = \frac{82 - 16}{\log_e \left(\frac{28}{16}\right)} = \frac{12}{\log_e (1.75)} = 21.4$$

$$A = \frac{3220}{21.4} = 150.4 \text{ ft}^2$$

$$\text{New installed cost} = \left(\frac{\text{area new}}{\text{area old}}\right)^{0.6} \times \text{old price}$$

$$= \left(\frac{150.4}{50}\right)^{0.6} \times \$9000 = \$17{,}400.$$

With an annual energy savings of $18,237, the investment in additional heat exchanger capacity is desirable.

7

Reducing Building Energy Losses

ENERGY LOSSES DUE TO HEAT LOSS AND HEAT GAIN

Depending on the time of year, a heat loss or a heat gain wastes energy. For example, a heat loss during the winter means wasted energy in heating the building. Similarly, during the summer months, a heat gain means wasted energy in cooling the building. The building construction affects the heat loss and heat gain. Fig. 7-1 illustrates the total heat gain of the building. The flow of heat is always from one temperature to a colder temperature. The heat loss of a building is illustrated by Fig. 7-2. In this case, the building is considered the "hot body."

In the context of this book, heat loss refers to heating loads, while heat gain refers to cooling loads. By considering building materials and constructions, the associated heat loss and heat gains can be reduced.

In this chapter, you will see: how to apply handy Building Construction Tables to solve most heat transfer problems, how substitutions of building materials saves energy, and how to apply different types of glass to save energy.

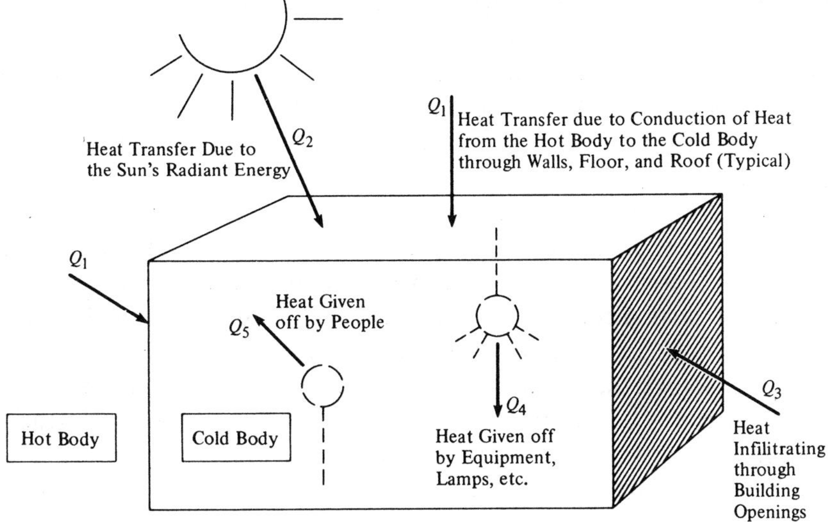

$$\text{Heat Gain} = Q_1 + Q_2 + Q_3 + Q_4 + Q_5$$

Fig. 7-1 Heat gain of a building.

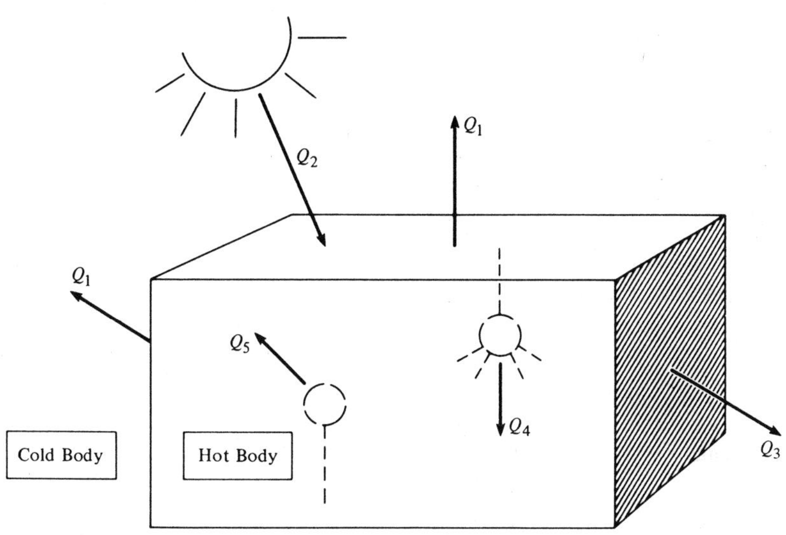

$$\text{Heat Loss} = Q_1 + Q_3 - Q_2 - Q_4 - Q_5$$

Fig. 7-2 Heat loss of a building.

CONDUCTIVITY THROUGH BUILDING MATERIALS

The best conductors of heat are metals. Insulations such as wood, asbestos, cork, and felt are poor conductors. Conductance is widely used because many materials used in the construction of buildings are non-homogeneous.

Table 7-1 should be used when the surface conductance, which is the transfer of heat from air to a surface or from a surface to air, is required. Table 7-1 also lists air space conductances for various positions and heat directions. Note that the conductance will change from the heat gain case to the heat loss case. The data for a 3/4 in. air space shows the insulating properties of air. When one surface of the air space is covered with aluminum foil, the resistance increases; this, in turn, will reduce the heat loss.

The heat transfer coefficients for various building materials are given in Table 7-2.

TABLE 7-1 Thermal Resistances for Surface Films and Air Spaces.

Medium & Position	Direction of Heat Flow	Building Materials: wood, paper, glass, masonry $\epsilon = 0.90$	Galvanized Steel $\epsilon = 0.20$	Aluminum Foil $\epsilon = 0.05$
A. Surface Films		*R*	*R*	*R*
1. Still Air				
a) Horizontal	Up	0.61	1.10	1.32
b) Horizontal	Down	0.92	2.70	4.55
c) Vertical	Horizontal	0.68	1.85	1.70
2. Moving Air				
a) 15 mph Wind (Winter)	Any	0.17	–	–
b) 7.5 mph Wind (Summer)	Any	0.25	–	–
B. Air Space 3/4″		$\epsilon = 0.82$	$\epsilon = 0.20$	$\epsilon = 0.05$
1. Air Mean Temp. 90°F/0°F.**		*R*	*R*	*R*
a) Horizontal	Up	0.76/1.02*	1.63/1.78	2.26/2.16
b) Horizontal	Down	0.84/1.31	2.08/2.88	3.25/4.04
c) Slope 45°	Up	0.81/1.13	1.90/2.13	2.81/2.71
d) Slope 45°	Down	0.84/1.31	2.09/2.88	3.24/4.04
e) Vertical	Horizontal	0.84/1.28	2.10/2.73	3.28/3.76

*Assume an average ΔT of 10°F.

**For resistance at temperatures other than 90°F or 0°F, interpolate between the two values–typical.

Reprinted by permission from *ASHRAE Handbook of Fundamentals* 1972.

TABLE 7-2 Heat Transfer Coefficients of Building Materials.*

MATERIAL	DESCRIPTION	CONDUCTIVITY K#	CONDUCTANCE C +
BUILDING BOARDS	ASBESTOS-CEMENT BOARD	4.0	
	GYPSUM OR PLASTER BOARD...1/2 IN.		2.25
	PLYWOOD	0.80	
	PLYWOOD...3/4 IN.		1.07
	SHEATHING (IMPREGNATED OR COATED)	0.38	
	SHEATHING (IMPREGNATED OR COATED) 25/32 IN.		0.49
	WOOD FIBER—HARDBOARD TYPE	1.40	
INSULATING MATERIALS	BLANKET AND BATT:		
	MINERAL WOOL FIBERS (ROCK, SLAG, OR GLASS)	0.27	
	WOOD FIBER	0.25	
	BOARDS AND SLABS:		
	CELLULAR GLASS	0.39	
	CORKBOARD	0.27	
	GLASS FIBER	0.25	
	INSULATING ROOF DECK...2 IN.		0.18
MASONRY MATERIALS	LOOSE FILL:		
	MINERAL WOOL (GLASS, SLAG, OR ROCK)	0.27	
	VERMICULITE (EXPANDED)	0.46	
	CONCRETE:		
	CEMENT MORTAR	5.0	
	LIGHTWEIGHT AGGREGATES, EXPANDED SHALE, CLAY, SLATE, SLAGS; CINDER; PUMICE; PERLITE; VERMICULITE	1.7	
	SAND AND GRAVEL OR STONE AGGREGATE	12.0	
	STUCCO	5.0	
	BRICK, TILE, BLOCK, AND STONE:		
	BRICK, COMMON	5.0	
	BRICK, FACE	9.0	
	TILE, HOLLOW CLAY, 1 CELL DEEP, 4 IN.		0.90
	TILE, HOLLOW CLAY, 2 CELLS, 8 IN.		0.54
	BLOCK, CONCRETE, 3 OVAL CORE:		
	SAND & GRAVEL AGGREGATE...4 IN.		1.40
	SAND & GRAVEL AGGREGATE...8 IN.		0.90
	CINDER AGGREGATE.............4 IN.		0.90
	CINDER AGGREGATE.............8 IN.		0.58
	STONE, LIME OR SAND	12.50	
PLASTERING MATERIALS	CEMENT PLASTER, SAND AGGREGATE	5.0	
	GYPSUM PLASTER:		
	LIGHTWEIGHT AGGREGATE...1/2 IN.		3.12
	LT. WT. AGG. ON METAL LATH...3/4 IN.		2.13
	PERLITE AGGREGATE	1.5	
	SAND AGGREGATE	5.6	
	SAND AGGREGATE ON METAL LATH 3/4 IN.		7.70
	VERMICULITE AGGREGATE	1.7	
ROOFING	ASPHALT ROLL ROOFING		6.50
	BUILT-UP ROOFING...3/8 IN.		3.00
SIDING MATERIALS	ASBESTOS-CEMENT, 1/4 IN. LAPPED		4.76
	ASPHALT INSULATING (1/2 IN. BOARD)		0.69
	WOOD, BEVEL, 1/2 X 8, LAPPED		1.23
WOODS	MAPLE, OAK, AND SIMILAR HARDWOODS	1.10	
	FIR, PINE, AND SIMILAR SOFTWOODS	0.80	
	FIR, PINE & SIM. SOFTWOODS 25/32 IN.		1.02

*Extracted with permission from ASHRAE Guide and Data Book, 1965.
#Conductivity given in Btu in. per hr sq ft F
+Conductance given in Btu per hr sq ft F

Courtesy of the Trane Company.

SIM 7-1

Calculate the heat loss through 10,000 ft^2 of building wall, as indicated by Fig. 7-3.

Assume a temperature differential of 17°F. When using Table 7-2, remember: $R = d/K$

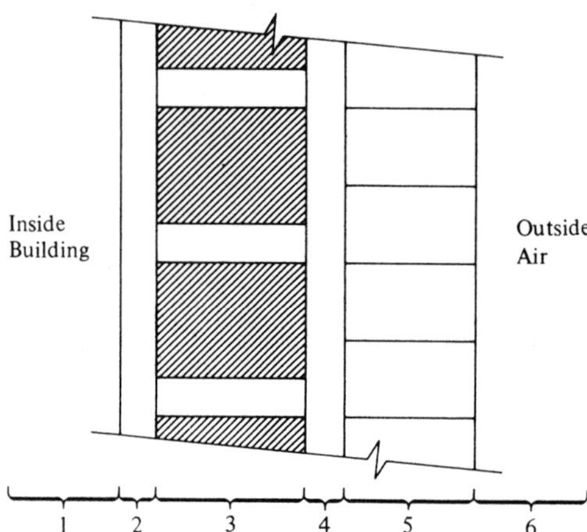

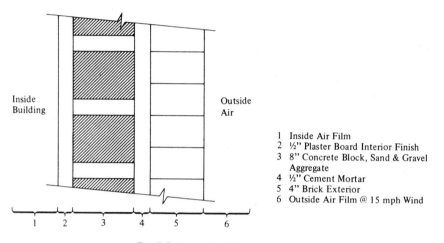

1 Inside Air Film
2 ½" Plaster Board Interior Finish
3 8" Concrete Block, Sand & Gravel Aggregate
4 ½" Cement Mortar
5 4" Brick Exterior
6 Outside Air Film @ 15 mph Wind

Fig. 7-3 Typical wall construction.

ANSWER

Item	*Description*	*Resistance*	*Reference*
1	Outside air film at 15 mph	0.17	Table 7-1
2	4″ brick	0.44	Table 7-2 $R = 4/9$
3	Mortar	0.10	Table 7-2 $R = ½/5$
4	Block	1.11	Table 7-2 $R = 1/.9$
5	Gypsum	0.44	Table 7-2 $R = 1/2.25$
6	Inside film	0.68	Table 7-1
	Total resistance	2.94	

$U = 1/R = 0.34$

$q = U A \Delta T$

$= 0.34 \times 10{,}000 \times 17 = 57{,}800$ Btu/h.

SIM 7-2

In order to reduce the heat loss through the building wall, the engineer is recommending substituting cider aggregate for the 8″ concrete block wall. (Assume the structural considerations are the same.) Comment on the recommendation.

ANSWER

The conductance of the cinder aggregate concrete block wall is 0.58, as compared to 0.90, thus the resistance will increase from 1.11 to 1.72. The effect is to increase the overall resistance of the wall to 3.54.

$U = 1/R$ - 0.28
q = 0.28 × 10,000 × 17 = 47,600 Btuh or 10,200 Btuh are saved with the change in wall construction. The total system should be evaluated from an energy conservation viewpoint.

SIM 7-3

Calculate the heat gain during the summer for the construction of SIM 7-1.

ANSWER

The only item which changes is the outside air film resistance:

Item	*Description*	*Resistance*
1	Outside air film at 7½ mps	0.25
Thus		
Total resistance =		3.02

$$U = \frac{1}{3.02} = 0.33$$

q = 56,100 Btuh.

Handy Tables to Save Time

Energy analysis of composite building constructions can be simplified by using Tables 7-3 through 7-13. These tables have been reprinted with the permission of ASHRAE and the Trane Company. To use them simply find the column and row which describes the type of construction. The corresponding coefficient of transmission (conductance) is then found.

Table 7-14 is another useful table which illustrates design temperatures for various cities in the United States.

TABLE 7-3 Coefficients of Transmission (U) of Solid Masonry Walls.

Coefficients are expressed in Btu per (hour) (square foot) (Fahrenheit degree difference in temperature between the air on the two sides), and are based on an outside wind velocity of 15 mph

EXTERIOR CONSTRUCTION[3]		NONE	INTERIOR FINISH										NO.
			PLAS. 5/8 IN. ON WALL		METAL LATH AND 3/4 IN. PLAS. ON FURRING		GYPSUM LATH (3/8 IN.) AND 1/2 IN. PLAS. ON FURRING			INSUL. BD. LATH (1/2 IN.) AND 1/2 IN. PLAS. ON FURRING		WOOD LATH AND 1/2 IN. PLAS.	
			(SAND AGG.)	(LT. WT. AGG.)	(SAND AGG.)	(LT. WT. AGG.)	NO PLAS.	(SAND AGG.)	(LT. WT. AGG.)	NO PLAS.	(SAND AGG.)	(SAND AGG.)	
RESISTANCE →			0.11	0.39	0.13	0.47	0.32	0.41	0.64	1.43	1.52	0.40	
MATERIAL	R	U	U	U	U	U	U	U	U	U	U	U	
		A	B	C	D	E	F	G	H	I	J	K	
BRICK (FACE AND COMMON)[4]													
(6 IN.)	0.61	0.68	0.64	0.54	0.39	0.34	0.36	0.35	0.33	0.26	0.25	0.35	1
(8 IN.)	1.24	0.48	0.45	0.41	0.31	0.28	0.30	0.29	0.27	0.22	0.22	0.29	2
(12 IN.)	2.04	0.35	0.33	0.30	0.25	0.23	0.24	0.23	0.22	0.19	0.19	0.23	3
(16 IN.)	2.84	0.27	0.26	0.25	0.21	0.19	0.20	0.20	0.19	0.16	0.16	0.20	4
BRICK (COMMON ONLY)													
(8 IN.)	1.60	0.41	0.39	0.35	0.28	0.26	0.27	0.26	0.25	0.21	0.20	0.26	5
(12 IN.)	2.40	0.31	0.30	0.27	0.23	0.21	0.22	0.22	0.21	0.18	0.17	0.22	6
(16 IN.)	3.20	0.25	0.24	0.23	0.19	0.18	0.19	0.18	0.18	0.16	0.15	0.18	7
STONE (LIME AND SAND)													
(8 IN.)	0.64	0.67	0.63	0.53	0.39	0.34	0.36	0.35	0.32	0.26	0.25	0.35	8
(12 IN.)	0.96	0.55	0.52	0.45	0.34	0.31	0.32	0.31	0.29	0.24	0.23	0.31	9
(16 IN.)	1.28	0.47	0.45	0.40	0.31	0.28	0.29	0.28	0.27	0.22	0.22	0.29	10
(24 IN.)	1.92	0.36	0.35	0.32	0.26	0.24	0.25	0.24	0.23	0.19	0.19	0.24	11
HOLLOW CLAY TILE													
(8 IN.)	1.85	0.36	0.36	0.32	0.26	0.24	0.25	0.25	0.23	0.20	0.19	0.25	12
(10 IN.)	2.22	0.33	0.31	0.29	0.24	0.22	0.23	0.22	0.21	0.18	0.18	0.23	13
(12 IN.)	2.50	0.30	0.29	0.27	0.22	0.21	0.22	0.21	0.20	0.17	0.17	0.21	14

POURED CONCRETE													
30 LB. PER CU FT													
(4 IN.)	4.44	0.19	0.19	0.18	0.16	0.15	0.15	0.15	0.14	0.13	0.13	0.15	15
(6 IN.)	6.66	0.13	0.13	0.13	0.12	0.11	0.11	0.11	0.11	0.10	0.10	0.11	16
(8 IN.)	8.88	0.10	0.10	0.10	0.09	0.09	0.09	0.09	0.09	0.08	0.08	0.09	17
(10 IN.)	11.10	0.08	0.08	0.08	0.08	0.07	0.08	0.08	0.07	0.07	0.07	0.08	18
80 LB. PER CU FT													
(6 IN.)	2.40	0.31	0.30	0.27	0.23	0.21	0.22	0.22	0.21	0.18	0.17	0.22	19
(8 IN.)	3.20	0.25	0.24	0.23	0.19	0.18	0.19	0.18	0.18	0.16	0.15	0.18	20
(10 IN.)	4.00	0.21	0.20	0.19	0.17	0.16	0.16	0.16	0.15	0.14	0.14	0.16	21
(12 IN.)	4.80	0.18	0.17	0.17	0.15	0.14	0.14	0.14	0.14	0.12	0.12	0.14	22
140 LB. PER CU FT													
(6 IN.)	0.48	0.75	0.69	0.58	0.41	0.36	0.38	0.37	0.34	0.27	0.26	0.37	23
(8 IN.)	0.64	0.67	0.63	0.53	0.39	0.34	0.36	0.35	0.32	0.26	0.25	0.35	24
(10 IN.)	0.80	0.61	0.57	0.49	0.36	0.32	0.34	0.33	0.31	0.25	0.24	0.33	25
(12 IN.)	0.96	0.55	0.52	0.45	0.34	0.31	0.32	0.31	0.29	0.24	0.23	0.31	26
CONCRETE BLOCK													
(GRAVEL AGG.) (8 IN.)	1.11	0.52	0.48	0.43	0.33	0.29	0.31	0.30	0.28	0.23	0.22	0.30	27
(12 IN.)	1.28	0.47	0.45	0.40	0.31	0.28	0.29	0.28	0.27	0.22	0.22	0.29	28
(CINDER AGG.) (8 IN.)	1.72	0.39	0.37	0.34	0.27	0.25	0.26	0.25	0.24	0.20	0.20	0.25	29
(12 IN.)	1.89	0.36	0.35	0.32	0.26	0.24	0.25	0.24	0.23	0.19	0.19	0.24	30
(LT. WT. AGG.) (8 IN.)	2.00	0.35	0.34	0.31	0.26	0.23	0.24	0.24	0.22	0.19	0.19	0.24	31
(12 IN.)	2.27	0.32	0.31	0.28	0.24	0.22	0.23	0.22	0.21	0.18	0.18	0.22	32

[3]If stucco or structural glass is applied to the exterior, the additional resistance value of 0.10 would have a negligible effect on the U value.
[4]Brick, 6 in. (5½ in. actual) is assumed to have no backing. Walls 8, 12 and 16 in. have 4 in. of face brick and balance of common brick.

Extracted with permission from 1965 ASHRAE Guide and Data Book.

Courtesy of the Trane Company.

TABLE 7-4 Coefficients of Transmission (U) of Masonry Partitions.

Coefficients are expressed in Btu per (hour) (square foot) (Fahrenheit degree difference in temperature between the air on the two sides), and are based on still air (no wind) conditions on both sides

TYPE OF PARTITION		SURFACE FINISH					
		NONE	PLAS. (LT. WT. AGG.) 5/8 IN.		PLAS. (SAND AGG.) 5/8 IN;		
RESISTANCE →			ONE SIDE 0.39	TWO SIDES 0.78	ONE SIDE 0.11	TWO SIDES 0.22	NUMBER
MATERIAL	R	U	U	U	U	U	
		A	B	C	D	E	
HOLLOW CONCRETE BLOCK							
(CINDER AGG.)							
(3 IN.)	0.86	0.45	0.38	0.33	0.43	0.41	1
(4 IN.)	1.11	0.40	0.35	0.31	0.39	0.37	2
(8 IN.)	1.72	0.32	0.29	0.26	0.31	0.30	3
(12 IN.)	1.89	0.31	0.27	0.25	0.30	0.29	4
(LT. WT. AGG.)							
(3 IN.)	1.27	0.38	0.33	0.30	0.36	0.35	5
(4 IN.)	1.50	0.35	0.31	0.27	0.34	0.32	6
(8 IN.)	2.00	0.30	0.27	0.24	0.29	0.28	7
(12 IN.)	2.27	0.28	0.25	0.23	0.27	0.26	8
(GRAVEL AGG.)							
(8 IN.)	1.11	0.40	0.35	0.31	0.39	0.37	9
(12 IN.)	1.28	0.38	0.33	0.29	0.36	0.35	10

HOLLOW CLAY TILE							
(3 IN.)	0.80	0.46	0.39	0.34	0.44	0.42	11
(4 IN.)	1.11	0.40	0.35	0.31	0.39	0.37	12
(6 IN.)	1.52	0.35	0.31	0.27	0.33	0.32	13
(8 IN.)	1.85	0.31	0.28	0.25	0.30	0.29	14
HOLLOW GYPSUM TILE							
(3 IN.)	1.35	0.37	0.32	0.29	0.35	0.34	15
(4 IN.)	1.67	0.33	0.29	0.26	0.32	0.31	16
SOLID PLASTER WALLS							
GYPSUM LATH (1/2 IN.) AND PLAS.							
3/4 IN. EACH SIDE							
(LT. WT. AGG.)	1.39	0.36	-	-	-	-	17
(SAND AGG.)	0.71	0.48	-	-	-	-	18
1 IN. EACH SIDE							
(LT. WT. AGG.)	1.73	0.32	-	-	-	-	19
(SAND AGG.)	0.81	0.46	-	-	-	-	20
METAL LATH AND PLAS.[1]							
2 IN. TOTAL THICKNESS							
(LT. WT. AGG.)	1.28	0.38	-	-	-	-	21
(SAND AGG.)	0.36	0.58	-	-	-	-	22
2 1/2 IN. TOTAL THICKNESS							
(LT. WT. AGG.)	1.60	0.34	-	-	-	-	23
(SAND AGG.)	0.45	0.55	-	-	-	-	24
GLASS AND GLASS BLOCKS	SEE TABLE 3-2						

[1]Metal Core and Supports disregarded. Plaster troweled smooth both sides.

Extracted with permission from 1965 ASHRAE Guide and Date Book.

Courtesy of the Trane Company.

TABLE 7-5 Coefficients of Transmission (U) of Frame Walls.

These coefficients are expressed in Btu per (hour) (square foot) (Fahrenheit degree difference in temperature between the air on the two sides), and are based on an outside wind velocity of 15 mph

EXTERIOR[1]			INTERIOR FINISH		TYPE OF SHEATHING[3]						NUMBER
					NONE, BUILDING PAPER	GYPSUM BOARD 1/2 IN.	PLYWOOD 5/16 IN.	WOOD 25/32 IN AND BUILDING PAPER	INSULATION BOARD SHEATHING 1/2 IN.	INSULATION BOARD SHEATHING 25/32 IN.	
			RESISTANCE →		0.06	0.45	0.39	1.04	1.32	2.06	
MATERIAL	R	AV. R	MATERIAL	R	U A	U B	U C	U D	U E	U F	
WOOD SHINGLES OVER INSUL.: BACKER BD. (5/16 IN.)	1.40	1.42[2]	NONE	—	0.43	0.37	0.38	0.30	0.28	0.23	23
ASPHALT INSUL. SIDING	1.45		GYPSUM BD. (3/8 IN.)	0.32	0.28	0.25	0.25	0.22	0.20	0.18	24
			GYPSUM LATH (3/8 IN.) AND 1/2 IN. PLAS. (LT. WT. AGG.)	0.64	0.25	0.23	0.23	0.20	0.19	0.17	25
			GYPSUM LATH (3/8 IN.) AND 1/2 IN. PLAS. (SAND AGG.)	0.41	0.27	0.24	0.25	0.21	0.20	0.18	26
			METAL LATH AND 3/4 IN. PLAS. (LT. WT. AGG.)	0.47	0.27	0.24	0.24	0.21	0.20	0.17	27
			METAL LATH AND 3/4 IN. PLAS. (SAND AGG.)	0.13	0.29	0.26	0.27	0.23	0.21	0.18	28
			INSUL. BD. (1/2 IN.)	1.43	0.21	0.20	0.20	0.18	0.17	0.15	29
			INSUL. BD. LATH (1/2 IN.) AND 1/2 IN. PLAS. (SAND AGG.)	1.52	0.21	0.19	0.19	0.17	0.16	0.15	30
			PLYWOOD (1/4 IN.)	0.31	0.28	0.25	0.25	0.22	0.20	0.18	31
			WOOD PANELS (3/4 IN.)	0.94	0.24	0.22	0.22	0.19	0.18	0.16	32
			WOOD LATH AND 1/2 IN. PLAS. (SAND AGG.)	0.40	0.27	0.24	0.25	0.21	0.20	0.18	33

			NONE	—	0.91	0.67	0.70	0.48	0.42	0.32	34
			GYPSUM BD. (3/8 IN.)	0.32	0.42	0.36	0.37	0.30	0.27	0.23	35
			GYPSUM LATH (3/8 IN.) AND 1/2 IN. PLAS. (LT. WT. AGG.)	0.64	0.37	0.32	0.33	0.27	0.25	0.21	36
			GYPSUM LATH (3/8 IN.) AND 1/2 IN. PLAS. (SAND AGG.)	0.41	0.40	0.35	0.36	0.29	0.27	0.22	37
ASBESTOS-CEMENT SIDING	0.21										
STUCCO[5] 1 IN.	0.20	0.19[2]	METAL LATH AND 3/4 IN. PLAS. (LT. WT. AGG.)	0.47	0.39	0.34	0.35	0.28	0.26	0.22	38
			METAL LATH AND 3/4 IN. PLAS. (SAND AGG.)	0.13	0.45	0.39	0.40	0.31	0.29	0.24	39
ASPHALT ROLL SIDING	0.15		INSUL. BD. (1/2 IN.)	1.43	0.29	0.26	0.26	0.22	0.21	0.18	40
			INSUL. BD. LATH (1/2 IN.) AND 1/2 IN. PLAS. (SAND AGG.)	1.52	0.28	0.25	0.26	0.22	0.21	0.18	41
			PLYWOOD (1/4 IN.)	0.31	0.42	0.36	0.37	0.30	0.27	0.23	42
			WOOD PANELS (3/4 IN.)	0.94	0.33	0.29	0.30	0.25	0.23	0.20	43
			WOOD LATH AND 1/2 IN. PLAS. (SAND AGG.)	0.40	0.40	0.35	0.36	0.29	0.27	0.22	44

[1]Note that although several types of exterior finish may be grouped because they have approximately the same thermal resistance value, it is not implied that all types may be suitable for application over all types of sheathing listed.
[2]Average resistance of items listed. This average was used in computation of U values shown.
[3]Building paper is not included except where noted.
[4]Small air space between building paper and brick veneer neglected.
[5]Where stucco is applied over insulating board or gypsum sheathing, building paper is generally required, but the change in U value is negligible.

Extracted with permission from 1965 ASHRAE Guide and Data Book.

Courtesy of the Trane Company.

TABLE 7-6 Coefficients of Transmission (U) of Masonry Walls.

Coefficients are expressed in Btu per (hour) (square foot) (Fahrenheit degree difference in temperature between the air on the two sides), and are based on an outside wind velocity of 15 mph

EXTERIOR FACING			BACKING		NONE	INTERIOR FINISH											NUMBER
						PLAS. 5/8 IN. ON WALL		METAL LATH AND 3/4 IN. PLAS. ON FURRING		GYPSUM LATH (3/8 IN.) AND 1/2 IN. PLAS. ON FURRING			INSUL. BD. LATH (1/2 IN.) and 1/2 IN. PLAS. ON FURRING		WOOD LATH AND 1/2 IN. PLAS.		
			RESISTANCE → ↓			(SAND AGG.) 0.11	(LT. WT. AGG.) 0.39	(SAND AGG.) 0.13	(LT. WT. AGG.) 0.47	NO PLAS. 0.32	(SAND AGG.) 0.41	(LT. WT. AGG.) 0.64	NO PLAS. 1.43	(SAND AGG.) 1.52	(SAND AGG.) 0.40		
MATERIAL	R	AV. R	MATERIAL	R	U	U	U	U	U	U	U	U	U	U	U		
					A	B	C	D	E	F	G	H	I	J	K		
			CONCRETE BLOCK														
			(CINDER AGG.)														
			(4 IN.)	1.11	0.41	0.39	0.35	0.28	0.26	0.27	0.26	0.25	0.21	0.20	0.26		1
			(8 IN.)	1.72	0.33	0.32	0.29	0.24	0.22	0.23	0.23	0.21	0.18	0.18	0.23		2
			(12 IN.)	1.89	0.31	0.30	0.28	0.23	0.21	0.22	0.22	0.21	0.18	0.17	0.22		3
			(LT. WT. AGG.)														
FACE BRICK			(4 IN.)	1.50	0.35	0.34	0.31	0.25	0.23	0.24	0.24	0.22	0.19	0.19	0.24		4
4 IN.	0.44		(8 IN.)	2.00	0.30	0.29	0.27	0.23	0.21	0.22	0.21	0.20	0.17	0.17	0.21		5
STONE			(12 IN.)	2.27	0.28	0.27	0.25	0.21	0.20	0.20	0.20	0.19	0.17	0.16	0.20		6
4 IN.	0.32		(SAND AGG.)														
PRECAST			(4 IN.)	0.71	0.49	0.46	0.41	0.32	0.29	0.30	0.29	0.27	0.22	0.22	0.29		7
CONCRETE		.39	(8 IN.)	1.11	0.41	0.39	0.35	0.28	0.26	0.27	0.26	0.25	0.21	0.20	0.26		8
(SAND AGG.)			(12 IN.)	1.28	0.38	0.37	0.33	0.27	0.25	0.26	0.25	0.24	0.20	0.20	0.25		9
4 IN.	0.32																
6 IN.	0.48		HOLLOW CLAY TILE														
			(4 IN.)	1.11	0.41	0.39	0.35	0.28	0.26	0.27	0.26	0.25	0.21	0.20	0.26		10
			(8 IN.)	1.85	0.31	0.30	0.28	0.23	0.22	0.22	0.22	0.21	0.18	0.18	0.22		11
			(12 IN.)	2.50	0.26	0.25	0.24	0.20	0.19	0.19	0.19	0.18	0.16	0.16	0.19		12
			CONCRETE														
			(SAND AGG.)														
			(4 IN.)	0.32	0.60	0.56	0.49	0.36	0.32	0.34	0.33	0.31	0.25	0.24	0.33		13
			(6 IN.)	0.48	0.55	0.52	0.45	0.34	0.31	0.32	0.31	0.29	0.24	0.23	0.31		14
			(8 IN.)	0.64	0.51	0.48	0.42	0.32	0.29	0.31	0.30	0.28	0.23	0.22	0.30		15

			CONCRETE BLOCK													
			(CINDER AGG.)													
			(4 IN.)	1.11	0.36	0.35	0.32	0.26	0.24	0.25	0.24	0.23	0.19	0.19	0.24	16
			(8 IN.)	1.72	0.29	0.29	0.26	0.22	0.21	0.21	0.21	0.20	0.17	0.17	0.21	17
			(12 IN.)	1.89	0.28	0.27	0.25	0.21	0.20	0.21	0.20	0.19	0.17	0.17	0.20	18
			(LT. WT. AGG.)													
			(4 IN.)	1.50	0.32	0.30	0.28	0.23	0.22	0.22	0.22	0.21	0.18	0.18	0.22	19
			(8 IN.)	2.00	0.27	0.26	0.25	0.21	0.20	0.20	0.20	0.19	0.16	0.16	0.20	20
COMMON BRICK			(12 IN.)	2.27	0.25	0.25	0.23	0.20	0.19	0.19	0.19	0.18	0.16	0.16	0.19	21
4 IN.	0.80		(SAND AGG.)													
PRECAST			(4 IN.)	0.71	0.42	0.40	0.36	0.29	0.26	0.27	0.27	0.25	0.21	0.21	0.27	22
CONCRETE		.72	(8 IN;)	1.11	0.36	0.35	0.32	0.26	0.24	0.25	0.24	0.23	0.19	0.19	0.24	23
(SAND AGG.			(12 IN.)	1.28	0.34	0.33	0.30	0.25	0.23	0.24	0.23	0.22	0.19	0.18	0.23	24
8 IN.	0.64															
			HOLLOW CLAY TILE													
			(4 IN.)	1.11	0.36	0.35	0.32	0.26	0.24	0.25	0.24	0.23	0.19	0.19	0.24	25
			(8 IN.)	1.85	0.28	0.28	0.26	0.22	0.20	0.21	0.20	0.19	0.17	0.17	0.20	26
			(12 IN.)	2.50	0.24	0.23	0.22	0.19	0.18	0.18	0.18	0.17	0.15	0.15	0.18	27
			CONCRETE													
			(SAND AGG.)													
			(4 IN.)	0.32	0.50	0.48	0.42	0.32	0.29	0.30	0.30	0.28	0.23	0.22	0.30	28
			(6 IN.)	0.48	0.47	0.44	0.39	0.31	0.28	0.29	0.28	0.27	0.22	0.22	0.28	29
			(8 IN.)	0.64	0.43	0.41	0.37	0.29	0.27	0.28	0.27	0.26	0.21	0.21	0.27	30

Extracted with permission from 1965 ASHRAE Guide and Data Book.

Courtesy of the Trane Company.

TABLE 7-7 Coefficients of Transmission (U) of Masonry Cavity Walls.

Coefficients are expressed in Btu per (hour) (square foot) (Fahrenheit degree difference in temperature between the air on the two sides), and are based on an outside wind velocity of 15 mph

EXTERIOR CONSTRUCTION			INNER SECTION		INTERIOR FINISH											NUMBER
					NONE	PLAS. 5/8 IN. ON WALL		METAL LATH AND 3/4 IN. PLAS. ON FURRING		GYPSUM LATH (3/8 IN.) AND 1/2 IN. PLAS. ON FURRING			INSUL. BD. LATH (1/2 IN.) AND 1/2 IN. PLAS. ON FURRING		WOOD LATH AND 1/2 IN. PLAS.	
			RESISTANCE → ↓			(SAND AGG.) 0.11	(LT. WT. AGG.) 0.39	(SAND AGG.) 0.13	(LT. WT AGG.) 0.47	NO PLAS. 0.32	(SAND AGG.) 0.41	(LT. WT AGG.) 0.64	NO PLAS. 1.43	(SAND AGG.) 1.52	(SAND AGG.) 0.40	
MATERIAL	R	AV. R	MATERIAL	R	U A	U B	U C	U D	U E	U F	U G	U H	U I	U J	U K	
FACE BRICK (4 IN.)		0.44	CONCRETE BLOCK (4 IN.)													
			(GRAVEL AGG.)...	0.71	0.34	0.32	0.30	0.25	0.23	0.23	0.23	0.22	0.19	0.18	0.23	1
			(CINDER AGG.)...	1.11	0.30	0.29	0.27	0.22	0.21	0.21	0.21	0.20	0.17	0.17	0.21	2
			(LT. WT. AGG.)...	1.50	0.27	0.26	0.24	0.21	0.19	0.20	0.19	0.19	0.16	0.16	0.19	3
			COMMON BRICK (4 IN.)............	0.80	0.33	0.32	0.29	0.24	0.22	0.23	0.23	0.21	0.18	0.18	0.23	4
			CLAY TILE (4 IN.)...	1.11	0.30	0.29	0.27	0.22	0.21	0.21	0.21	0.20	0.17	0.17	0.21	5

COMMON BRICK (4 IN.)	0.80		CONCRETE BLOCK (4 IN.) (GRAVEL AGG.)...	0.71	0.30	0.29	0.27	0.23	0.21	0.22	0.21	0.20	0.18	0.17	0.21	6
			(CINDER AGG.)...	1.11	0.27	0.26	0.25	0.21	0.19	0.20	0.20	0.19	0.16	0.16	0.20	7
CONCRETE BLOCK (GRAVEL AGG.) (4 IN.)	0.71	0.76	(LT. WT. AGG.)...	1.50	0.25	0.24	0.22	0.19	0.18	0.19	0.18	0.18	0.15	0.15	0.18	8
			COMMON BRICK (4 IN.)............	0.80	0.30	0.29	0.27	0.22	0.21	0.21	0.21	0.20	0.17	0.17	0.21	9
			CLAY TILE (4 IN.)...	1.11	0.27	0.26	0.25	0.21	0.19	0.20	0.20	0.19	0.16	0.16	0.20	10
			CONCRETE BLOCK (4 IN.) (GRAVEL AGG.)...	0.71	0.27	0.27	0.25	0.21	0.20	0.20	0.20	0.19	0.17	0.16	0.20	11
CONCRETE BLOCK			(CINDER AGG.)...	1.11	0.25	0.24	0.23	0.19	0.18	0.19	0.18	0.18	0.16	0.15	0.18	12
			(LT. WT. AGG.)...	1.50	0.23	0.22	0.21	0.18	0.17	0.17	0.17	0.17	0.15	0.14	0.17	13
(CINDER AGG.) (4 IN.)		1.11	COMMON BRICK (4 IN.)............	0.80	0.27	0.26	0.24	0.21	0.19	0.20	0.20	0.19	0.16	0.16	0.20	14
			CLAY TILE (4 IN.)...	1.11	0.25	0.24	0.23	0.19	0.18	0.19	0.18	0.18	0.16	0.15	0.18	15

Extracted with permission from 1965 ASHRAE Guide and Data Book.

Courtesy of the Trane Company.

TABLE 7-8 Coefficients of Transmission (U) of Frame Partitions or Interior Walls.

Coefficients are expressed in Btu per (hour) (square foot) (Fahrenheit degree difference in temperature between the air on the two sides), and are based on still air (no wind) conditions on both sides

TYPE OF INTERIOR FINISH MATERIAL	R	SINGLE PARTITION (FINISH ON ONLY ONE SIDE OF STUDS) U A	DOUBLE PARTITION (FINISH ON BOTH SIDES OF STUDS) U B	NUMBER
GYPSUM BD. (3/8 IN.)	0.32	0.60	0.34	1
GYPSUM LATH (3/8 IN.) AND 1/2 IN. PLAS. (LT. WT. AGG.)	0.64	0.50	0.28	2
GYPSUM LATH (3/8 IN.) AND 1/2 IN. PLAS. (SAND AGG.)	0.41	0.56	0.32	3
METAL LATH AND 3/4 IN. PLAS. (LT. WT. AGG.)	0.47	0.55	0.31	4
METAL LATH AND 3/4 IN. PLAS. (SAND AGG.)	0.13	0.67	0.39	5
INSUL. BD. (1/2 IN.)	1.43	0.36	0.19	6
INSUL. BD. LATH (1/2 IN.) AND 1/2 PLAS. (SAND AGG.)	1.52	0.35	0.19	7
PLYWOOD: (1/4 IN.)	0.31	0.60	0.34	8
(3/8 IN.)	0.47	0.55	0.31	9
(1/2 IN.)	0.63	0.50	0.28	10
WOOD PANELS (3/4 IN.)	0.94	0.43	0.24	11
WOOD-LATH AND 1/2 IN. PLAS. (SAND AGG.)	0.40	0.57	0.32	12
SHEET-METAL PANELS ADHERED TO WOOD (FRAMING)	0	0.74	0.43	13
GLASS AND GLASS BLOCKS	SEE TABLE 3-2			

Extracted with permission from 1965 ASHRAE Guide and Data Book.

Courtesy of the Trane Company.

TABLE 7-9 Coefficients of Transmission (U) of Frame Walls.

These coefficients are expressed in Btu per (hour) (square foot) (Fahrenheit degree difference in temperature between the air on the two sides), and are based on an outside wind velocity of 15 mph

EXTERIOR[1]			INTERIOR FINISH		TYPE OF SHEATHING[3]						NUMBER
					NONE, BUILDING PAPER	GYPSUM BOARD 1/2 IN.	PLYWOOD 5/16 IN.	WOOD 25/32 IN. AND BUILDING PAPER	INSULATION BOARD SHEATHING 1/2 IN.	INSULATION BOARD SHEATHING 25/32 IN.	
			RESISTANCE		0.06	0.45	0.39	1.04	1.32	2.06	
MATERIAL	R	AV. R	MATERIAL	R	U A	U B	U C	U D	U E	U F	
WOOD SIDING		0.85[2]	NONE	-	0.57	0.47	0.48	0.36	0.33	0.27	1
DROP-(1 IN. X 8 IN.)	0.79		GYPSUM BD. (3/8 IN.)	0.32	0.33	0.29	0.30	0.25	0.23	0.20	2
BEVEL (1/2 IN. X 8 IN.)	0.81		GYPSUM LATH (3/8 IN.) AND 1/2 IN. PLAS. (LT. WT. AGG.)	0.64	0.30	0.27	0.27	0.23	0.22	0.19	3
WOOD SHINGLES			GYPSUM LATH (3/8 IN.) AND 1/2 IN. PLAS. (SAND AGG.)	0.41	0.32	0.28	0.29	0.24	0.23	0.19	4
7 1/2 IN. EXPOSURE	0.87		METAL LATH AND 3/4 IN. PLAS. (LT. WT. AGG.)	0.47	0.31	0.28	0.28	0.24	0.22	0.19	5
WOOD PANELS (3/4 IN.)	0.94		METAL LATH AND 3/4 IN. PLAS. (SAND AGG.)	0.13	0.35	0.31	0.31	0.26	0.24	0.21	6
			INSUL. BD. (1/2 IN.)	1.43	0.24	0.22	0.22	0.19	0.18	0.16	7
			INSUL. BD. LATH (1/2 IN.) AND 1/2 IN. PLAS. (SAND AGG.)	1.52	0.24	0.22	0.22	0.19	0.18	0.16	8
			PLYWOOD (1/4 IN.)	0.31	0.33	0.29	0.30	0.25	0.23	0.20	9
			WOOD PANELS (3/4 IN.)	0.94	0.27	0.25	0.25	0.22	0.20	0.18	10
			WOOD LATH AND 1/2 IN. PLAS. (SAND AGG.)	0.40	0.32	0.28	0.29	0.24	0.23	0.19	11
FACE-BRICK VENEER[4]	0.44	0.45[2]	NONE	–	0.73	0.56	0.58	0.42	0.38	0.30	12
PLYWOOD (3/8 IN.)	0.47		GYPSUM BD. (3/8 IN.)	0.32	0.37	0.33	0.33	0.27	0.25	0.21	13
			GYPSUM LATH (3/8 IN.) AND 1/2 IN. PLAS. (LT. WT. AGG.)	0.64	0.33	0.30	0.30	0.25	0.24	0.20	14
			GYPSUM LATH (3/8 IN.) AND 1/2 IN. PLAS. (SAND AGG.)	0.41	0.36	0.32	0.32	0.27	0.25	0.21	15
			METAL LATH AND 3/4 IN. PLAS. (LT. WT. AGG.)	0.47	0.35	0.31	0.32	0.26	0.25	0.21	16
			METAL LATH AND 3/4 IN. PLAS. (SAND AGG.)	0.13	0.40	0.35	0.36	0.29	0.27	0.22	17
			INSUL. BD. (1/2 IN.)	1.43	0.26	0.24	0.24	0.21	0.20	0.17	18
			INSUL. BD. LATH (1/2 IN.) AND 1/2 IN. PLAS. (SAND AGG.)	1.52	0.26	0.23	0.24	0.21	0.19	0.17	19
			PLYWOOD (1/4 IN.)	0.31	0.38	0.33	0.33	0.27	0.26	0.21	20
			WOOD PANELS (3/4 IN.)	0.94	0.30	0.27	0.28	0.23	0.22	0.19	21
			WOOD LATH AND 1/2 IN. PLAS. (SAND AGG.)	0.40	0.36	0.32	0.32	0.27	0.25	0.21	22

[1]Note that although several types of exterior finish may be grouped because they have approximately the same thermal resistance value, it is not implied that all types may be suitable for application over all types of sheathing listed.
[2]Average resistance of items listed. This average was used in computation of U values shown.
[3]Building paper is not included except where noted.
[4]Small air space between building paper and brick veneer neglected.
[5]Where stucco is applied over insulating board or gypsum sheathing, building paper is generally required, but the change in U value is negligible.

Extracted with permission from 1965 ASHRAE Guide and Data Book.

Courtesy of the Trane Company.

TABLE 7-10 Coefficients of Transmission (U) of Frame Construction Ceilings and Floors.

Coefficients are expressed in Btu per (hour) (square foot) (Fahrenheit degree difference between the air on the two sides) and are based on still air (no wind) conditions on both sides

DIRECTION OF HEAT	HEAT FLOW UPWARD (WINTER CONDITIONS)						HEAT FLOW DOWNWARD (SUMMER CONDITIONS)						
	TYPE OF FLOOR												
			WOOD SUBFLOOR (25/32 IN.), FELT, AND—						WOOD SUBFLOOR (25/32 IN.), FELT, AND—				
TYPE OF CEILING	NONE	WOOD SUBFLOOR (25/32 IN.)	CEMENT (1 1/2 IN.) AND CERAMIC TILE (1/2 IN.)	HARDWOOD FLOOR (3/4 IN.)	PLYWOOD (5/8 IN.) AND FLOOR TILE OR LINOLEUM (1/8 IN.)	INSUL. BD. (3/8 IN.) AND HARD BD. (1/4 IN.) AND FLOOR TILE OR LINOLEUM (1/8 IN.)	NONE	WOOD SUBFLOOR (25/32 IN.)	CEMENT (1 1/2 IN.) AND CERAMIC TILE (1/2 IN.)	HARDWOOD FLOOR (3/4 IN.)	PLYWOOD (5/8 IN.) AND FLOOR TILE OR LINOLEUM (1/8 IN.)	INSUL. BD. (3/8 IN.) AND HARD BD. (1/4 IN.) AND FLOOR TILE OR LINOLEUM (1/8 IN.)	NUMBER
RESISTANCE ↑↓	—	0.98	1.38	1.72	1.87	2.26	—	0.98	1.38	1.72	1.87	2.26	

MATERIAL	R	U	U	U	U	U	U	U	U	U	U	U	U	
		A	B	C	D	E	F	G	H	I	J	K	L	
NONE	—	—	0.45	0.38	0.34	0.32	0.29	—	0.35	0.31	0.28	0.26	0.24	1
GYPSUM BD. (3/8 IN.)	0.32	0.65	0.30	0.27	0.24	0.23	0.22	0.46	0.24	0.22	0.21	0.20	0.19	2
GYPSUM LATH (3/8 IN.) AND 1/2 IN. PLAS. (LT. WT. AGG.)	0.64	0.54	0.27	0.24	0.23	0.22	0.20	0.40	0.22	0.21	0.19	0.19	0.17	3
GYPSUM LATH (3/8 IN.) AND 1/2 IN. PLAS. (SAND AGG.)	0.41	0.61	0.29	0.26	0.24	0.23	0.21	0.44	0.24	0.22	0.20	0.20	0.18	4
METAL LATH AND 3/8 IN. PLAS. (LT. WT. AGG.)	0.47	0.59	0.28	0.26	0.23	0.23	0.21	0.43	0.23	0.21	0.20	0.19	0.18	5
METAL LATH AND 3/8 IN. PLAS. (SAND AGG.)	0.13	0.74	0.31	0.28	0.26	0.25	0.22	0.51	0.25	0.23	0.21	0.21	0.19	6
INSUL BD. (1/2 IN.)	1.43	0.38	0.22	0.20	0.19	0.19	0.17	0.31	0.19	0.18	0.17	0.16	0.15	7
INSUL. BD. LATH (1/2 IN.) AND 1/2 IN. PLAS. (SAND AGG.)	1.52	0.36	0.22	0.20	0.19	0.18	0.17	0.30	0.19	0.17	0.17	0.16	0.15	8
ACOUSTICAL TILE														
(1/2 IN.) ON GYPSUM BD. (3/8 IN.)	1.51[2]	0.37	0.22	0.20	0.19	0.18	0.17	0.30	0.19	0.17	0.17	0.16	0.15	9
(1/2 IN.) ON FURRING	1.19	0.41	0.24	0.22	0.20	0.19	0.18	0.33	0.20	0.19	0.17	0.17	0.16	10
(3/4 IN.) ON GYPSUM BD. (3/8 IN.)	2.10[2]	0.30	0.19	0.18	0.17	0.17	0.15	0.25	0.17	0.16	0.15	0.15	0.14	11
(3/4 IN.) ON FURRING	1.78	0.33	0.21	0.19	0.18	0.17	0.16	0.28	0.18	0.17	0.16	0.15	0.15	12
WOOD LATH AND 1/2 IN. PLAS. (SAND AGG.)	0.40	0.62	0.29	0.26	0.24	0.23	0.21	0.45	0.24	0.22	0.20	0.20	0.18	13

[1]Includes asphalt, rubber, and plastic tile (½ in.), ceramic tile, or terrazzo (1 in.).
[2]Includes thermal resistance of ⅜ in. gypsum wall board.

Extracted with permission from 1965 ASHRAE Guide and Data Book.

Courtesy of the Trane Company.

TABLE 7-11 Coefficients of Transmission (U) of Concrete Floor–Ceiling Constructions (Summer Conditions, Downward Flow).

Coefficients are expressed in Btu per (hour) (square foot) (Fahrenheit degree difference in temperature between the air on the two sides), and are based on still air (no wind) conditions on both sides

TYPE OF DECK			TYPE OF FINISH FLOOR	↑↓	TYPE OF CEILING														NUMBER
					CEILING APPLIED DIRECTLY TO SLAB					SUSPENDED CEILING									
										GYPSUM BD. (3/8 IN.) AND PLAS.			METAL LATH AND PLAS.		ACOUSTICAL TILE				
					NONE	PLAS.		ACOUSTICAL TILE-GLUED							ON FURRING OR CHANNELS		ON GYPSUM BD. (3/8 IN.)		
						(LT. WT. AGG.) 1/8 IN.	(SAND AGG.) 1/8 IN.	1/2 IN.	3/4 IN.	NO PLAS.	(LT. WT. AGG.) 1/2 IN.	(SAND AGG.) 1/2 IN.	(LT. WT. AGG.) 3/4 IN.	(SAND AGG.) 3/4 IN.	1/2 IN.	3/4 IN.	1/2 IN.	3/4 IN.	
			RESISTANCE		-	0.08	0.02	1.19	1.78	0.32	0.64	0.41	0.47	0.13	1.19	1.78	1.51	2.10	
MATERIAL	R	AV. R	MATERIAL	AV. R	U	U	U	U	U	U	U	U	U	U	U	U	U	U	
					O	P	Q	R	S	T	U	V	W	X	Y	Z	Z′	Z″	
CONCRETE[4] (SAND AGG.) (4 IN.) 0.32 (6 IN.) 0.48		0.40	NONE	-----	0.45	0.43	0.44	0.29	0.25	0.28	0.26	0.27	0.27	0.30	0.23	0.20	0.21	0.19	1
			FLOOR TILE[5] OR LINOLEUM (1/8 IN.)	0.05	0.44	0.42	0.43	0.29	0.25	0.28	0.26	0.27	0.27	0.29	0.22	0.20	0.21	0.19	2
			WOOD BLOCK (13/16 IN.) ON SLAB	0.74	0.34	0.33	0.33	0.24	0.21	0.23	0.22	0.23	0.23	0.24	0.19	0.17	0.18	0.17	3
			FLOOR ON SLEEPERS PLYWOOD SUBFLOOR (5/8 IN.), FELT AND FLOOR TILE[5] OR LINOLEUM (1/8 IN.)	0.89	0.23	0.23	0.23	0.18	0.17	0.19	0.18	0.19	0.18	0.20	0.16	0.15	0.15	0.14	4
			WOOD SUBFLOOR (25/32 IN.), FELT AND HARD-WOOD (3/4 IN.)	1.72	0.20	0.19	0.20	0.16	0.15	0.16	0.15	0.16	0.16	0.17	0.14	0.13	0.14	0.13	5

CONCRETE[4] (SAND AGG.) (8 IN.) 0.64 (10 IN.) 0.80	0.72	NONE	-----	0.39	0.38	0.39	0.27	0.23	0.26	0.24	0.25	0.25	0.27	0.21	0.19	0.20	0.18	6
		FLOOR TILE[5] OR LINOLEUM (1/8 IN.)	0.05	0.38	0.37	0.38	0.26	0.23	0.26	0.24	0.25	0.25	0.27	0.21	0.19	0.20	0.18	7
		WOOD BLOCK (13/16 IN.) ON SLAB	0.74	0.30	0.30	0.30	0.22	0.20	0.22	0.20	0.21	0.21	0.23	0.18	0.16	0.17	0.16	8
		FLOOR ON SLEEPERS PLYWOOD SUBFLOOR (5/8 IN.), FELT AND FLOOR TILE[5] OR LINOLEUM (1/8 IN.)	0.89	0.22	0.21	0.22	0.17	0.16	0.18	0.17	0.18	0.17	0.18	0.15	0.14	0.15	0.14	9
		WOOD SUBFLOOR (25/32 IN.), FELT AND HARD-WOOD (3/4 IN.)	1.72	0.19	0.18	0.18	0.15	0.14	0.16	0.15	0.15	0.15	0.16	0.14	0.13	0.13	0.12	10

[4] Concrete is assumed to have a thermal conductivity k of 12.0.
[5] Includes asphalt, rubber, and plastic tile (⅛ in.), ceramic tile on terrazzo (1 in.).

Extracted with permission from 1965 ASHRAE Guide and Data Book.

Courtesy of the Trane Company.

TABLE 7-12 Coefficients of Transmission (U) of Flat Masonry Roofs with Built-up Roofing with and without Suspended Ceiling (Summer Conditions, Downward Flow).

These coefficients are expressed in Btu per (hour) (square foot) (Fahrenheit degree difference in temperature between the air on the two sides), and are based on an outside wind velocity of 7.5 mph

TYPE OF DECK		TYPE OF FORM		TYPE OF CEILING																NUMBER
				ROOF INSULATION—NO CEILING							SUSPENDED CEILING									
				C VALUE OF ROOF INSULATION							GYPSUM BD. (3/8 IN.) AND PLAS.			METAL LATH AND PLAS.		ACOUSTICAL TILE				
				NONE	0.72	0.36	0.24	0.19	0.15	0.12	NO PLAS.	LT. WT. AGG. (1/2 IN.)	SAND AGG. (1/2 IN.)	LT. WT. AGG. (3/4 IN.)	SAND AGG. (3/4 IN.)	ON FURRING OR CHANNELS 1/2 IN.	ON FURRING OR CHANNELS 3/4 IN.	ON GYPSUM BD. (3/8 IN.) 1/2 IN.	ON GYPSUM BD. (3/8 IN.) 3/4 IN.	
		RESISTANCE →		—	1.39	2.78	4.17	5.26	6.67	8.33	0.32	0.64	0.41	0.47	0.13	1.19	1.78	1.51	2.10	
MATERIAL	R	MATERIAL	R	U	U	U	U	U	U	U	U	U	U	U	U	U	U	U	U	
				A′	B′	C′	D′	E′	F′	G′	H′	I′	J′	K′	L′	M′	N′	O′	P′	
CONCRETE SLAB[4] (GRAVEL AGG.)																				
(4 IN.)	0.32	TEMPORARY	—	0.55	0.31	0.22	0.17	0.14	0.12	0.10	0.32	0.29	0.31	0.30	0.34	0.25	0.22	0.23	0.20	1
(6 IN.)	0.48	TEMPORARY	—	0.51	0.30	0.21	0.16	0.14	0.12	0.10	0.30	0.28	0.30	0.29	0.32	0.24	0.21	0.22	0.20	2
(8 IN.)	0.64	TEMPORARY	—	0.47	0.28	0.20	0.16	0.14	0.11	0.10	0.29	0.27	0.28	0.28	0.31	0.23	0.20	0.22	0.19	3
LT. WT. AGG[5]																				
(2 IN.)	2.22	CORRUGATED METAL[3]	0	0.27	0.20	0.15	0.13	0.11	0.10	0.08	0.20	0.19	0.20	0.19	0.21	0.17	0.15	0.16	0.15	4
		INSUL. BD. (1 IN.)	2.78	0.15	0.13	0.11	0.09	0.09	0.08	0.07	0.13	0.12	0.13	0.13	0.13	0.12	0.11	0.11	0.10	5
		INSUL. BD. (1 1/2 IN.)	4.17	0.13	0.11	0.09	0.08	0.08	0.07	0.06	0.11	0.11	0.11	0.11	0.11	0.10	0.09	0.10	0.09	6
		GLASS FIB. BD. (1 IN.)	4.00	0.13	0.11	0.10	0.08	0.08	0.07	0.06	0.11	0.11	0.11	0.11	0.11	0.10	0.10	0.10	0.09	7

(3 IN.)...3.33	CORRUGATED METAL[3]	0	0.21	0.16	0.13	0.11	0.10	0.09	0.08	0.16	0.16	0.16	0.16	0.17	0.14	0.13	0.14	0.13	8
	INSUL. BD. (1 IN.)......	2.78	0.13	0.11	0.10	0.08	0.08	0.07	0.06	0.11	0.11	0.11	0.11	0.11	0.10	0.10	0.10	0.09	9
	INSUL. BD. (1 1/2 IN.)..	4.17	0.11	0.10	0.08	0.08	0.07	0.06	0.06	0.10	0.09	0.10	0.10	0.10	0.09	0.09	0.09	0.08	10
	GLASS FIB. BD. (1 IN.)	4.00	0.11	0.10	0.09	0.08	0.07	0.06	0.06	0.10	0.10	0.10	0.10	0.10	0.09	0.09	0.09	0.08	11
(4 IN.)...4.44	CORRUGATED METAL[3]	0	0.17	0.14	0.11	0.10	0.09	0.08	0.07	0.14	0.13	0.14	0.14	0.14	0.12	0.12	0.12	0.11	12
	INSUL. BD. (1 IN.)	2.78	0.11	0.10	0.09	0.08	0.07	0.06	0.06	0.10	0.10	0.10	0.10	0.10	0.09	0.09	0.09	0.09	13
	INSUL. BD. (1 1/2 IN.)..	4.17	0.10	0.09	0.08	0.07	0.07	0.06	0.05	0.09	0.09	0.09	0.09	0.09	0.08	0.08	0.08	0.08	14
	GLASS FIB. BD. (1 IN.)	4.00	0.10	0.09	0.08	0.07	0.07	0.06	0.05	0.09	0.09	0.09	0.09	0.09	0.08	0.08	0.08	0.08	15
GYPSUM SLAB[7] (2 IN.)...1.20	GYPSUM BD. (1/2 IN.)	0.45	0.32	0.22	0.17	0.14	0.12	0.10	0.09	0.22	0.21	0.22	0.22	0.23	0.19	0.17	0.18	0.16	16
	INSUL. BD. (1 IN.)......	2.78	0.18	0.15	0.12	0.10	0.09	0.08	0.07	0.15	0.14	0.15	0.14	0.15	0.13	0.12	0.13	0.12	17
	INSUL. BD. (1 1/2 IN.)..	4.17	0.15	0.12	0.10	0.09	0.08	0.07	0.07	0.12	0.12	0.12	0.12	0.13	0.11	0.10	0.11	0.10	18
	ASBESTOS-CEMENT BD.[6] (1/4 IN.)..........	0.06	0.34	0.23	0.18	0.14	0.12	0.10	0.09	0.25	0.23	0.24	0.24	0.26	0.20	0.18	0.19	0.17	19
	GLASS FIB. BD. (1 IN.)	4.00	0.15	0.12	0.11	0.09	0.08	0.07	0.07	0.13	0.12	0.12	0.12	0.13	0.11	0.11	0.11	0.10	20
(3 IN.)...1.80	GYPSUM BD. (1/2 IN.)	0.45	0.27	0.19	0.15	0.13	0.11	0.10	0.08	0.20	0.19	0.19	0.19	0.21	0.17	0.15	0.16	0.15	21
	INSUL. BD. (1 IN.)......	2.78	0.16	0.13	0.11	0.10	0.09	0.08	0.07	0.14	0.13	0.13	0.13	0.14	0.12	0.11	0.12	0.11	22
	INSUL. BD. (1 1/2 IN.)..	4.17	0.13	0.11	0.10	0.09	0.08	0.07	0.06	0.11	0.11	0.11	0.11	0.12	0.10	0.11	0.10	0.10	23
	ASBESTOS-CEMENT BD (1/4 IN.)................	0.06	0.30	0.21	0.16	0.13	0.12	0.10	0.09	0.21	0.20	0.21	0.21	0.22	0.18	0.16	0.17	0.16	24
	GLASS FIB. BD. (1 IN.)	4.00	0.14	0.12	0.10	0.09	0.08	0.07	0.06	0.12	0.11	0.12	0.11	0.12	0.11	0.10	0.10	0.10	25
(4 IN.)...2.40	GYPSUM BD. (1/2 IN.)	0.45	0.23	0.17	0.14	0.12	0.10	0.09	0.08	0.18	0.17	0.17	0.17	0.18	0.15	0.14	0.15	0.13	26
	INSUL. BD. (1 IN.)......	2.78	0.15	0.12	0.11	0.09	0.08	0.07	0.07	0.13	0.12	0.12	0.12	0.13	0.11	0.11	0.11	0.10	27
	INSUL. BD. (1 1/2 IN.)..	4.17	0.12	0.11	0.09	0.08	0.08	0.07	0.06	0.11	0.10	0.11	0.11	0.11	0.10	0.09	0.10	0.09	28
	ASBESTOS-CEMENT BD.[3] 1/4 IN.)..........	0.06	0.25	0.19	0.15	0.12	0.11	0.09	0.08	0.19	0.18	0.19	0.19	0.20	0.16	0.15	0.16	0.14	29
	GLASS FIB. BD. (1 IN.)	4.00	0.13	0.11	0.09	0.08	0.08	0.07	0.06	0.11	0.11	0.11	0.11	0.11	0.10	0.09	0.10	0.09	30

[3]U values would also apply if slab were poured on metal lath, paper-backed wire, fabric, or asbestos-cement board (¼ in.)
[4]Concrete assumed to have a thermal conductivity k of 12.0 and a density of 140 lb per cu ft.
[5]Concrete assumed to have a thermal conductivity k of 0.90 and a density of 30 lb per cu ft.
[6]Gypsum slab 2¼ in. thick since this is recommended practice.
[7]Gypsum fiber concrete with 12½ percent wood chips (thermal conductivity k = 1.66).

Extracted with permission from 1965 ASHRAE Guide and Data Book.

Courtesy of the Trane Company.

TABLE 7-13 Coefficients of Transmission (U) of Wood or Metal Construction Flat Roofs and Ceilings (Summer Conditions, Downward Flow).

Coefficients are expressed in Btu per (hour) (square foot) (Fahrenheit degree difference in temperature between the air on the two sides), and are based upon an outside wind velocity of 7.5 mph

TYPE OF DECK (BUILT-UP ROOF IN ALL CASES)		INSULATION ADDED ON TOP OF DECK[4]		TYPE OF CEILING											NUMBER
				NONE	GYPSUM BD. (3/8 IN.) AND PLAS.			METAL LATH AND PLAS.		INSUL. BD. (1/2 IN.)	ACOUSTICAL TILE				
											ON FURRING		ON GYPSUM DB. (1/2 IN.)		
		CON-DUCT-ANCE INSUL.	RE-SIST-ANCE		NONE	LT. WT. AGG. 1/2 IN.	SAND AGG. 1/2 IN.	LT. WT. AGG. 3/4 IN.	SAND AGG. 3/4 IN.	PLAIN (1.43) OR 1/2 IN. PLAS. SAND AGG. 1.52	1/2 IN.	3/4 IN.	1/2 IN	3/4 IN.	
RESISTANCE → ↓					0.32	0.64	0.41	0.47	0.13	1.47	1.19	1.78	1.51	2.10	
MATERIAL	R	C	R	U A′	U B′	U C′	U D′	U E′	U F′	U G′	U H′	U I′	U J′	U K′	
WOOD[3] 1 IN.	0.98	NONE	-----	0.40	0.26	0.24	0.26	0.25	0.28	0.20	0.22	0.19	0.20	0.18	1
		0.72	1.39	0.26	0.19	0.18	0.19	0.19	0.20	0.16	0.17	0.15	0.16	0.14	2
		0.36	2.78	0.19	0.15	0.15	0.15	0.15	0.16	0.13	0.14	0.13	0.13	0.12	3
		0.24	4.17	0.15	0.13	0.12	0.12	0.12	0.13	0.11	0.11	0.11	0.11	0.10	4
		0.19	5.26	0.13	0.11	0.11	0.11	0.11	0.11	0.10	0.10	0.10	0.10	0.09	5
		0.15	6.67	0.11	0.10	0.09	0.10	0.09	0.10	0.09	0.09	0.08	0.09	0.08	6
		0.12	8.33	0.09	0.08	0.08	0.08	0.08	0.08	0.08	0.08	0.07	0.08	0.07	7
WOOD[3] 2 IN.	2.03	NONE	-----	0.28	0.21	0.19	0.20	0.20	0.22	0.17	0.18	0.16	0.17	0.15	8
		0.72	1.39	0.20	0.16	0.15	0.16	0.16	0.17	0.14	0.14	0.13	0.14	0.13	9
		0.36	2.78	0.16	0.13	0.13	0.13	0.13	0.13	0.11	0.12	0.11	0.11	0.11	10
		0.24	4.17	0.13	0.11	0.11	0.11	0.11	0.11	0.10	0.10	0.10	0.10	0.09	11
		0.19	5.26	0.11	0.10	0.10	0.10	0.10	0.10	0.09	0.09	0.09	0.09	0.08	12
		0.15	6.67	0.10	0.09	0.09	0.09	0.09	0.09	0.08	0.08	0.08	0.08	0.08	13
		0.12	8.33	0.08	0.08	0.07	0.07	0.08	0.08	0.07	0.07	0.07	0.07	0.07	14

WOOD[3] 3 IN.	3.23	NONE	-----	0.21	0.17	0.16	0.16	0.16	0.17	0.14	0.15	0.13	0.14	0.13	15
		0.72	1.39	0.16	0.13	0.13	0.13	0.13	0.14	0.12	0.12	0.11	0.12	0.11	16
		0.36	2.78	0.13	0.11	0.11	0.11	0.11	0.12	0.10	0.10	0.10	0.10	0.09	17
		0.24	4.17	0.11	0.10	0.10	0.10	0.10	0.10	0.09	0.09	0.09	0.09	0.08	18
		0.19	5.26	0.10	0.09	0.09	0.09	0.09	0.09	0.08	0.08	0.09	0.08	0.08	19
		0.15	6.67	0.09	0.08	0.08	0.08	0.08	0.08	0.07	0.07	0.08	0.07	0.07	20
		0.12	8.33	0.08	0.07	0.07	0.07	0.07	0.07	0.07	0.07	0.07	0.06	0.06	21
PREFORMED SLABS-WOOD FIBER AND CEMENT BINDER 2 IN.	3.60	NONE	-----	0.20	0.16	0.15	0.15	0.15	0.16	0.13	0.14	0.15	0.13	0.12	22
3 IN.	5.40	NONE	-----	0.14	0.12	0.12	0.12	0.12	0.13	0.11	0.11	0.12	0.11	0.10	23
FLAT METAL ROOF DECK	0	NONE	-----	0.67	0.36	0.32	0.34	0.34	0.38	0.25	0.27	0.23	0.25	0.22	24
		0.72	1.39	0.35	0.24	0.22	0.23	0.23	0.25	0.19	0.20	0.18	0.19	0.17	25
		0.36	2.78	0.23	0.18	0.17	0.18	0.17	0.19	0.15	0.16	0.14	0.15	0.14	26
		0.24	4.17	0.18	0.14	0.14	0.14	0.14	0.15	0.12	0.13	0.12	0.12	0.11	27
		0.19	5.26	0.15	0.12	0.12	0.12	0.12	0.13	0.11	0.11	0.11	0.11	0.10	28
		0.15	6.67	0.12	0.11	0.10	0.10	0.10	0.11	0.09	0.10	0.09	0.09	0.09	29
		0.12	8.33	0.10	0.09	0.09	0.09	0.09	0.09	0.08	0.08	0.08	0.08	0.08	30

[3] Wood deck 1, 2, and 3 in. is assumed to be 25/32, 1⅝, and 2⅝, in. thick, respectively. The thermal conductivity k is assumed to be 0.80.
[4] If a vapor barrier is used beneath roof insulation it will have a negligible effect on the U value.

Extracted with permission from 1965 ASHRAE Guide and Data Book.

Courtesy of the Trane Company.

TABLE 7-14 Outside Design Temperatures and Latitudes for Various Cities in the United States.

STATE	CITY	DESIGN TEMP.		NORTH LATITUDE, DEGREES
		DB	WB	
ALABAMA	ANNISTON	95	75	33
	BIRMINGHAM	95	78	33
	MOBILE	95	80	31
ALASKA	JUNEAU	65	52	58
ARIZONA	PHOENIX	105	76	33
	TUCSON	105	72	32
ARKANSAS	LITTLE ROCK	95	78	35
CALIFORNIA	FRESNO	105	74	37
	LOS ANGELES	90	70	34
	SACRAMENTO	100	72	38
	SAN FRANCISCO	85	65	38
COLORADO	DENVER	95	64	40
	PUEBLO	95	65	38
CONNECTICUT	HARTFORD	93	75	42
DELAWARE	WILMINGTON	95	78	40
DISTRICT OF COLUMBIA	WASHINGTON	95	78	39
FLORIDA	JACKSONVILLE	95	78	30
	MIAMI	91	79	26
GEORGIA	ATLANTA	95	76	34
	SAVANNAH	95	78	32
HAWAII	HONOLULU	83	73	21
IDAHO	BOISE	95	65	44
	POCATELLO	95	65	43
ILLINOIS	CHICAGO	95	75	43
	PEORIA	96	76	43
	SPRINGFIELD	98	77	42
INDIANA	EVANSVILLE	95	78	38
	FORT WAYNE	95	75	41
	INDIANAPOLIS	95	76	40
IOWA	DES MOINES	95	78	42
	DUBUQUE	95	78	42
	SIOUX CITY	95	78	42
KANSAS	KANSAS CITY	100	76	39
	WICHITA	100	75	38

KENTUCKY	ASHLAND	95	76	38
	LOUISVILLE	95	78	38
LOUISIANA	NEW ORLEANS	95	80	30
	SHREVEPORT	100	78	32
MAINE	PORTLAND	90	73	44
MARYLAND	BALTIMORE	95	78	39
MASSACHUSETTS	BOSTON	92	75	42
	HOLYOKE	93	75	42
MICHIGAN	DETROIT	95	75	42
	GRAND RAPIDS	95	75	43
MINNESOTA	DULUTH	93	73	47
	MINNEAPOLIS	95	75	45
MISSISSIPPI	JACKSON	95	78	32
MISSOURI	KANSAS CITY	100	76	39
	SPRINGFIELD	100	75	37
	ST. LOUIS	95	78	39
MONTANA	BILLINGS	90	66	46
	HELENA	95	67	46
NEBRASKA	LINCOLN	95	78	41
	OMAHA	95	78	41
NEVADA	RENO	95	65	39
NEW HAMPSHIRE	CONCORD	90	73	43
NEW JERSEY	TRENTON	95	78	40
NEW MEXICO	ALBUQUERQUE	95	70	35
	SANTA FE	95	65	36
NEW YORK	ALBANY	93	75	43
	BUFFALO	93	73	43
	NEW YORK	95	75	41
NORTH CAROLINA	ASHEVILLE	93	75	36
	GREENSBORO	95	78	36
NORTH DAKOTA	BISMARCK	95	73	47
OHIO	CINCINNATI	95	78	39
	CLEVELAND	95	75	42
OKLAHOMA	OKLAHOMA CITY	101	77	35
	TULSA	101	77	36

TABLE 7-14 (Cont.) Outside Design Temperatures and Latitudes for Various Cities in the United States.

STATE	CITY	DESIGN TEMP.		NORTH LATITUDE, DEGREES
		DB	WB	
OREGON	PORTLAND	90	68	45
PENNSYLVANIA	PHILADELPHIA	95	78	40
	PITTSBURGH	95	75	40
RHODE ISLAND	PROVIDENCE	93	75	42
SOUTH CAROLINA	CHARLESTON	95	78	33
	GREENVILLE	95	75	35
SOUTH DAKOTA	RAPID CITY	95	70	44
TENNESSEE	CHATTANOOGA	95	76	35
	MEMPHIS	95	78	35
TEXAS	DALLAS	100	78	33
	EL PASO	100	69	32
	GALVESTON	95	80	29
	HOUSTON	95	80	30
	SAN ANTONIO	100	78	29
UTAH	SALT LAKE CITY	95	64	41
VIRGINIA	NORFOLK	95	78	37
	RICHMOND	95	78	38
	ROANOKE	95	78	37
WASHINGTON	SEATTLE	85	65	48
	SPOKANE	95	65	48
WEST VIRGINIA	CHARLESTON	95	75	42
WISCONSIN	EAU CLAIRE	95	75	45
	MADISON	95	75	43
	MILWAUKEE	95	75	43
WYOMING	CHEYENNE	95	65	41

Courtesy of the Trane Company.

SIM 7-4

The engineer is evaluating a 10,000 ft^2, 4 in. common brick exterior wall with 4 in. cinder aggregate concrete block backing with a 5/8 in. sand aggregate plaster on interior walls.

a. Comment on adding a 1 inch glass fiber insulating board prior to the interior finish.
b. Instead of insulating board, comment on providing a 3/4 inch air space with aluminum foil on one side. (Use the data for 0°F.)

ANSWER

From Table 7-6,

a. $U = 0.35$ (Coordinate B-16)

or

$R = 1/0.35 = 2.85$

The effect of the insulating material can be seen by using Table 7-2 with

$$R = \frac{1}{0.25} = 4$$

Thus, Total $R = 2.85 + 4 = 6.85$

$$U = 1/6.85 = 0.14.$$

Energy Savings

$$q = (0.35 - 0.14) \times 10{,}000 \times 17 = 35{,}700 \text{ Btuh}$$

b. From Table 7-1,

The effect of the air space at 0°F is 3.76

Thus, R Total = 2.85 + 3.76 = 6.61

Energy Savings

$$q = \left(0.35 - \frac{1}{6.61}\right) \times 10{,}000 \times 17 = 33{,}781 \text{ Btuh.}$$

THE EFFECT OF SUNLIGHT

Heat from the sun's rays greatly increases heat gain of a building. If the building energy requirements were mainly due to cooling, then this gain should be minimized. Solar energy affects a building in the following ways:

1. *Raises the surface temperature:* Thus a greater temperature differential will exist at roofs than at walls.
2. A large percentage of direct solar radiation and diffuse sky radiation *passes through* transparent materials, such as glass.

Surface Temperatures

The temperature of a wall or roof depends upon:

(a) the angle of the sun's rays
(b) the color and roughness of the surface
(c) the reflectivity of the surface
(d) the type of construction.

When an engineer is specifying building materials, he should consider the above factors. A simple example is color. The darker the surface, the more Solar radiation will be absorbed. Obviously, white surfaces have a lower temperature than black surfaces after the same period of solar heating. Another factor is that smooth surfaces reflect more radiant heat than do rough ones.

In order to properly take solar energy into account, the angle of the sun's rays must be known. If the latitude of the plant is known, the angle can be determined.

Tables 7-15 and 7-16 illustrates typical temperature differentials for the roof and walls, based on different locations in the United States and various material. Table 7-16 is simplified to indicate Dark (D) and Light (L) color materials only. A full set of tables can be found in the ASHRAE Guide and Data Book.

SIM 7-5

An engineer is evaluating three designs for a roof. In the first case the construction is as follows: a 2 in. lightweight aggregate deck with no insulation. The second case is the same as the first, except a 1 in. glass fibreboard form with roof insulation of $R = 4.17$ is used. The third case is the same as case two, except a 4 in. light weight aggregate is used instead of the 2 inch used in case one. The roof area is 5000 ft^2. What is the maximum heat gain for August 24, if the plant was located in Indianapolis, Indiana?

ANSWER

From Table 7-12, the conductance for each case is as follows:

Case 1: $U = 0.27\ (A' - 4)$
Case 2: $U = 0.13\ (D' - 4)$
Case 3: $U = 0.10\ (D' - 12)$

To find the latitude of the city, Table 7-14 is used. From Table 7-14, Indianapolis is at 40 degrees N. Latitude. Thus, Table 7-15 can be used to find the temperature differentials. For other latitudes, refer to the ASHRAE Guide and Data Book.

From Table 7-15 at 2:00 P.M. Case 1 and 2: $\Delta T = 60$
From Table 7-15 at 4:00 P.M. Case 3: $\Delta T = 54$

Thus,
Case 1: q max = 0.27 × 5000 × 60 = 81,000 Btuh

Case 2: q max = 0.13 × 5000 × 60 = 39,000 Btuh

Case 3: q max = 0.10 × 5000 × 52 = 27,000 Btuh

Case 3 reduces the heat gain through the roof by 54,000 Btuh.

Sunlight and Glass Considerations

A danger in the energy conservation movement is to take steps backward. A simple example would be to exclude glass from building designs because of the poor conductance and solar heat gain factors of clear glass. The engineer needs to evaluate various alternate glass constructions and coatings in order to maintain and improve the aesthetic qualities of good design while minimizing energy inefficiencies. Table 7-17 illustrates several categories of glass which are presently available. It should be noted that the method to reduce heat gain of glass due to conductance is to provide an insulating air space.

To reduce the solar radiation that passes through glass, several techniques are available. Heat absorbing glass (tinted glass) is very popular. Reflective glass is gaining popularity, as it greatly reduces solar heat gains. Fig. 7-4 illustrates a building utilizing reflective glass.

To calculate the relative heat gain through glass, a simple method is illustrated below:

$$Q = UA\,(t_0 - t_1) + A \times S_1 \times S_2 \qquad (48)$$

where

Q is the total heat gain for each glass orientation (Btuh).
U is the conductance of the glass (Btu/h-ft^2-°F)
A is the area of glass; The area used should include framing, since it will generally have a poor conductance as compared with the surrounding material. (ft^2)
$t_0 - t_1$ is the temperature difference between the inside temperature and outside ambient. (°F)
S_1 is the shading coefficient; S_1 takes into account external shades,

Fig. 7-4 Reflective glass installation Mountain Bell Plaza, Phoenix, Arizona. (*Courtesy of PPG Industries*.)

such as venetian blinds and draperies, and the qualities of the glass, such as tinting and reflective coatings.

S_2 is the solar heat gain factor and is determined from Table 7-18; This factor takes into account direct and diffused radiation from the sun. Diffused radiation is basically caused by reflections from dust particles and moisture in the air.

When using Table 7-18, determine the data and hour which gives the maximum heat gain when the contributions of all walls are considered. To use the table for directions corresponding to N, NW, W, SW, and S use the hours at the bottom of the table.

CAUTION: When using reflective glass always insure that the reflections from the building do not increase the solar radiation on adjacent buildings.

TABLE 7-15 Total Equivalent Temperature Differentials for Roofs for April 20 and August 24#.

DESCRIPTION OF ROOF CONSTRUCTION*	SUN TIME								
	A.M.			P.M.					
	8	10	12	2	4	6	8	10	12
LIGHT CONSTRUCTION ROOFS-EXPOSED TO SUN									
1" WOOD**OR 1" WOOD + 1" OR 2" INSULATION	14.0	40.0	56.0	64.0	52.0	21.0	12.0	6.0	2.0
MEDIUM CONSTRUCTION ROOFS-EXPOSED TO SUN									
2" CONCRETE OR 2" CONCRETE + 1" OR 2" INSULATION OR 2" WOOD	8.0	32.0	50.0	60.0	52.0	34.0	16.0	8.0	4.0
2" GYPSUM OR 2" GYPSUM + 1" INSULATION 1" WOOD OR 2" WOOD OR } + 4" ROCK WOOL 2" CONCRETE OR } IN FURRED CEILING 2" GYPSUM	2.0	22.0	42.0	54.0	56.0	44.0	22.0	12.0	8.0
4" CONCRETE OR 4" CONCRETE WITH 2" INSULATION	2.0	22.0	40.0	52.0	54.0	42.0	24.0	14.0	8.0
HEAVY CONSTRUCTION ROOFS-EXPOSED TO SUN									
6" CONCRETE	6.0	8.0	26.0	40.0	48.0	46.0	34.0	20.0	14.0
6" CONCRETE + 2" INSULATION	8.0	8.0	22.0	36.0	44.0	46.0	36.0	22.0	16.0
ROOFS COVERED WITH WATER-EXPOSED TO SUN									
LIGHT CONSTRUCTION ROOF WITH 1" WATER	2.0	6.0	18.0	24.0	20.0	16.0	12.0	4.0	2.0
HEAVY CONSTRUCTION ROOF WITH 1" WATER	0	0	-2.0	12.0	16.0	18.0	16.0	12.0	8.0
ANY ROOF WITH 6" WATER	0	2.0	2.0	8.0	12.0	12.0	10.0	6.0	2.0
ROOFS WITH ROOF SPRAYS-EXPOSED TO SUN									
LIGHT CONSTRUCTION	2.0	6.0	14.0	20.0	18.0	16.0	12.0	4.0	2.0
HEAVY CONSTRUCTION	0	0	4.0	10.0	14.0	16.0	14.0	12.0	8.0
ROOFS IN SHADE									
LIGHT CONSTRUCTION	-2.0	2.0	8.0	14.0	16.0	14.0	10.0	4.0	2.0
MEDIUM CONSTRUCTION	-2.0	0	4.0	10.0	14.0	14.0	12.0	8.0	4.0
HEAVY CONSTRUCTION	0	0	2.0	6.0	10.0	12.0	12.0	10.0	6.0

*Includes 3/8" felt roofing with or without slag. May also be used for shingle roof.
**Nominal thickness of wood.
#Table 7-15 is for 40 degrees North latitude. It may also be used for 40 degrees South latitude for February 20 and October 23.

Courtesy of the Trane Company.

TABLE 7-16 Total Equivalent Temperature Differentials for Walls for April 20 and August 24*.

NORTH LATITUDE		SUN TIME																		SOUTH LATITUDE
		A.M.						P.M.												
		8		10		12		2		4		6		8		10		12		
		EXTERIOR COLOR D = DARK L = LIGHT																		
		D	L	D	L	D	L	D	L	D	L	D	L	D	L	D	L	D	L	
		FRAME																		
	NE	19	9	21	11	14	11	14	12	16	16	16	16	12	12	8	6	4	4	SE
	E	32	16	38	20	34	18	14	14	16	16	16	16	12	12	8	8	4	4	E
	SE	18	10	33	21	34	22	28	19	18	16	16	16	12	12	8	6	4	4	NE
	S	-2	-2	9	3	32	18	41	26	33	25	20	17	13	13	9	9	5	5	N
	SW	-2	-2	2	0	8	6	31	26	46	32	49	33	29	24	8	6	4	4	NW
	W	-2	-2	2	2	8	8	22	14	42	30	50	36	24	24	10	10	4	4	W
	NW	-2	-2	2	0	8	6	14	12	24	21	37	26	31	23	8	6	4	4	SW
(SHADE)	N	-2	-2	0	0	6	6	12	12	16	16	14	14	10	10	6	6	2	2	S (SHADE)
		4″ BRICK OR STONE VENEER + FRAME																		
	NE	0	-2	21	11	18	10	11	8	14	12	16	16	14	14	12	12	8	6	SE
	E	4	2	32	16	33	19	16	16	14	14	16	16	14	14	12	10	8	8	E
	SE	5	0	26	14	35	21	31	20	21	17	16	16	14	14	12	10	8	8	NE
	S	-2	-2	0	0	19	11	34	22	35	24	26	20	14	14	10	10	6	6	N
	SW	3	0	2	0	4	4	15	10	38	26	42	30	40	28	12	10	8	8	NW
	W	2	0	2	2	6	4	12	10	28	20	42	30	44	30	18	16	8	8	W
	NW	-2	-2	0	0	4	4	10	8	14	14	29	22	32	24	13	12	8	8	SW
(SHADE)	N	-2	-2	0	0	2	2	8	8	12	12	14	14	14	14	10	10	6	6	S (SHADE)
		8″ HOLLOW TILE OR 8″ CINDER BLOCK																		
	NE	2	2	2	2	18	10	15	10	11	8	14	12	15	14	14	12	10	10	SE
	E	6	4	14	6	26	14	28	16	22	14	14	12	16	14	16	12	12	10	E
	SE	5	2	5	2	21	12	25	16	24	17	17	14	17	14	14	12	10	8	NE
	S	3	3	3	3	6	3	19	11	34	20	35	21	26	18	15	12	11	8	N
	SW	5	2	5	2	5	2	9	7	15	13	31	21	35	24	31	21	10	8	NW
	W	6	4	6	4	6	4	8	6	12	10	20	16	32	24	34	24	20	16	W
	NW	2	2	2	2	3	2	5	4	10	8	14	12	22	19	28	22	11	10	SW
(SHADE)	N	0	0	0	0	0	0	2	2	8	8	12	12	12	12	12	12	8	8	S (SHADE)

8" BRICK OR 12" HOLLOW TILE OR 12" CINDER BLOCK																				
	NE	4	4	4	4	10	4	15	9	14	9	11	8	12	10	12	12	11	10	SE
	E	10	8	10	8	16	10	20	12	20	12	16	10	16	12	16	12	14	12	E
	SE	11	7	9	7	9	7	18	14	23	16	20	15	15	12	15	12	15	13	NE
	S	8	5	8	5	8	5	8	5	16	10	22	14	22	16	16	13	14	11	N
	SW	11	7	9	7	9	7	11	7	13	9	15	10	24	15	29	19	24	17	NW
	W	10	6	8	6	8	8	10	8	12	8	16	10	22	18	26	18	26	18	W
	NW	4	4	4	4	4	4	5	4	7	6	10	8	12	10	17	15	18	15	SW
(SHADE)	N	2	2	2	2	2	2	2	2	4	4	8	8	10	10	10	10	8	8	S (SHADE)
12" BRICK																				
	NE	9	8	9	7	9	6	9	6	11	6	12	7	12	7	11	8	11	8	SE
	E	14	10	14	10	14	10	12	8	14	10	16	12	16	12	16	10	16	10	E
	SE	13	8	13	9	13	9	13	9	13	9	16	11	18	13	18	13	15	10	N
	S	12	9	13	10	10	7	10	7	10	7	13	7	16	10	18	12	17	11	NE
	SW	13	8	13	9	13	9	13	9	13	9	13	11	13	11	15	11	17	13	NW
	W	14	10	14	10	14	10	12	8	12	8	12	8	12	8	14	10	18	12	W
	NW	9	8	9	7	9	6	9	6	9	6	9	6	9	7	11	8	11	8	SW
(SHADE)	N	6	6	4	4	4	4	4	4	4	4	4	4	4	4	6	6	8	8	S (SHADE)
8" CONCRETE OR STONE OR 6" OR 8" CONCRETE BLOCK																				
	NE	7	5	5	4	5	4	6	4	11	8	15	11	17	13	15	11	11	9	SE
	E	8	6	16	10	26	14	26	14	20	12	16	12	16	12	14	12	12	10	E
	SE	9	4	9	7	21	14	23	16	22	15	17	15	15	12	15	13	13	11	N
	S	5	3	5	3	8	3	18	10	23	18	25	17	19	16	14	11	12	9	NE
	SW	9	4	7	4	9	4	11	6	18	13	27	20	29	19	27	20	13	11	NW
	W	8	6	8	6	8	6	10	8	14	10	22	16	30	20	28	20	16	12	W
	NW	5	4	5	2	5	4	6	6	8	8	13	11	20	15	21	16	9	8	SW
(SHADE)	N	2	2	2	2	2	2	4	4	6	6	8	8	10	10	8	8	6	6	S (SHADE)
12" CONCRETE OR STONE																				
	NE	7	5	7	4	7	4	13	9	14	9	11	9	11	10	13	12	11	10	SE
	E	12	8	10	8	12	8	20	12	20	14	18	12	14	12	16	12	16	12	E
	SE	11	7	11	7	9	7	18	11	20	13	20	13	17	13	15	12	15	13	N
	S	11	8	8	5	8	5	8	5	16	10	20	15	22	17	19	13	14	11	NE
	SW	11	7	11	7	9	7	9	7	11	9	13	11	22	17	24	17	22	15	NW
	W	12	8	10	8	10	8	12	8	12	8	14	10	18	12	26	16	24	16	W
	NW	7	5	7	4	7	4	7	5	7	6	9	8	11	10	18	13	20	15	SW
(SHADE)	N	2	2	2	2	2	2	2	2	4	4	6	6	9	8	10	10	8	8	S (SHADE)

*Table 7-16 is for 40 degrees latitude. It may be used for south latitude for February 20 and October 23.

Courtesy of the Trane Company.

TABLE 7-17 Typical Glass Characteristics.

MONOLITHIC GLASS CLEAR AND TINTED

GLASS	THICKNESS in	THICKNESS mm	TRANSMITTANCE Average Daylight %	TRANSMITTANCE Total Solar %	REFLECTANCE Average Daylight %	RELATIVE HEAT GAIN Btu/hr-sq ft	RELATIVE HEAT GAIN W/m²	U VALUE Btu/hr-sq ft / °F	U VALUE W/m² °k	SHADING COEFFICIENTS No Shade	Venetian Blinds Light	Venetian Blinds Med	Draperies Light	Draperies Med	Draperies Dark
SHEET	SS	2.5	91	87	8	215	678	1.10	6.3	1.00	.55	.64	.56	.61	.70
	DS	3	90	86											
	3/16	5	90	83		211	665			.98					
CLEAR	1/8	3	90	83		215	678			1.00	.55	.64	.56	.61	.70
	3/16	5	89	79		205	646			.95					
	1/4	6	88	77		201	634			.93					
	5/16	8	88	77		201	634			.93					
CLEAR HEAVY DUTY	3/8	10	87	75		199	627			.92					
	1/2	12	86	71		191	602			.88	.54	.62	.52	.56	.66
	5/8	15	85	67		185	583			.85					
	3/4	19	83	63		179	564			.82	.54	.60	.50	.53	.62
	7/8	22	81	59		171	539			.78					
BLUE-GREEN	1/8	3	83	63	7	179	564			.82	.54	.60	.50	.53	.62
	3/16	5	79	55		161	508			.73	.53	.57	.45	.47	.54
	1/4	6	75	47		155	489			.70					
GREY	3/16	5	51	53	6	163	514			.74	.53	.58	.47	.49	.51
	1/4	6	44	46	5	151	476			.68	.52	.56	.44	.47	.52
	5/16	8	35	38		137	432			.61	.47	.50	.42	.44	.49
	3/8	10	28	31		125	394			.55	.43	.45	.40	.42	.45
	1/2	12	19	22	4	111	350			.48	.38	.40	.30	.38	.40

BRONZE	1/8	3	68	65	6	183	577	.84	.54	.61	.51	.54	.64
	3/16	5	59	55		165	520	.75	.53	.58	.47	.49	.57
	1/4	6	52	49		155	489	.70	.53	.57	.45	.47	.54
	5/16	8	44	40	5	141	444	.63	.48	.51	.43	.45	.50
	3/8	10	38	34		131	413	.58	.45	.48	.41	.43	.47
	1/2	12	28	24		113	356	.49	.38	.40	.30	.38	.40

THERMOPANE INSULATING GLASS (inboard light clear)

						AIR SPACE		AIR SPACE									
						1/2″	12mm	1/4″ 6mm		1/2″ 12mm							
CLEAR	1/8	3	80	69	15	182	574					.87	.51	.57	.50	.53	.58
	3/16	5	79	62	14	172	542					.82	.51	.56	.48	.52	.57
	1/4	6	77	59	14	166	523					.79	.49	.54	.41	.50	.56
BLUE-GREEN	1/8	3	75	52	13	148	466					.70	.44	.48	.44	.47	.52
	3/16	5	70	43	12	132	416					.62	.39	.43	.41	.42	.43
	1/4	6	66	36	12	120	378	.60	3.4	.55	3.1	.56	.36	.39	.38	.41	.44
GREY	3/16	5	45	42	8	128	403					.60	.38	.42	.40	.43	.47
	1/4	6	39	35	7	120	370					.54	.35	.38	.37	.40	.43
BRONZE	1/8	3	61	54	11	150	473					.71	.44	.48	.44	.47	.52
	3/16	5	53	43	9	122	385					.62	.34	.43	.41	.43	.48
	1/4	6	46	38	8	120	378					.56	.36	.39	.38	.41	.44

THERMOPANE Xi

							AIR SPACE							
							3/16	5mm						
SHEET	SS&DS	81	75	15	188	593	.58	3.4	.90	.51	.57	.50	.53	.58

Courtesy of Libbey-Owens-Ford Company.

TABLE 7-17 (Cont.) Typical Glass Characteristics.

LAMINATED VARI-TRAN

GLASS	THICKNESS in	THICKNESS mm	VARI-TRAN COATING Color	VARI-TRAN COATING Number	TRANSMITTANCE Average Daylight %	TRANSMITTANCE Average Daylight Tol	TRANSMITTANCE Total Solar %	REFLECTANCE Average Daylight %	REFLECTANCE Average Daylight Tol	RELATIVE HEAT GAIN [1] Btu/hr-sq ft	RELATIVE HEAT GAIN [1] W/m²	U VALUE Btu/hr-sq ft/°F	U VALUE W/m²°k	SHADING COEFFICIENTS No Shade	Venetian Blinds Light	Venetian Blinds Med	Draperies Light	Draperies Med	Draperies Dark
CLEAR	¼	6	Silver	1-108	8	±1.5	9	43	±3.0	71	224	1.05	6.0	.28	.21	.23	.21	.22	.23
				1-114	14	±2.0	15	33		89	281			.37	.27	.31	.28	.29	.31
				1-120	20	±2.5	19	27		95	299			.40	.29	.33	.30	.32	.34
			Golden	1-208	8	±1.5	8	27	±3.0	73	230			.29	.22	.24	.22	.22	.24
				1-214	14	±2.0	13	24		85	268			.35	.26	.29	.26	.27	.29
				1-220	20	±2.5	20	21		99	312			.42	.31	.35	.32	.33	.36

MONOLITHIC VARI-TRAN

GLASS	THICKNESS in	THICKNESS mm	VARI-TRAN COATING Color	VARI-TRAN COATING Number	TRANSMITTANCE Average Daylight %	TRANSMITTANCE Average Daylight Tol	TRANSMITTANCE Total Solar %	REFLECTANCE Average Daylight %	REFLECTANCE Average Daylight Tol	RELATIVE HEAT GAIN [1] Btu/hr-sq ft	RELATIVE HEAT GAIN [1] W/m²	U VALUE Btu/hr-sq ft/°F	U VALUE W/m²°k	SHADING COEFFICIENTS No Shade	Venetian Blinds Light	Venetian Blinds Med	Draperies Light	Draperies Med	Draperies Dark
CLEAR	¼	6	Silver	1-108	8	±1.5	9	44	±3.0	57	180	.80	4.5	.23	.18	.19	.18	.18	.19
				1-114	14	±2.0	16	33		82	258	.85	4.8	.35	.26	.29	.26	.28	.29
				1-120	20	±2.5	20	27		91	287	.90	5.1	.39	.28	.32	.29	.31	.33
			Golden	1-208	8	±1.5	9	28	±3.0	63	199	.80	4.5	.26	.20	.22	.20	.20	.21
				1-214	14	±2.0	14	26		76	240	.85	4.8	.32	.24	.27	.24	.25	.26
				1-220	20	±2.5	21	24		93	293	.90	5.1	.40	.29	.33	.30	.32	.34
B/GREEN			Blue	2-350	50	±5.0	35	18	±3.0	129	407	1.05	6.0	.57	.42	.48	.42	.44	.48
GREY TUF-FLEX			Grey	3-108 [2]	8	±1.5	11	11	±2.0	78	246	.85	4.8	.33	.25	.28	.25	.26	.27
				3-114 [2]	14	±2.0	17	9		93	293	.90	5.1	.40	.29	.33	.30	.32	.34
				3-120 [2]	20	±2.5	24	7		106	334	1.00	5.7	.46	.34	.38	.35	.36	.39
GREY				3-134	34	±4.0	36	6		135	426	1.05	6.0	.60	.44	.50	.43	.46	.50
BRONZE TUF-FLEX			Bronze	4-108 [2]	8	±1.5	9	14	±2.0	74	233	.85	4.8	.31	.24	.26	.24	.24	.25
				4-114 [2]	14	±2.0	14	11		89	281	.90	5.1	.38	.28	.32	.29	.30	.32
				4-120 [2]	20	±2.5	20	9		99	312	.95	5.4	.43	.32	.35	.32	.34	.37
BRONZE				4-134	34	±4.0	31	6		127	400	1.05	6.0	.56	.42	.47	.41	.43	.47

THERMOPANE VARI-TRAN (inboard light clear except Vari-Tran 2-350-2)

CLEAR	1	25	Silver	1-108	7	±1.5	7	44	±3.0	41	129	.50	2.8	.17	.16	.16	.15	.15	.15
				1-114	13	±2.0	14	33		59	186			.26	.23	.23	.22	.22	.23
				1-120	18	±2.5	16	27		67	211			.30	.26	.27	.26	.26	.27
			Golden	1-208	7	±1.5	7	28	±3.0	43	136			.18	.17	.17	.16	.16	.16
				1-214	13	±2.0	12	26		55	173			.24	.21	.22	.21	.21	.22
				1-220	18	±2.5	17	24		69	218			.31	.27	.28	.27	.27	.28
BLUE-GREEN			Blue	2-350	45	±5.0	28	20	±3.0	98	309	.55	3.1	.45	.38	.39	.36	.38	.41
				2-350-2 ③	38	±5.0	20	20		96	303			.44	.37	.38	.35	.37	.40
GREY TUF-FLEX			Grey	3-108 ②	7	±1.5	9	11	±2.0	53	167	.50	2.8	.23	.20	.21	.20	.20	.21
				3-114 ②	13	±2.0	14	9		65	205			.29	.25	.26	.25	.25	.26
				3-120 ②	18	±2.5	20	7		76	240	.55	3.1	.34	.29	.30	.28	.29	.31
GREY				3-134	30	±4.0	29	7		102	322			.47	.39	.41	.38	.40	.43
BRONZE TUF-FLEX			Bronze	4-108 ②	7	±1.5	7	14	±2.0	49	154	.50	2.8	.21	.19	.20	.18	.18	.19
				4-114 ②	13	±2.0	11	11		61	192			.27	.24	.24	.23	.23	.24
				4-120 ②	18	±2.5	15	9		70	221	.55	3.1	.31	.27	.28	.27	.27	.28
BRONZE				4-134	30	±4.0	25	7		94	296			.43	.36	.37	.35	.37	.40

① When ASHRAE Solar Heat Gain Factor is 200 Btu/hr-sq. ft. and outdoor air is 14°F. warmer than indoor air, with no indoor shading.
② Tempered only May be furnished in 5⁄16″ (8 mm) thickness in some sizes.
③ May require both lights be tempered

ADDITIONAL TECHNICAL DATA UPON APPLICATION

The satisfactory performance of all LOF products requires selection of the appropriate type, size and thickness of glass, and proper glazing in adequate sash. The information contained in this catalog is intended as a general guide in planning your glass requirements. Libbey-Owens-Ford assumes no responsibility for its use or application.

TABLE 7-18 Solar Heat Gain Factors for 40 North Latitude.

Solar Heat Gain Factors

June 21										
	AM						PM			
	8	**9**	**10**	**11**	**12**	**1**	**2**	**3**	**4**	
N	29	33	35	37	38	37	35	33	29	**N**
NE	156	113	62	40	38	37	35	31	26	**NW**
E	215	192	145	80	41	37	35	31	26	**W**
SE	152	161	148	116	71	41	36	31	26	**SW**
S	29	45	69	88	95	88	69	45	29	**S**
	4	**3**	**2**	**1**	**12**	**11**	**10**	**9**	**8**	
	PM						AM			

September 21										
	AM						PM			
	8	**9**	**10**	**11**	**12**	**1**	**2**	**3**	**4**	
N	16	22	26	29	30	29	26	22	16	**N**
NE	87	47	28	29	30	29	26	22	16	**NW**
E	205	195	148	77	32	29	26	22	16	**W**
SE	199	226	221	192	141	77	30	23	16	**SW**
S	71	124	165	191	200	191	165	124	71	**S**
	4	**3**	**2**	**1**	**12**	**11**	**10**	**9**	**8**	
	PM						AM			

Courtesy of Libbey-Owens-Ford Company.

Applying Insulation

As illustrated in this chapter, insulation offers one of the best methods to improve the efficiency of building designs. Table 7-19 offers a comparison of various insulation materials. In practice, the thermal conductivities of insulation should be based on the manufacturer's data.

TABLE 7-19 Thermal Conductivity (k) of Industrial Insulation (Design Values)[a] (For Mean Temperatures Indicated).

Expressed in Btu per (hour) (square foot) (Fahrenheit degree temperature difference per in.)

Form	Material (Composition)	Accepted Max Temp for Use,* F	Typical Density (lb/cu ft)	Typical Conductivity k at Mean Temp F													
				-100	-75	-50	-25	0	25	50	75	100	200	300	500	700	900
BLANKETS and FELTS	MINERAL FIBER (Rock, Slag, or Glass)																
	Blanket, Metal Reinforced	1200	6–12									0.26	0.32	0.39	0.54		
		1000	2.5–6									0.24	0.32	0.40	0.61		
	Mineral Fiber, Glass																
	Blanket, Flexible, Fine-Fiber Organic Bonded	350	0.65				0.25	0.26	0.28	0.30	0.33	0.36	0.53				
			0.75				0.24	0.25	0.27	0.29	0.32	0.34	0.48				
			1.0				0.23	0.24	0.25	0.27	0.29	0.32	0.43				
			1.5				0.21	0.22	0.23	0.25	0.27	0.28	0.37				
			2.0				0.20	0.21	0.22	0.23	0.25	0.26	0.33				
			3.0				0.19	0.20	0.21	0.22	0.23	0.24	0.31				
	Blanket, Flexible, Textile-Fiber Organic Bonded	350	0.65				0.27	0.28	0.29	0.30	0.31	0.32	0.50	0.68			
			0.75				0.26	0.27	0.28	0.29	0.31	0.32	0.48	0.66			
			1.0				0.24	0.25	0.26	0.27	0.29	0.31	0.45	0.60			
			1.5				0.22	0.23	0.24	0.25	0.27	0.29	0.39	0.51			
			3.0				0.20	0.21	0.22	0.23	0.24	0.25	0.32	0.41			
	Felt, Semi-Rigid Organic	400	3–8						0.24	0.25	0.26	0.27	0.35	0.44			
	Bonded	850	3	0.16	0.17	0.18	0.19	0.20	0.21	0.22	0.23	0.24	0.35	0.55			
	Laminated & Felted Without Binder	1200	7.5											0.35	0.45	0.60	
	VEGETABLE and ANIMAL FIBER																
	Hair Felt or Hair Felt plus Jute	180	10						0.26	0.28	0.29	0.30					
BLOCKS, BOARDS and PIPE INSULATION	ASBESTOS																
	Laminated Asbestos Paper	700	30									0.40	0.45	0.50	0.60		
	Corrugated & Laminated Asbestor Paper																
	4-ply	300	11–13								0.54	0.57	0.68				
	6-ply	300	15–17								0.49	0.51	0.59				
	8-ply	300	18–20								0.47	0.49	0.57				

TABLE 7-19 (Continued)

Form	Material (Composition)	Accepted Max Temp for Use,* F	Typical Density (lb/cu ft)	Typical Conductivity k at Mean Temp F													
				-100	-75	-50	-25	0	25	50	75	100	200	300	500	700	900
	MOLDED AMOSITE and																
	BINDER	1500	15–18									0.32	0.37	0.42	0.52	0.62	0.72
	85% MAGNESIA	600	11–12									0.35	0.38	0.42			
	CALCIUM SILICATE	1200	11–13									0.35	0.41	0.44	0.52	0.62	0.72
		1800	12–15												0.63	0.74	0.95
	CELLULAR GLASS	800	9			0.32	0.33	0.35	0.36	0.38	0.40	0.42	0.48	0.55			
	DIATOMACEOUS SILICA	1600	21–22												0.64	0.68	0.72
		1900	23–25												0.70	0.75	0.80
	MINERAL FIBER																
	Glass, Organic Bonded,																
	Block and Boards	400	3–10	0.16	0.17	0.18	0.19	0.20	0.22	0.24	0.25	0.26	0.33	0.40			
	Non-Punking Binder	1000	3–10									0.26	0.31	0.35	0.52		
	Pipe Insulation, slag or																
	or glass	350	3–4					0.20	0.21	0.22	0.23	0.24	0.29				
		500	3–10					0.20	0.22	0.24	0.25	0.26	0.33	0.40			
	Inorganic Bonded Block	1000	10–15									0.33	0.38	0.45	0.55		
		1800	15–24									0.32	0.37	0.42	0.52	0.62	0.74
	Pipe Insulation slag																
	or glass	1000	10–15									0.33	0.35	0.45	0.55		
	MINERAL FIBER																
	Resin Binder		15			0.23	0.24	0.25	0.26	0.28	0.29						
	Rigid Polystyrene																
	Extruded, R-12 exp	170	3.5	0.16	0.16	0.15	0.16	0.16	0.17	0.18	0.19	0.20					
	Extruded, R-12 exp	170	2.2	0.16	0.16	0.17	0.16	0.17	0.18	0.19	0.20						
	Extruded	170	1.8	0.17	0.18	0.19	0.20	0.21	0.23	0.24	0.25	0.27					
	Molded Beads	170	1	0.18	0.20	0.21	0.23	0.24	0.25	0.26	0.28						
	Polyurethane**																
	R-11 exp	210	1.5–2.5	0.16	0.17	0.18	0.18	0.18	0.17	0.16	0.16	0.17					
	RUBBER, Rigid Foamed	150	4.5						0.20	0.21	0.22	0.23					
	VEGETABLE and ANIMAL																
	FIBER																
	Wool Felt (Pipe Insulation)	180	20						0.28	0.30	0.31	0.33					

INSULATING CEMENTS	MINERAL FIBER (Rock, Slag, or Glass) With Colloidal Clay Binder	1800	24–30									0.49	0.55	0.61	0.73	0.85	
	With Hydraulic Setting Binder	1200	30–40									0.75	0.80	0.85	0.95		
LOOSE FILL	Cellulose insulation (Milled pulverized paper or wood pulp)		2.5–3							0.26	0.27	0.29					
	Mineral fiber, slag, rock or glass		2–5			0.19	0.21	0.23	0.25	0.26	0.28	0.31					
	Perlite (expanded		5–8	0.25	0.27	0.29	0.30	0.32	0.34	0.35	0.37	0.39					
	Silica aerogel		7.6			0.13	0.14	0.15	0.15	0.16	0.17	0.18					
	Vermiculite (expanded)		7–8.2			0.39	0.40	0.42	0.44	0.45	0.47	0.49					
			4–6			0.34	0.35	0.38	0.40	0.42	0.44	0.46					

[a]Representative values for dry materials as selected by the ASHRAE Technical Committee 2.4 on Insulation. They are intended as design (not specification) values for materials of building construction for normal use. For the thermal resistance of a particular product, the user may obtain the value supplied by the manufacturer or secure the results of unbiased tests.

*These temperatures are generally accepted as maximum. When operating temperature approaches these limits the manufacturer's recommendations should be followed.

**These are values for aged board stock.

Reprinted by permission from *ASHRAE Handbook of Fundamentals*, 1972.

WINDOW TREATMENTS

Several types of window treatments to reduce losses have become available. This section describes some of the products on the market based on information supplied by manufacturers. No claims are made concerning the validity or completeness described. The summary is based on "Windows For Energy Efficient Buildings" as prepared by the Lawrence Berkeley Laboratory for U.S. DOE under contract W-7405-ENG-48.

Solar Control

Solar Control Films. A range of tinted and reflective polyester films are available to adhere to inner window surfaces to provide solar control for existing clear glazing. Films are typically two- or three-layer laminates composed of metalized, transparent and/or tinted layers. Films are available with a wide range of solar and visible light transmittance values, resulting in shading coefficients as low as 0.24. Most films are adhered with precoated pressure sensitive adhesives. Reflective films will reduce winter U values by about 20%. (Note that a new solar control film, which provides a U value of 0.68, is described in the Thermal Barriers section below). Films adhered to glass improve the shatter resistance of glazing and reduce transmission, thus reducing fading of furnishings.

Fiberglass Solar Control Screens. Solar control screen provides sun and glare control as well as some reduction in winter heat loss. Screens are woven from vinyl-coated glass strands and are available in a variety of colors. Depending on color and weave, shading coefficients of 0.3 to 0.5 are achieved. Screens are durable, maintenance-free, and provide impact resistance. They are usually applied on the exterior of windows and may (1) be attached to mounting rails and stretched over windows, (2) mounted in rigid frames and installed over windows, or (3) made into roller shades which can be retracted and stored as desired.

Motorized Window Shading System. A variety of plastic and fabric shades is available for use with a motorized window shading system. Reversible motor is located within the shade tube roller and contains a brake mechanism to stop and hold in any position. Motor controls may be gauged and operated locally or from a master station. Automatic photoelectric controls are available that (1) monitor sun intensity and angle and adjust shade position to provide solar control and (2) employ an internal light sensor and provide a preset level of internal ambient light.

Exterior Sun Control Louvers. Operable external horizontal and vertical louver systems are offered for a variety of building sun control applications. Louvers are hinged together and can be rotated in unison to provide the desired degree of shading for any sun position. Operation may be manual or electric; electrical operation may be initiated by manual switches, time clock, or sun sensors. Louvers may be closed to reduce night thermal losses. Sun control elements are available in several basic shapes and in a wide range of sizes.

External Venetian Blinds. Externally mounted, all-weather venetian blinds may be manually operated from within a building or electrically operated and controlled by means of automatic sun sensors, time controls, manual switches, etc. Aluminum slats are held in position with side guides and controlled by weatherproof lifting tapes. Slats can be tilted to modulate solar gain, closed completely, or restricted to admit full light and heat. Blinds have been in use in Europe for many years and have been tested for resistance to storms and high winds.

Adjustable Louvered Windows. Windows incorporating adjustable external louvered shading devices are available. Louvers are extruded aluminum or redwood, 3 to 5 inches wide, and are manually controlled. Louvers may be specified on double-hung, hinged, or louvered-glass windows. When open, the louvers provide control of solar gain and glare; when closed, they provide privacy and security.

Solar Shutters. The shutter is composed of an array of aluminum slats set at 45° or $22\frac{1}{2}°$ from the vertical to block direct sunlight. Shutters are designed for external application and may be mounted vertically in front of window or projected outward from the bottom of the window. Other rolling and hinged shutters are stored beside the window and roll or swing into place for sun control, privacy, or security.

Thermal Barriers

Multilayer, Roll-up Insulating Window Shade. A multilayer window shade stores in a compact roll and utilizes spacers to separate the aluminized plastic layers in the deployed position, thereby creating a series of dead air spaces. A five-layer shade combined with insulated glass provides R8 thermal resistance.

Insulating Window Shade. A ThermoShade thermal barrier is a roll-up shade composed of hollow, lens-shaped, rigid, white PVC slats with virtually no air leakage through connecting joints. The side tracking system reduces window infiltration. Designed for interior installation and manual or automatic operation.

Insulating Window Shade. When added to a window, the roll-up insulating shade provides R4.5 for a single-glazed window or R5.5 for a double-glazed window. Quilt is composed of fabric outer surfaces and two polyester fiberfill layers sandwiched around a reflective vapor barrier. Quilt layers are ultrasonically welded. Shade edges are enclosed in a side track to reduce infiltration.

Reflective, Perforated Solar Control Laminate. Laminate of metalized weatherable polyester film and black vinyl is then perforated with 225 holes/in.^{2}, providing 36 percent open area. Available in a variety of metallized and nonmetallized colors, the shading coefficients vary from 0.30 to 0.35 for externally mounted screens and 0.37 to 0.45 for the material adhered to the inner glass surface. The laminate is typically mounted in aluminum screen frames which are hung externally, several inches from the window; it can also be utilized in a roll-up form. Some reduction in winter U value can be expected with external applications.

Semi-transparent Window Shades. Roll-up window shades made from a variety of tinted or reflective solar control film laminates. These shades provide most of the benefits of solar control film applied directly to glass but provide additional flexibility and may be retracted on overcast days or when solar heat gain is desired. Shades available with spring operated and gravity (cord and reel) operated rollers as well as motorized options. Shading coefficients as low as 0.13 are achieved and a tight fitting shade provides an additional air space and thus reduced U-value.

Louvered Metal Solar Screens. The solar screen consists of an array of tiny louvers which are formed from a sheet of thin aluminum. The louvered aluminum sheet is then installed in conventional screen frames and may be mounted against a window in place of a regular insect screen or mounted away from the building to provide free air circulation around the window. View to the outside is maintained while substantially reducing solar gain. Available in a light green or black finish with shading coefficients of 0.21 or 0.15, respectively.

Operable External Louvre Blinds. Solar control louvre blinds, mounted on the building exterior, can be controlled manually or automatically by sun and wind sensors. Slats can be tilted to modulate light, closed completely, or retracted to admit full light and heat. Developed and used extensively in Europe, they provide summer sun control, control of natural light, and reduction of winter heat loss.

Louvered Metal Solar Screens. Solar screen consists of an array of tiny fixed horizontal louvers which are woven in place. Louvers are tilted at 17° to provide sun control. Screen material is set in metal frames which may be permanently installed in a variety of configurations or designed for removal. Installed screens have considerable wind and impact resistance. Standard product (17 louvers/inch) has a shading coefficient of 0.23; low sun angle variant (23 louvers/inch) has a shading coefficient of 0.15. Modest reductions in winter U value have been measured.

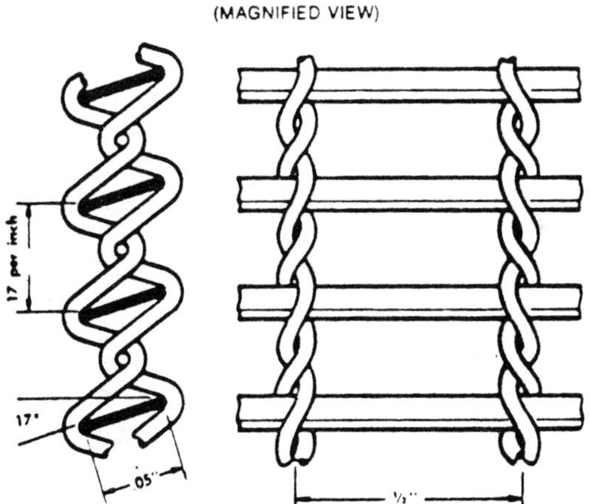

Insulating Solar Control Film. A modified solar control film designed to be adhered to the interior of windows provides conventional solar control function and has greatly improved insulating properties. Film emissivity is 0.23 to 0.25, resulting in a U value of 0.68 Btu/ft^2 hr-°F under winter conditions, compared to 0.87 for conventional solar control films and 1.1 for typical single-glazed windows.

Interior Storm Window. Low cost, do-it-yourself interior storm window with a rigid plastic glazing panel. Glazing panel may be removed for cleaning or summer storage. Reduces infiltration losses as well as conductive/convective heat transfer.

Retrofit Insulating Glass System. Single glazing is converted to double glazing by attaching an extra pane of glass with neoprene sealant. A dessicant-filled aluminum spacer absorbs moisture between the panes. An electric resistance wire embedded in the neoprene is heated with a special power source. This hermetically seals the window. New molding can then be applied if desired.

Infiltration

Weather-strip Tape. A polypropylene film scored along its centerline so that it can be easily formed into a "V" shape. It has a pressure sensitive adhesive on one leg of the "V" for application to seal cracks around doors and windows. On an average fitting, double-hung window, it will reduce infiltration by over 70 percent. It can be applied to rough or smooth surfaces.

BUILDING DESIGN CONSIDERATIONS

There are several factors influencing building design, namely local and state building codes, ASHRAE Standard 90, and Building Energy Performance Standards. To achieve optimum building performance, computer program simulations are usually required, which takes into account the system.

Computer applications are discussed in the next chapter.

ASHRAE Standard 90 is essentially a "prescriptive" standard. The Building Energy Performance Standards (BEPS), which were proposed in 1980, were met with great resistance by the engineering community. These standards would prescribe an energy budget of so many Btus per square foot for different types of buildings. Each Kw of electricity would be converted to a Btu by applying a Resource Utilization Factor (RUFS) that takes into account inefficiencies of transmission.

In order to meet the BEPS Standard, computer simulations would be essential.

8

Heating, Ventilation and Air-Conditioning System Optimization

EFFICIENT USE OF HEATING AND COOLING EQUIPMENT SAVES DOLLARS

Most people have probably experienced improper air conditioning or heating controls which waste energy. Energy is saved when efficient heating, ventillation, and air conditioning (HVAC) systems are used. In this chapter, you will learn how to compare the efficiency of various systems and apply the heat pump to save energy, see how various refrigeration systems can be used to save energy, learn the basics of air conditioning design from an energy conservation viewpoint, and begin to apply the computer approach for energy conservation.

Measuring System Efficiency by Using the Coefficient of Performance

The coefficient of performance (COP) is the basic parameter used to compare the performance of refrigeration and heating systems. COP for cooling and heating applications is defined as follows:

$$\text{COP (Cooling)} = \frac{\text{Rate of Net Heat Removal}}{\text{Total Energy Input}} \qquad (8\text{-}1)$$

$$\text{COP (Heating, Heat Pump*)} = \frac{\text{Rate of Useful Heat Delivered*}}{\text{Total Energy Input}} \qquad (8\text{-}2)$$

*For Heat Pump Applications, exclude supplemental heating.

APPLYING THE HEAT PUMP TO SAVE ENERGY

The heat pump has gained wide attention due to its high potential COP. The heat pump in its simplest form can be thought of as a window air conditioner. During the summer, the air on the room side is cooled while air is heated on the outside air side. If the window air conditioner is turned around in the winter, some heat will be pumped into the room. Instead of switching the air conditioner around, a cycle reversing valve is used to switch functions. This valve switches the function of the evaporator and condenser, and refrigeration flow is reversed through the device. *Thus, the heat pump is heat recovery through a refrigeration cycle.* Heat is removed from one space and placed in another. In Chapter 7, it was seen that the direction of heat flow is from hot to cold. Basically, energy or pumping power is needed to make heat flow "up hill". The mechanical refrigeration compressor "pumps" absorbed heat to a higher level for heat rejection. The refrigerant gas is compressed to a higher temperature level so that the heat absorbed by it, during the evaporation or cooling process, is rejected in the condensing or heating process. Thus, the heat pump provides cooling in the summer and heating in the winter. The source of heat for the heat pump can be from one of three elements: air, water or the ground.

Air to Air Heat Pumps

Heat exists in air down to 460°F below zero. Using outside air as a heat source has its limitations, since the efficiency of a heat pump drops off as the outside air level drops below 55°F. This is because the heat is more dispersed at lower temperatures, or more difficult to capture. Thus, heat pumps are generally sized on cooling load capacities. Supplemental heat is added to compensate for declining capacity of the heat pump. This approach allows for a realistic first cost and an economical operating cost.

Heat Pumps Do Save Energy

An average of 2 to 3 times as much heat can be moved for each Kw input compared to that produced by use of straight resistance heating. Heat pumps can have a COP of greater than 3 in industrial pro-

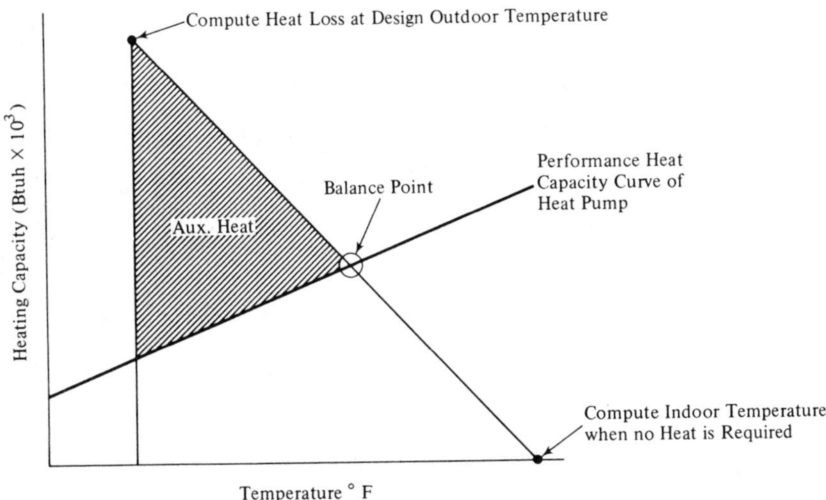

Fig. 8-1. Determining balance point of air to air heat pump.

cesses, depending on temperatures. Commercially available heat pumps range in size from two to three tons for residences to up to 40 tons for commercial and industrial users. Figure 8-1 illustrates a simple scheme for determining the supplemental heat required when using an air-air heat pump.

Hydronic Heat Pump

The hydronic heat pump is similar to the air to air unit, except the heat exchange is between water and refrigerant instead of air to refrigerant, as illustrated in Fig. 8-2. Depending on the position of the reversing valve, the air heat exchanger either cools or heats room air. In the case of cooling, heat is rejected through the water cooled condenser to the building water. In the case of heating, the reversing valve causes the water to refrigerant heat exchanger to become an evaporator. Heat is then absorbed from the water and discharged to the room air.

Imagine several hydronic heat pumps connected to the same building water supply. In this arrangement, it is conceivable that while one unit is providing cool air to one zone, another is providing hot air to another zone; the first heat pump is providing the heat source for the second unit, which is heating the room. This illustrates the principle of energy conservation. In practice, the heat rejected by

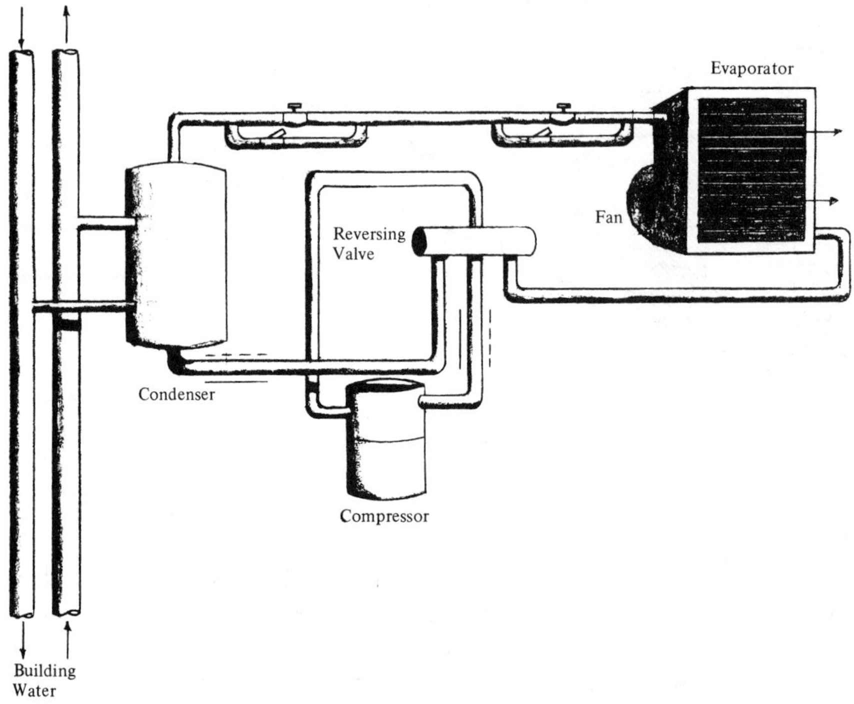

Fig. 8-2. Hydronic heat pump.

the cooling units does not equal the heat absorbed. An additional evaporative cooler is added to the system to help balance the loads. A better heat source would be the water from wells, lakes, or rivers which is thought of as a constant heat source. Care should be taken to insure that a heat pump connected to such a heat source does not violate ecological interests.

EFFICIENT APPLICATIONS OF REFRIGERATION EQUIPMENT

Liquid Chiller

A liquid chilling unit (mechanical refrigeration compressor) cools water, brine, or any other refrigeration liquid, for air conditioning or refrigeration purposes. The basic components include a compressor, liquid cooler, condenser, the compressor drive, and auxiliary components. A simple liquid chiller is illustrated in Fig. 8-3. The type of chiller usually depends on the capacity required. For example, small

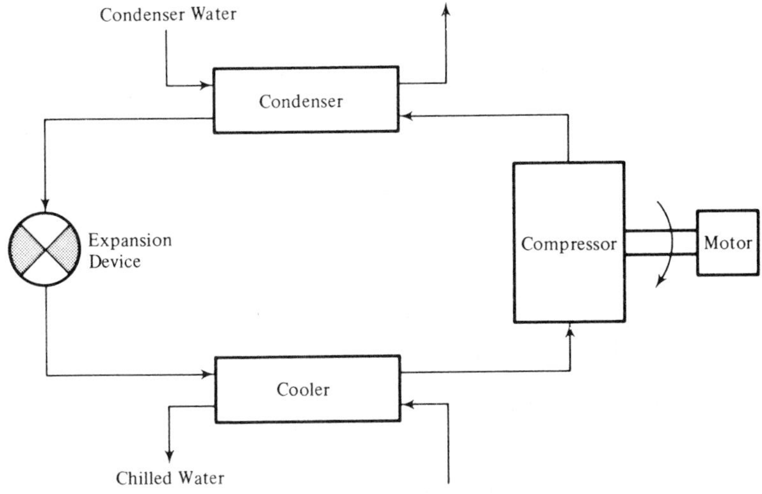

Fig. 8-3. Liquid chiller.

units below 80 tons are usually reciprocating, while units above 350 tons are usually centrifugal.

A factor which affects the power usage of liquid chillers is the percent load and the temperature of the condensing water. *A reduced condenser water temperature saves energy.* In Fig. 8-4, it can be seen that by reducing the original condenser water temperature by ten degrees, the power consumption of the chiller is reduced. Likewise, a chiller operating under part load consumes less power. The "ideal" coefficient of performance (COP) is used to relate the measure of cooling effectiveness. Approximately 0.8 Kw is required per ton of refrigeration (0.8 Kw is power consumption at full load, based on typical manufactures data).

Thus:

$$\text{COP} = \frac{1 \text{ Ton} \times 12000 \text{ Btu/ton}}{0.8 \text{ Kw} \times 3412 \text{ Btu/Kw}} = 4.4.$$

Chillers in Series and in Parallel

Multiple chillers are used to improve reliability, offer standby capacity, reduce inrush currents and decrease power costs at partial loads. Fig. 8-5 shows two common arrangements for chiller staging; namely, chillers in parallel and chillers in series.

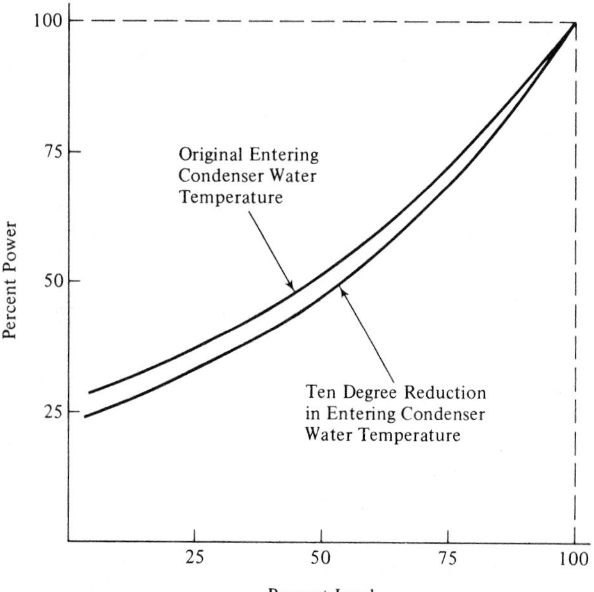

Fig. 8-4. Typical power consumption curve for centrifugal liquid chiller.

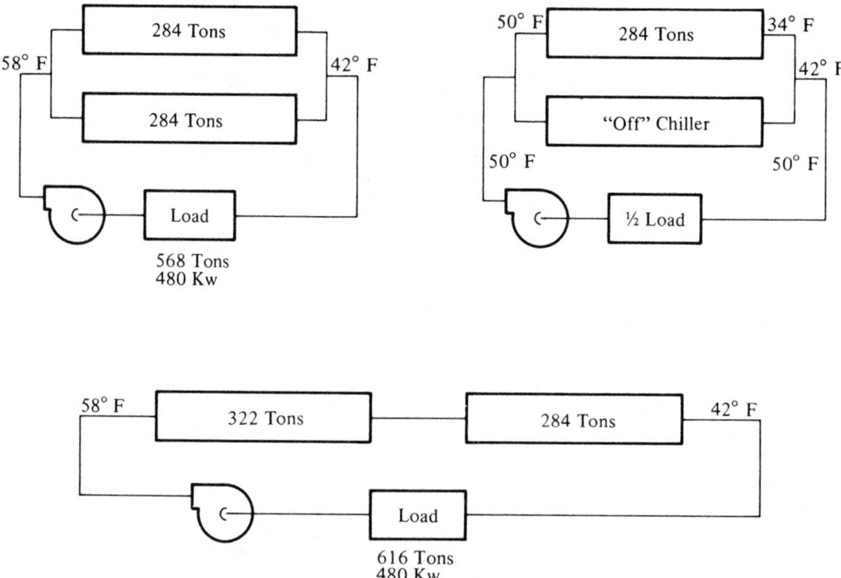

Fig. 8-5. Multiple chiller arrangements.

In the parallel chiller arrangement, liquid to be chilled is divided among the liquid chillers and the streams are combined after chilling. Under part load conditions, one unit must provide colder than designed chilled liquid so that when the streams combine, including the one from the off chiller, the supply temperature is provided. The parallel chillers have a lower first cost than the series chillers counterparts but usually consume more power.

In the series arrangement, a constant volume of flow of chilled water passes through the machines, producing better temperature control and better efficiency under part load operation; thus, the upstream chiller requires less Kw input per ton output. The waste of energy during the mixing aspect of the parallel chiller operation is avoided. The series chillers, in general, require higher pumping costs. The energy conservation engineer should evaluate the best arrangement, based on load required and the partial loading conditions.

The Absorption Refrigeration Unit

Any refrigeration system uses external energy to "pump" heat from a low temperature level to a high temperature. Mechanical refrigeration compressors pump absorbed heat to a higher temperature level for heat rejection. Similarly, absorption refrigeration changes the energy level of the refrigerant (water) by using lithium bromide to alternately absorb it at a low temperature level and reject it at a high level by means of a concentration-dilution cycle.

The single stage absorption refrigeration unit uses 10 to 12 psig steam as the driving force. Whenever users can be found for low pressure steam, energy savings will be realized. A second aspect for using adsorption chillers is that they are compatible for use with solar collector systems. Several manufacturers offer absorption refrigeration equipment which uses high temperature water (160°–200°F) as the driving force. Solar collectors are discussed in detail in Chapter 9.

A typical schematic for a single stage absorption unit is illustrated in Fig. 8-6. The basic components of the system are the evaporator, absorber, concentrator, and condenser. These components can be grouped in a single or double shell. Fig. 8-6 represents a single stage arrangement.

Evaporator. Refrigerant is sprayed over the top of the tube bundle

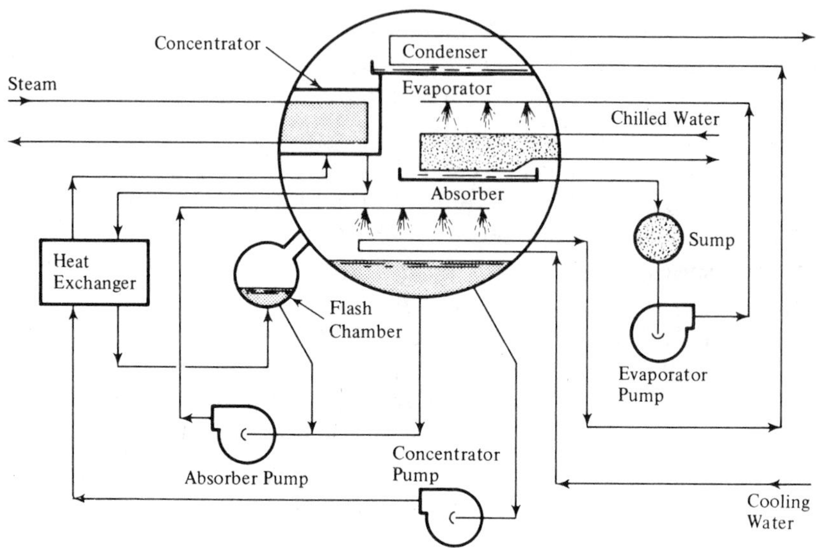

Fig. 8-6. One-shell lithium bromide cycle water chiller. (Source: Trane Air Conditioning Manual.)

to provide for a high rate of transfer between water in the tubes and the refrigerant on the outside of the tubes.

Absorber. The refrigerant vapor produced in the evaporator migrates to the bottom half of the shell where it is absorbed by a lithium bromide solution. Lithium bromide is basically a salt solution which exerts a strong attractive force on the molecules of refrigerant (water) vapor. The lithium bromide is sprayed into the absorber to speed up the condensing process. The mixture of lithium bromide and the refrigerant vapor collects in the bottom of the shell; this mixture is referred to as the dilute solution.

Concentrator. The dilute solution is then pumped through a heat exchanger where it is preheated by hot solution leaving the concentrator. The heat exchanger improves the efficiency of the cycle by reducing the amount of steam or hot water required to heat the dilute solution in the concentrator. The dilute solution enters the upper shell containing the concentrator. Steam coils supply heat to boil away the refrigerant from the solution. The absorbent left in the bottom of the concentrator has a higher percentage of absorbent than it does refrigerant, thus it is referred to as concentrated.

Condenser. The refrigerant vapor boiling from the solution in the concentrator flows upward to the condenser and is condensed. The condensed refrigerant vapor drops to the bottom of the condenser and from there flows to the evaporator through a regulating orifice. This completes the refrigerant cycle.

The single stage absorption unit consumes approximately 18.7 pounds of steam per ton of capacity (Steam consumption at full load based on typical manufacturers data.) For a single state absorption unit,

$$\text{COP} = \frac{1 \text{ ton} \times 12{,}000 \text{ Btu/ton}}{18.7 \text{ lb} \times 955 \text{ Btu/lb}} = 0.67.$$

The single stage absorption unit is not as efficient as the mechanical chiller. It is usually justified based on availability of low pressure steam, equipment considerations, or use with solar collector systems.

SIM 8-1

Compute the energy wasted when 15 psig steam is condensed prior to its return to the power plant. Comment on using the 15 psig steam directly for refrigeration.

ANSWER

From Steam Tables 12-20 for 30 psia steam, hfg is 945 Btu per pound of steam; thus, 945 Btu per pound of steam is wasted. In this case where *excess low pressure* steam cannot be used, absorption units should be considered in place of their electrical mechanical refrigeration counterparts.

SIM 8-2

2000 lb/hr of 15 psig steam is being wasted. Calculate the yearly (8000 hr/yr) energy savings if a portion of the centrifugal refrigeration system is replaced with single stage absorption. Assume 20 Kw additional energy is required for the pumping and cooling tower cost associated with the single stage absorption unit. Energy rate is \$.045 Kwh and the absorption unit consumes 18.7 lb of steam per ton of capacity.

The centrifugal chiller system consumes 0.8 Kwh per ton of refrigeration.

ANSWER

Tons of mechanical chiller capacity replaced. = 2000/18.7 = 106.95 tons.

Yearly energy savings = 2000/18.7 X 8000 X 0.8 X \$.045 −
20 Kw X 8000 X \$.045 = \$23,602

Two Stage Absorption Unit

The two stage absorption refrigeration unit as illustrated in Fig. 8-7 uses steam at 125 to 150 psig as the driving force. In situations

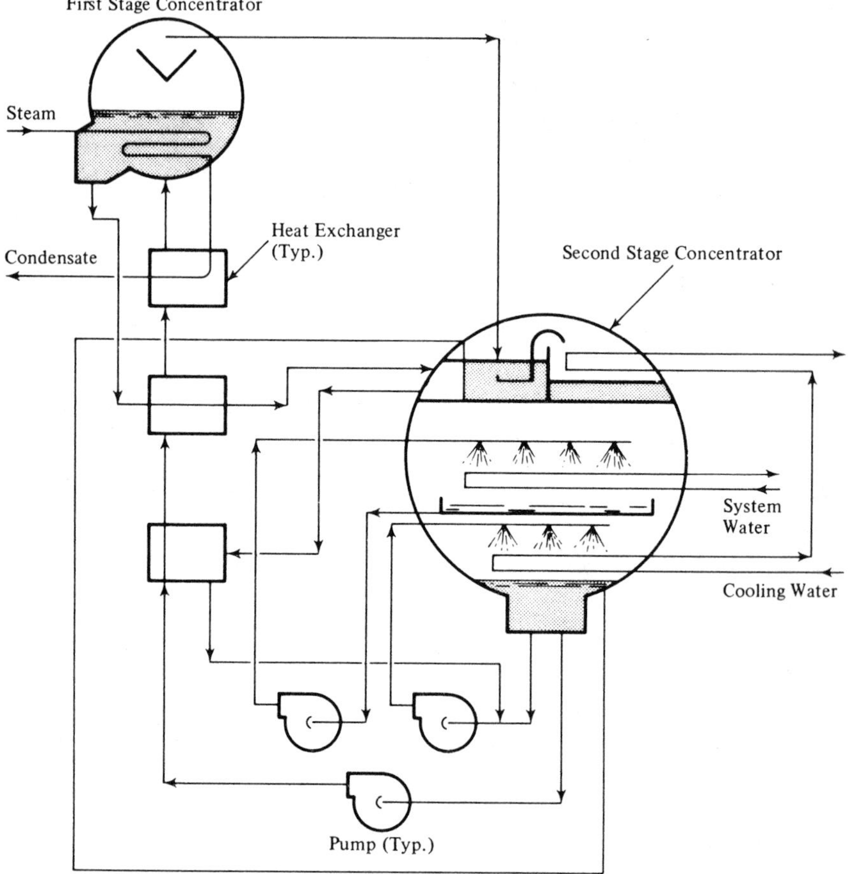

Fig. 8-7. Two stage absorption unit.

where excess medium pressure steam exists, this unit is extremely desirable. The unit is similar to the single stage absorption unit. The two stage absorption unit operates as follows:

Medium pressure steam is introduced into the first stage concentrator. This provides the heat required to boil out refrigerant from the dilute solution of water and lithium bromide salt. The liberated refrigerant vapor passes into the tubes of the second stage concentrator, where its temperature is utilized to again boil a lithium bromide solution, which in turn further concentrates the solution and liberates additional refrigerant. In effect, the concentrator frees an increased amount of refrigerant from solution with each unit of input energy.

The condensing refrigerant in the second stage concentrator is piped directly into the condenser section. The effect of this is to reduce the cooling water load. A reduced cooling water load decreases the size of the cooling tower which is used to cool the water. The remaining portions of the system are basically the same as the single stage unit.

The two stage absorption unit consumes approximately 12.2 pounds of steam per ton of capacity; thus, it is more efficient than its single stage counterpart. The associated COP is

$$\text{COP} = \frac{1 \text{ ton} \times 12000 \text{ Btu/ton}}{12.2 \text{ lb} \times 860 \text{ Btu/lb}} = 1.14.$$

Either type of absorption unit can be used in conjunction with centrifugal chillers when it is desirable to reduce the peak electrical demand of the plant, or to provide for a solar collector addition at a later date.

BASICS OF AIR CONDITIONING SYSTEM DESIGN FOR ENERGY CONSERVATION

Chapter 7 illustrated the fundamental aspects of heat gains and losses in buildings. It was seen that several building loads, such as lighting, and equipment losses remain fairly constant with outside temperature. Other loads, caused by the transmission of heat through building walls or by the ventilation of outside air into a building, vary with temperature. Solar radiation can be assumed to be constant with outside temperature.

The following summarizes heat gains and heat losses:

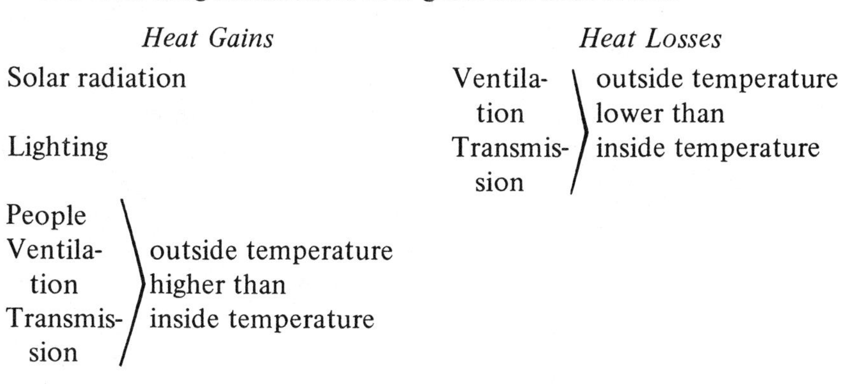

Heat Gains
Solar radiation
Lighting
People
Ventilation
Transmission
outside temperature higher than inside temperature
Heat Losses
Ventilation
Transmission
outside temperature lower than inside temperature

When heat gains are traded with heat losses, energy savings are realized. The central system is the basic air conditioning system used today. It allows for cooling, dehumidification, and heating equipment to be located in one area. As the central system developed, it became apparent that heat sources varied within the space; thus, the total area had to be divided into zones or sections.

The general categories of cooling systems most commonly used today are:

1. Central units serving individual zones
2. Multi-zone units
3. Double duct systems
4. Reheat systems
5. Variable volume systems
6. Individual room units

There are many variations and combinations, but the above are the basic options available. The major problem facing energy conservation engineering is to avoid generalizing. Each category has its application and economic advantage and should be evaluated based on the economics of the particular systems. In this section, particular applications of air conditioning systems will be discussed.

APPLYING VARIABLE AIR VOLUME SYSTEMS

The Variable Volume System has gained popularity in recent years, due to its potential of saving energy. This system is not a cure-all, but will save energy when *properly selected.* Variable Air Volume (VAV) Systems vary the quantity of air at a constant temperature, to match the system load requirements. The energy consumption closely parallels the load in air conditioning systems. This is in contrast to constant volume air systems which control building temperatures by varying the temperature of a constant flow of air. The most popular type of VAV System contains a terminal unit which modulates the air volume between maximum and a pre-determined value.

There are four main classes of VAV Systems which encompasses 80% of the VAV market. These classes combine the best features of several systems, to obtain maximum efficiency. Fig. 8-8 illustrates the four classes.

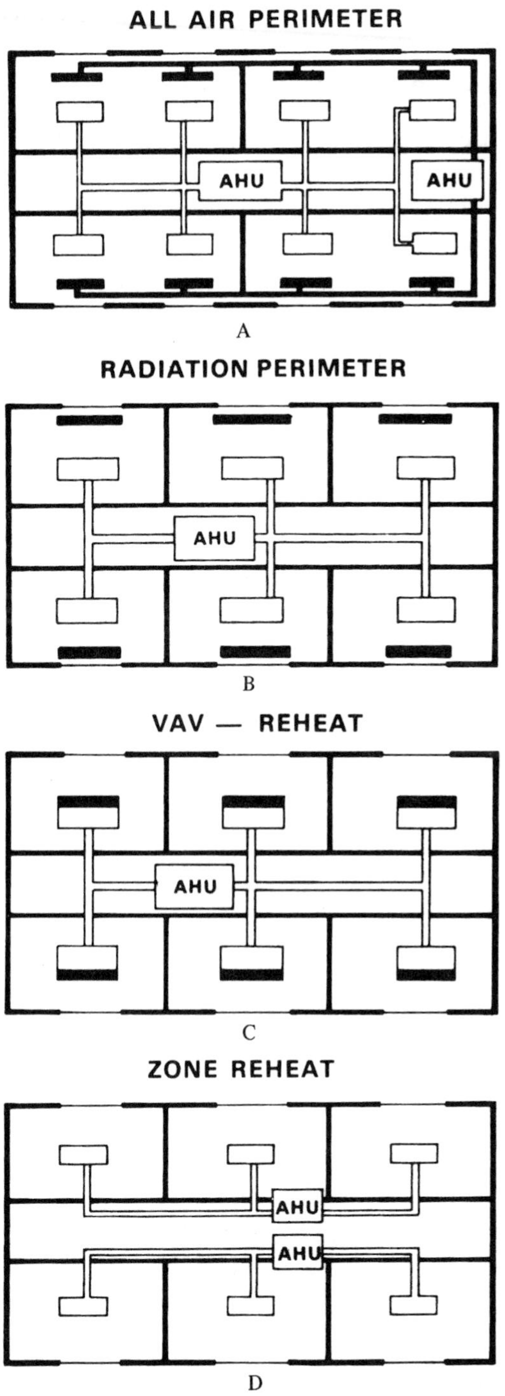

ALL AIR PERIMETER

AHU

AHU

A

RADIATION PERIMETER

AHU

B

VAV — REHEAT

AHU

C

ZONE REHEAT

AHU

AHU

D

Fig. 8-8. Typical variable air volume systems. (Courtesy of the Trane Company.)
Note: AHU is an air handling unit.

VAV with Independent Perimeter System (Fig. 8-8A)

This system uses a constant volume variable temperature system to offset the transmission losses or gains through the building walls. The *variable volume system* is for cooling only and is used for temperature control of occupied spaces.

The temperature of the constant volume air is controlled from an outside thermostat (shielded from direct sunlight) and the system operates on return air. During unoccupied hours, the larger variable volume system can be shut down, while the building temperature is maintained by the smaller constant volume system.

VAV with Radiation Perimeter (Fig. 8-8B)

This system is essentially similar to the previous class except the perimeter temperature is controlled via hydronic or electric heat. This system is mainly used in northern climates where under window heat is needed to offset down draft.

VAV with Zone Reheat (Fig. 8-8C)

This system provides heating as well as cooling using a VAV System. This system finds application in smaller buildings, mild climates, and buildings with good construction.

VAV with Reheat or with Dual Duct Systems (Fig. 8-8D)

This system modulates the volume to a fixed minimum before activating the reheat coils or mixing warm air at the individual zone. Once the minimum volume is reached, the volume remains constant and heat is added.

Energy Savings

Variable Air Volume Systems will be of greater benefit in buildings having the following characteristics:

1. High ratio of perimeter to interior area
2. Buildings which have high variations in loads, such as schools where the number of people change regularly
3. Small zone sizes

4. Building faces east and west and has a high percentage of glass (VAV reheat should be investigated for this case)

VAV Systems of the modulating type, which can be used for both perimeter and interior spaces and operating at medium static pressure, are usually the most economical. A true variable volume system has the following characteristic: system volume reduction corresponds to loads, which permits savings in annual fan Kwh and mechanical refrigeration.

APPLYING THE ECONOMIZER CYCLE

The basic concept of the economizer cycle is to use outside air as the cooling source when it is cold enough. There are several parameters which should be evaluated in order to determine if an economizer cycle is justified. These include:

1. Weather
2. Building occupancy
3. The zoning of the building
4. The compatibility of the economizer with other systems
5. The cost of the economizer.

What are the Costs of Using the Economizer Cycle?

In life, nothing comes free. Outside air cooling is accomplished at the expense of an additional return air fan, economizer control equipment, and an additional burden on the humidification equipment. Therefore, economizer cycles must be carefully evaluated based on the specific details of the application.

Maximum Savings

The economizer control system is the most significant aspect in achieving maximum savings. Outside air temperature alone does not insure that the economizer cycle is operating at its maximum potential. Ideally, the economizer controls should be based on Btus, not degrees. The enthalpy sensor offers the greatest potential for energy conservation, since it measures the heat content of the outside air. Unfortunately, enthalpy sensors have poor maintenance records and are many times replaced with thermometers.

APPLYING HEAT RECOVERY

Several examples of "heat recovery" have been indicated in this text. Basically, whenever wasted energy is recovered and used, the principle of "heat recovery" is being followed.

Coil Run-Around Cycle

The coil run-around cycle, as illustrated in Fig. 8-9A, is another example of heat recovery. The cycle essentially transfers energy from the exhuast stream to the make-up stream. The heat transfer is accomplished by installing coils in each stream and continuously circulating a heat transfer media, such as ethylene glycol fluid, between the two coils.

In winter, the warm exhaust air passes through the exhaust coils and transfers heat to the ethylene glycol fluid. The fluid is pumped to the make-up air coil where it preheats the incoming air. The system is most efficient in winter operation, but some recovery is possible during the summer.

Heat Wheels

If inlet and outlet exhaust ducts are close to one another, the heat wheel concept can be used. The heat wheel cycle accomplishes the

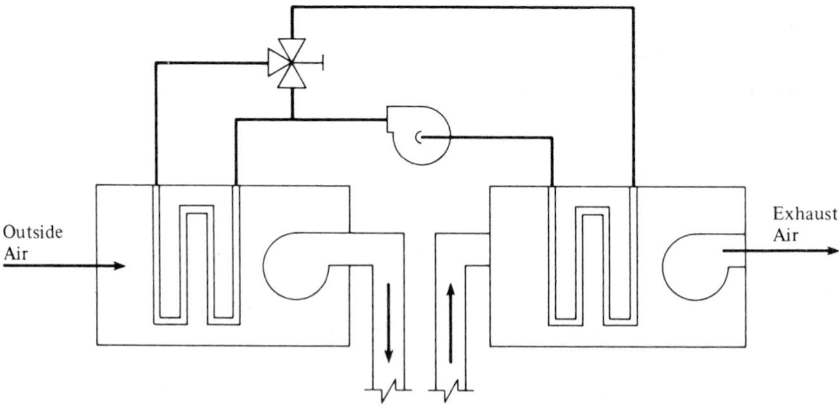

Fig. 8-9A. Coil-run-around cycle

Fig. 8-9. Heat recovery systems. (Courtesy of the Edison Electric Institute.)

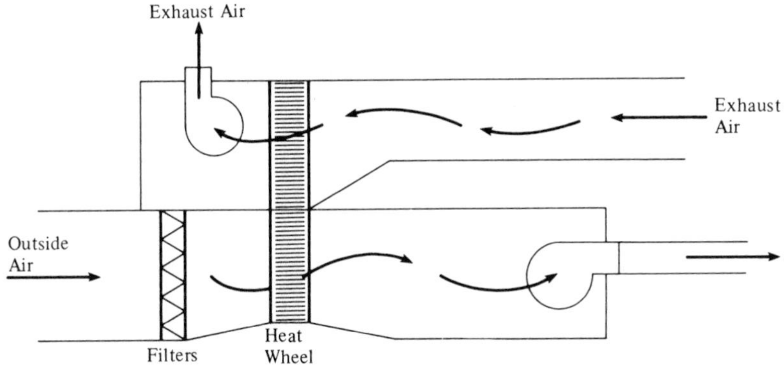

Fig. 8-9B. Heat wheels.

Fig. 8-9. *(Continued)*

same objectives of the coil run-around cycle, as illustrated in Fig. 8-9B. The heat wheel can be used to transfer sensible heat or latent heat. It consists of a motor driven wheel frame packed with a heat-absorbing material, such as aluminum, corrugated asbestos, or stainless steel mesh.

To optimize the system in the summer, lithium chloride impregnated asbestos can be used. Since lithium chloride absorbs moisture as well as heat, it can be used to cool and dehumidify make-up air.

Air to Air Heat Pipes and Exchangers

Another form of heat recovery is relatively simple, since no moving parts are involved. Either air to air heat pipes or exchangers can be used, as illustrated in Fig. 8-9C. A heat pipe is installed through adjacent walls of inlet and outlet ducts; it consists of a short length of copper tubing sealed at both ends. Inside is a porous cylindrical wick and a charge of refrigerant. Its operation is based on a temperature difference between the ends of the pipe, which causes the liquid in the wick to migrate to the warmer end to evaporate and absorb heat. When the refrigerant vapor returns through the hollow center of the wick to the cooler end, it gives up heat, condenses, and the cycle is repeated.

The air to air heat exchanger consists of an open ended steel box which is compartmentalized into multiple narrow channels. Each passage carries exhaust air alternating with make-up air. Energy is transmitted by means of conduction through the walls.

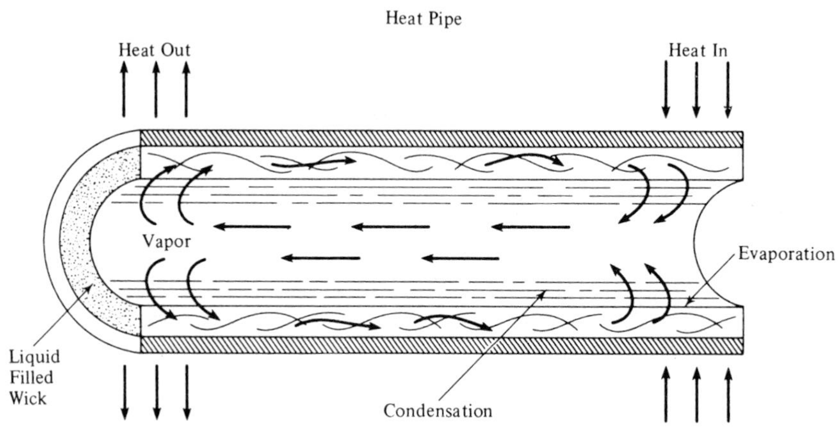

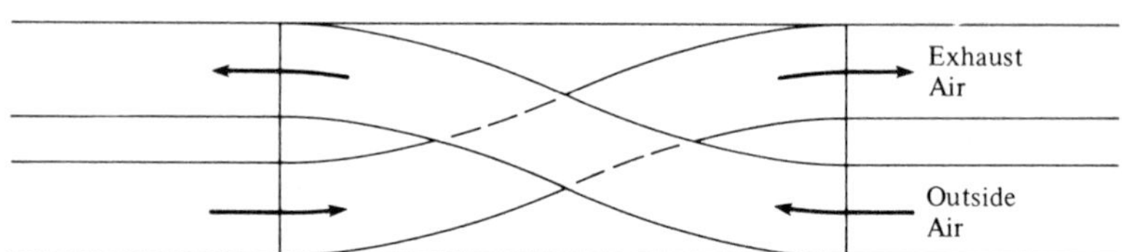

Fig. 8-9C. Air to air heat pipes and exchangers.

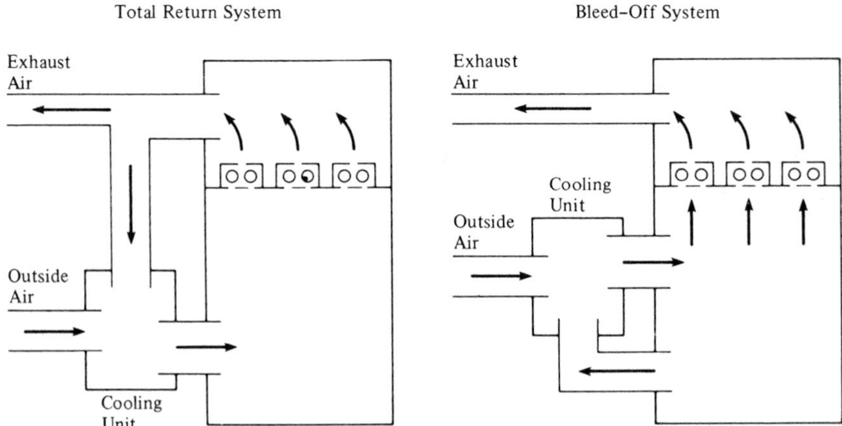

Fig. 8-9D. Recovery from lighting fixtures.

Fig. 8-9. *(Continued)*

Heat from Lighting Systems.

Heat dissipated by lighting fixtures which is recovered will reduce air conditioning loads, will produce up to 13 percent more light output for the same energy input, and can be used as a source of hot air. Two typical recovery schemes are illustrated in Fig. 8-9D. In the total return system, all of the air is returned through the luminaires. In the bleed off system, only a portion is drawn through the lighting fixtures. The system is usually used in applications requiring high ventilation rates.

Refrigeration Systems-Double Bundle Condenser

Refrigeration systems using double bundle condensers can be used for heat recovery and offer an economical way to prevent contamination of building water with cooling tower water. Fig. 8-9E illustrates the addition of a double bundle condenser to a refrigeration unit. The heat rejected by the compressor is now available to the building

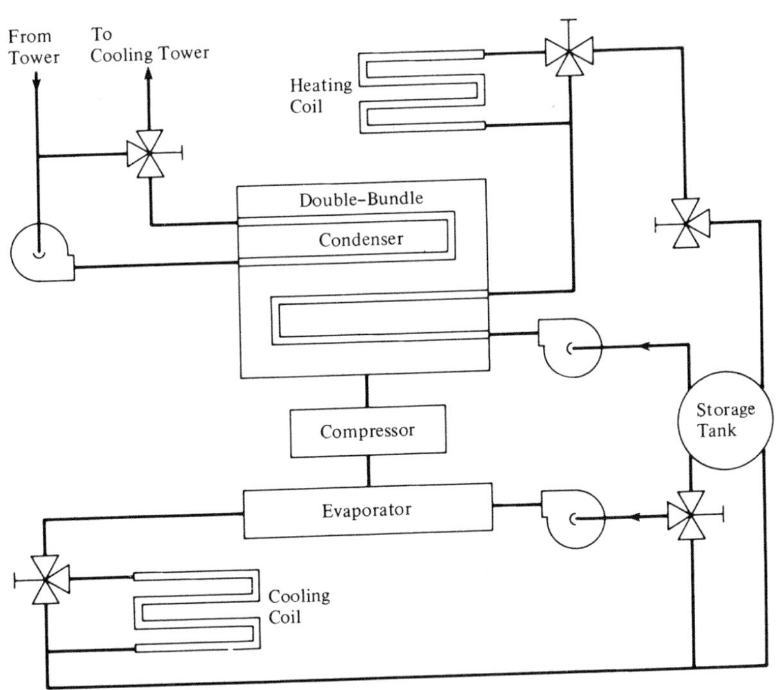

Fig. 8-9E. Recovery from refrigeration systems.

Fig. 8-9. *(Continued)*

water circuit. A heating coil has been added to provide supplemental heating. In cases where the amount of heat recovered during occupied hours exceeds the daytime heating requirements of perimeter zones, a heat storage tank can be added.

COOL STORAGE SYSTEM PERFORMANCE

Several utilities are offering substantial discounts to a number of their customers and even waiving demand charges on the use of power by offering time-of-day rates.

Time-of-day rates charge a premium during periods of high demand (on peak) and offer reduced rates during periods of low demand (off peak). The charge may be based on demand, usage, or both. Time-of-day rates have revived the concept of using thermal storage for cooling.

Thermal storage uses large volumes of water or ice in central or modular packages to store refrigeration produced during off peak periods for later use during periods of peak demand. Thermal storage systems can be placed in two general classifications: partial storage and total storage as indicated in Figure 8-10. Partial storage trims the peak refrigeration load only and requires some chiller operation during on peak periods. The storage volumes are manageable and the demand savings are attractive. Total storage systems must produce the entire daily refrigeration load the evening before. No chiller operation is allowed during on peak periods. Total storage systems require enormous thermal storage volumes and are usually attractive only where time-of-day rates offer substantial discounts for off peak power.

Chilled water has been the preferred choice of storage medium. Chilled water storage systems have lower energy requirements when compared to conventional systems. Producing and storing chilled water requires no unusual hardware. However, chilled water storage demands a fair amount of space. Approximately 10 cubic feet of water are required to store one tonhour of cooling. This storage volume can be reduced by a factor of two to three by using ice storage. Can this space savings justify the additional installed cost and higher energy cost of ice storage? Some recent innovations in packaging give the system designer some options when designing

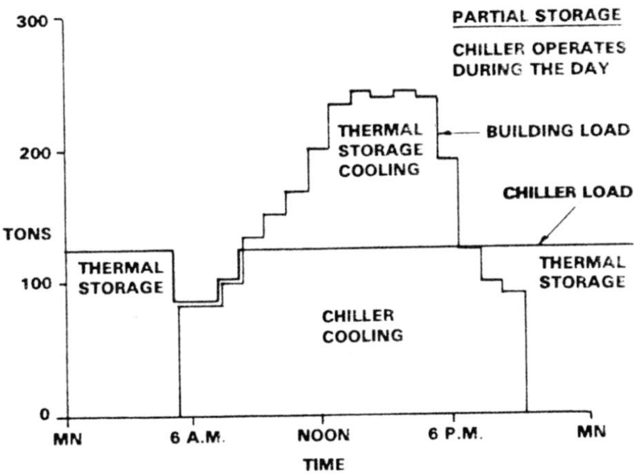

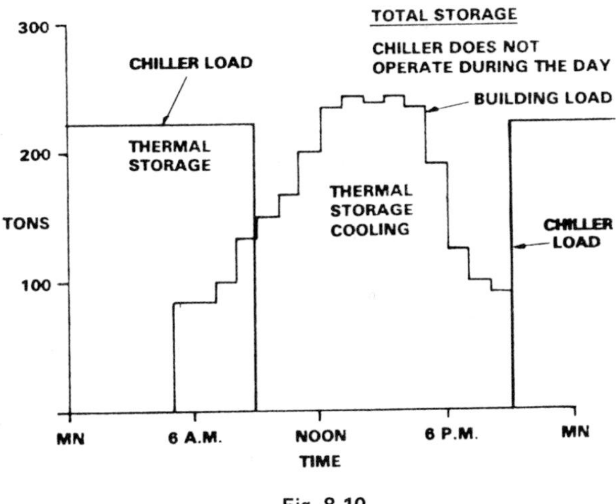

Fig. 8-10

ice storage systems. A review of ice storage systems past and present would be helpful.

Ice is stored on the heat transfer surface that produces the ice. The size of the ice tank is then a function of the heat transfer efficiency and the arrangement of the heat transfer surface as well as the amount of cooling the designer wishes to store. The ice tank is a large insulated tank containing several feet of steel pipe. The ice is built up on the outside of the pipe while the refrigerant (R22) is circulated inside the pipe. When fully charged, two and a half inches of ice surround the pipe or 13 pounds of ice for each lineal foot of pipe. This equates to five feet of steel pipe for every tonhour of cooling stored.

There are several methods to circulate the refrigerant within the tubes. Pumping large volumes of Refrigerant 22 is possible, but the pumping system must not allow the premature formation of flash gas in the system. Refrigerant 22 is circulated through the ice tank tubes by direct expansion, liquid recirculation or gravity (flooded systems).

Direct expansion employs the pressure difference between the high side receiver and the suction line accumulator to transport refrigerant through the ice tank. With direct expansion, proper sizing of the suction line will minimize oil return problems. From a design standpoint, direct expansion offers a simple and reliable system. However direct expansion suffers some inefficiencies compared to other systems. Fifteen to 20 percent of the tube surface must be used to provide superheat and is not available for making ice. This results in an installed cost penalty for additional surface area and/or lower efficiencies due to lower suction temperatures.

Liquid recirculation systems use either the compressor or a separate liquid refrigerant feed pump to circulate refrigerant as indicated in Figure 8-11. The recirculation rate is two to three times the evaporation rate. This causes liquid overfeed systems, as they are sometimes called, to return a two-phase mixture from the ice tank to a low pressure receiver. From the low pressure receiver, refrigerant vapor is returned to the compressor while the refrigerant liquid is available for recirculation to the ice tank. The liquid level control meters additional refrigerant from the high pressure receiver to the low pressure receiver, to make up for refrigerant returned to the compressor. The design of the feed pump is critical in liquid recirculation systems. Proper design of the pumping system must be followed to prevent flashing of the refrigerant at the pump suction or in the pump itself. Semihermetic pumps or open pumps with specially designed seals are used to keep refrigerant losses to a minimum.

Pumping drums use hot gas to recirculate refrigerant, thereby eliminating the need for the refrigerant feed pump. The "pumper"

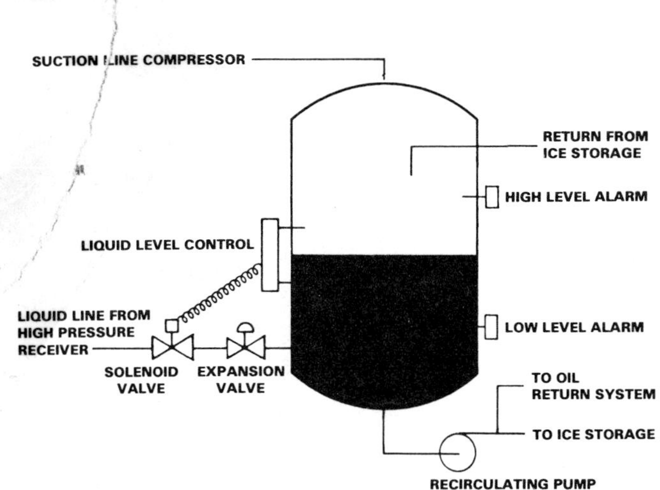

Fig. 8-11. Low Pressure Receiver

is a smaller receiver that is located below the low pressure receiver. When vented to the low pressure receiver, liquid refrigerant drains by gravity into the pumper drum. The pumper is then isolated from the low pressure receiver by means of solenoid valves. Hot gas is used to pressurize the pumper forcing refrigerant from the pumper through the ice tank. Two pumping drums are used so one is filling while the other is draining. The "double pumper" replaces the expensive and high maintenance liquid feed pump.

Gravity feed systems eliminate the need for pumping refrigerant altogether. With gravity systems, a low pressure receiver, or surge drum, is paired with each ice tank as a coil header. The vertical header carries the refrigerant to the ice tank inlet while a suction line returns refrigerant from the ice tank back to the surge drum. At the surge drum, vapor is returned to the compressor while liquid is recirculated to the ice tank. Due to the large refrigerant charge and the cost of multiple surge drums and controls, gravity flooded systems may be impractical in larger ice storage systems. An horizontal surge drum system is illustrated in Figure 8-12.

Oil return is a special concern with all liquid recirculation systems. At the low pressure receiver, refrigerant vapor is drawn off by the compressor, but the oil remains in the receiver. When a liquid feed pump or pumper drum is used, sufficient pressure drop is available

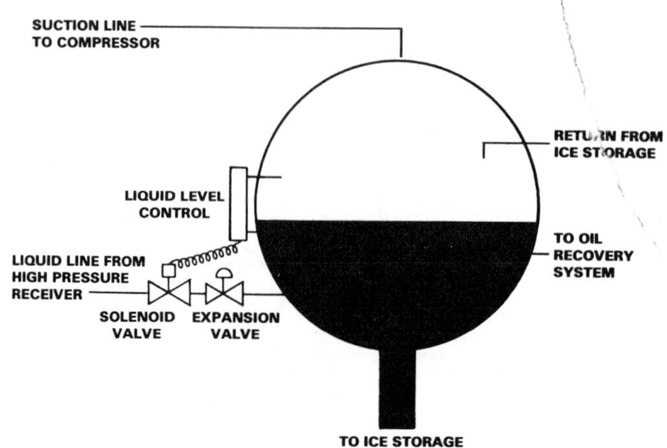

Figure 8-12. Horizontal Surge Drum

to employ an oil return system. The liquid line to the ice tank is tapped and a small amount of refrigerant is bled off to an expansion valve. The line downstream of the expansion valve is sized to maintain sufficient velocity to carry the entrained oil back to the compressor inlet. With a gravity flooded system, sufficient head is not available and a separate oil recovery system must be used. In addition to an oil return or oil recovery system, an oil separator in the compressor discharge is a must.

Finally, there are ice tank accessories. To maintain uniform ice thickness, the water in the ice tank must be thoroughly mixed. An air pump is used to agitate the ice tank water during both the freezing and melting cycles. During the ice melting cycle, chilled water must circulate around the ice covered pipes. An ice thickness probe is required to prevent the ice tank from overfreezing or "bridging." Bridges of ice forming between the ice covered pipes will impede the flow of chilled water through the ice bundle during the melting cycle, resulting in uneven ice melting.

Ice storage systems of this nature require very large refrigerant charges and incur substantial construction costs due to the number of field erected components. Due to these construction costs, many ice systems are difficult to justify except in larger tonnage applications. Recent innovations in the packaging of glycol type ice tanks may make the option of ice storage a greater possibility.

Glycol systems as indicated in Figure 8-13 use an ethylene glycol/water mixture as a low temperature heat transfer medium to transfer heat from the ice storage tanks to a packaged chiller and from the cooling coils to either the ice storage tanks or the chiller. The use of freeze protected chilled water eliminates the design time, field construction, large refrigerant charges and leaks formerly associated with ice systems.

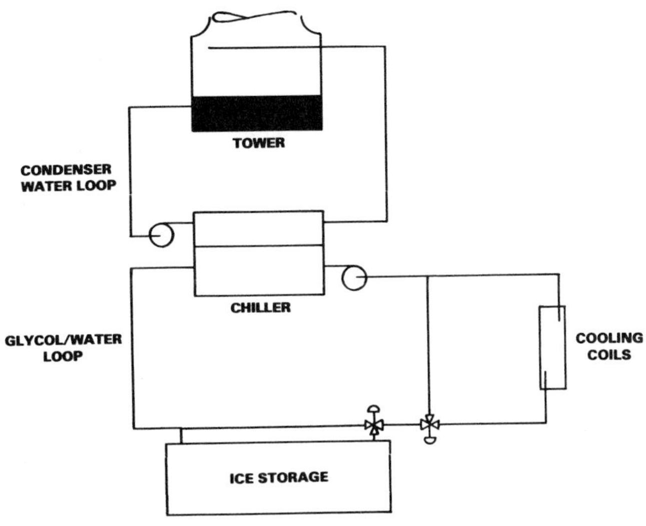

Fig. 8-13. Glycol System (Ethylene Glycol/Water)

Glycol ice storage tanks create ice by circulating the low temperature fluid through half-inch polyethylene tubing. The polyethylene tubing is coiled inside insulated polystyrene tanks. The ethylene glycol/water mixture is also used for cooling by circulating the warm fluid through the tubing, melting the ice. The water that experiences a phase change remains in the tank. Since no water circulates around the tubing, the ice tank may be frozen solid. The problems of ice bridging and an air pump for agitation are eliminated. In this configuration the glycol ice tank is a sealed system similar to a packaged chiller or a car battery.

The heat transfer surface of the glycol ice tank can be increased by four to five times the area used in refrigerant ice tanks due to the low cost of polyethylene. The extended heat transfer area decreases the approach temperature required to make ice. Centrifugal or reciprocating chillers producing 23 to 26 degree glycol are well suited for this application. Centrifugal chillers have an excellent track record in low temperature applications including food processing, cosmetics, pharmaceuticals, clean rooms, other industrial applications and, of course, ice rinks.

THERMAL STORAGE CONTROL SYSTEMS

The control of thermal storage systems and equipment mandates the uses of automated control systems. The complexity of controls should not exceed the complexity of the system. However, more than a time clock and a few extra thermostats are required to insure the maximum benefit of the thermal storage system. Control concepts can range from simple to comprehensive.

The simplest control is chiller priority. This control scheme allows the chiller to accept all of the cooling load until the load exceeds chiller capacity. Excess load is then met by melting ice. Control defaults to limiting chiller capacity by chiller selection or some form of demand limiting. Control is simple, but use of thermal storage is limited to cooling in excess of chiller capacity. This control may not provide the maximum financial return on the thermal storage investment.

There are several variations of ice priority control schemes. The intent of ice priority control is to gain maximum benefit of time-of-day rates by maximizing ice usage each day. During intermediate seasons when the daily cooling load is less than thermal storage capacity, chiller operation can be limited to off peak hours. When daily cooling load exceeds thermal storage capacity, chiller and ice must share the on peak cooling load. The task of assigning cooling load to either ice or chiller is well suited for building automation systems with equation processing capabilities. If ice is used too early or too rapidly, the ice will be depleted before the end of the daytime cooling load, leaving several hours of building load in excess of chiller capacity. On the other hand, if the ice is not used to its maximum potential, the operating cost benefits of thermal storage are not totally realized.

How does the automation system predict the daily cooling load the day before? Most of the required tools are already in place. The automation system has the ability to monitor several building or load producing trends. What the automation system does not possess is the ability to equate these trends to building cooling load.

Fortunately, several HVAC system simulation programs are available for this task. Simulation programs can isolate single load components and plot their dependence to an easily monitored variable. Internal loads are a function of time and calendar. Ventilation and solar loads can be equated to monitored weather variables. A daily cooling load profile can now be constructed by measurement of the monitored variables, profiling those variables for the period of prediction, and summing the results. Automation systems that can produce an expected daily load profile provide the building operator with a very powerful management tool. Only informed and prudent operation of thermal storage systems can guarantee their success.

THE VENTILATION AUDIT

Several existing codes require outside air requirements in excess of those required to dilute carbon dioxide and odors. Some localities have adopted ASHRAE Standard 62-1981 for outside air and exhaust requirements. This standard establishes a basic SCFM per person outside air requirement. The meaning of the word ventilation has also changed as a result of this standard. Ventilation in effect means "an outside (outdoor) air supply plus any recirculated air that has been treated to maintain the desired quality of air within a designated space."

To accomplish an energy audit of the ventilation system the following steps can be followed.

1. Measure volume of air at the outdoor air intakes of the ventilation system. Record ventilation and fan motor nameplate data.
2. Determine local code requirements and compare against measurements.
3. Check if code requirements exceed "Recommended Ventilation Standards."

4. Apply for code variance if it is determined that existing standards are higher than required.

To decrease CFM, the fan pulley can be changed. Two savings are derived from this change, namely:

- Brake horsepower of fan motor is reduced.
- Reduced heat loss during heating season.

To compute the savings, Equations 8-3 and 8-4 are used. Figure 8-14 can also be used to compute fan power savings as a result of air flow reduction.

$$\text{HP (New)} = \text{HP} \times \left(\frac{\text{CFM (New)}}{\text{CFM (Old)}}\right)^3 \tag{8-3}$$

$$Q\text{(Saved)} = \frac{1.08 \text{ BTU}}{\text{HR - CFM - °F}} \times \text{CFM (Saved)} \times \Delta T \tag{8-4}$$

$$\text{KW} = \text{HP} \times 0.746/\eta \tag{8-5}$$

where

HP = motor horsepower
CFM = cubic feet per minute

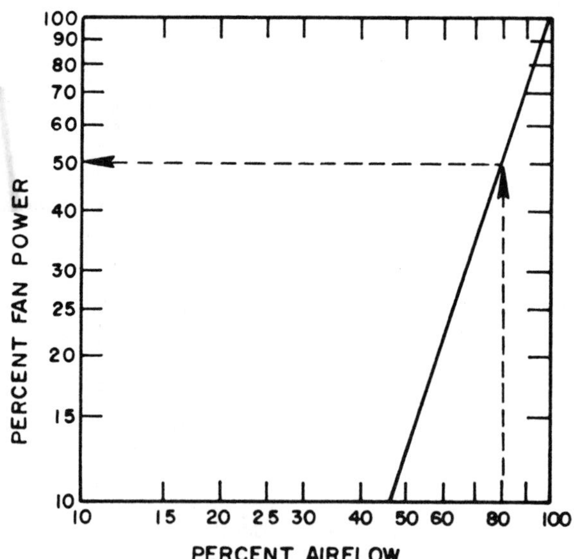

Fig. 8-14 Decrease in horsepower accomplished by reducing fan speed (based on laws of fan performance). (Source: *NBS Handbook 115 Supplement 1.*)

ΔT = average temperature gradient
KW = motor kilowatts (K = 1000)
η = motor efficiency

In addition to reducing air flow during occupied periods, consideration should be given to shutting the system down during unoccupied hours.

If the space was cooled, additional savings will be achieved. The quantity of energy required to cool and dehumidify the ventilated air to indoor conditions is determined by the enthalpy difference between outdoor and indoor air.

ENERGY ANALYSIS UTILIZING SIMULATION PROGRAMS

There are a number of software programs which have been successfully used to simulate how a system will perform various load and weather conditions. Before the microcomputer revolution of the 1980s, two of the most widely used public domain programs were DOE-II developed by Lawrence Berkeley Laboratory for the Department of Energy and Building Loads Analysis and Steam Thermodynamics better known as BLAST developed for the Army and Air Force. In the private sector the TRACE program developed by the Trane Company and Energy Analysis developed by Ross R. Meriwether and Associates dominated the market.

The microcomputer revolution has made available high capacity, low cost computers and numerous software options. As indicated in Chapter 2, the Alliance to Save Energy "ENVEST" program allows for complete energy investment decision making.

Carrier's E20-II Hourly Analysis Program combines the base and hourly load calculation logarithms of the E20-II Commercial Load Estimating Program with the sophisticated equipment simulation of the E20-II Operating Cost Analysis Program.

The below summaries some other commonly used microcomputer energy analysis programs.

Simplified Energy Analysis (SEA) – developed by Ferreira & Kalosineky Associates, P.O. Box L-6, New Bedford, Massachusetts 02745.

COGEN – Performs thermodynamic and financial analysis for turbine generators cogeneration systems. Software programs for combustion efficiency, steam calculations and pipe heat loss are also available from Software Systems, 5766 Balcone Drive, Suite 203, Austin, Texas 78731. In addition, spreadsheet programs such as Microsoft "Chart" can be used by the designers in analyzing energy management information.

The microcomputer provides a cost effective way to analyze energy consumption, make detailed building energy analysis, and implement life cycle costing in the decision making process. There are numerous software options available which can help the plant engineer make the best possible choice.

TEST AND BALANCE CONSIDERATIONS

Probably the biggest overlooked low-cost energy audit requirement is a thorough test, balance, and adjust Program. In essence, the audit should include the following steps.

1. ***Test.*** Quantitative determination of conditions within the system boundary, including flow rates, temperature and humidity measurements, pressures, etc.
2. ***Balance.*** Balance the system for required distribution of flows by manipulation of dampers and valves.
3. ***Adjust.*** Control instrument settings, regulating devices, and control sequences should be adjusted for required flow patterns.

In essence, the above program checks the designer's intent against actual performance and balances and adjusts the system for peak performance.

Several sources outlining Test and Balance Procedures are:

- Construction Specifications Institute (CSI), which offers a specification series that includes a guide specification Document 15050 entitled, "Testing and Balancing of Environmental Systems." Reprints of this paper are available. It explains factors to be considered in using the guide specification for project specifications.

- Associated Air Balance Council (AABC), the certifying body of independent agencies.
- National Environmental Balancing Bureau (NEBB), sponsored jointly by the Mechanical Contractors Association of America and the Sheet Metal and Air Conditioning Contractors National Association as the certifying body of the installing contractors' subsidiaries.

9

Cogeneration

Cogeneration is the combined production of electrical or mechanical energy and heat. The heat can be in the form of hot gases, hot liquids, or process steam. The combined generation of work (electrical or mechanical) and process heat provides a better overall utilization of the fuel used in a plant. For example, if you have a plant that simultaneously needs electricity and low-temperature heat, you could fill these needs in a variety of ways. You could

1. buy or make electricity and use a portion to fill the thermal requirements through resistance heating.
2. buy or make electricity and use oil, gas, wood, or coal to fill the thermal requirements, or
3. make electricity and use the rejected heat to fill the thermal requirement

If you choose option 3, the fuel consumption will be less than that of either option 1 or option 2. For example, if the energy for heat is the same for all three options, the incremental fuel for electricity in option 3 is about 50 percent less than what would have been burned at a central utility plant to generate the same amount of electricity.

In order for option 3 to be available, the electricity and heat needs of the plant must be matched. Unfortunately, many industries do not have a good match between these needs, and therefore they cannot combine their production. This potential problem will be discussed later.

Cogeneration includes topping units (in which primary energy is used to produce electricity and rejected heat is used to satisfy the needs of a relatively low-temperature thermal process) and bottom-

---

Appreciation is expressed to John Belding who authored this original work as part of Manual 15, Industrial-Energy Conservation published by MIT Press. Special permission to reprint portions, granted by publisher.

ing units (in which energy is used first to satisfy the thermal demands of a high-temperature process and rejected heat is then used to produce power). In the near term, topping units offer more opportunity for energy savings because of the ready availability of appropriate technologies and because low-temperature processes account for the majority of total thermal demands.

Technologies currently available for cogeneration include the following.

- *Extraction steam turbine-generator.* High pressure, high-temperature steam is raised in a boiler, fed to a turbogenerator, and extracted from the turbine at temperature and pressure suitable for process thermal needs. The major advantage of this system is that it can use various fuels, including coal and industrial wastes. At process-steam pressures up to a few hundred psi, the electricity generation is about 50-70 kWh per million Btu of energy delivered to the process.

- *Combustion turbine system with waste-heat boiler.* This system uses the high-temperature (800°F) turbine exhaust to generate steam in a waste-heat boiler. The hot exhaust may also be used directly in some industrial processes. In comparison with the steam-turbine method, the power generated per unit of process thermal energy is higher. The chief disadvantage is the fuel inflexibility of current turbine technology, which is limited to natural gas or petroleum distillates. Major utilization of gas turbines for cogeneration would cause a shift from coal and nuclear fuels used in central stations to gas and distillate. Development work is progressing on the use of residual oils and coal-derived fuels, but reliability and maintainability are major concerns.

- *Diesel with waste-heat recovery.* This system employs a water-jacket heat exchanger and an exhaust-heat boiler to raise process steam using the rejected heat of the diesel engine. Like the gas-turbine system, the system achieves a very high power generation per unit of thermal energy delivered to the process. The major drawback of the diesel is the same as that of the turbine: at present, industrial diesels burn only natural gas or distillate. However, the adaptation of heavy-duty residual-oil-burning marine diesel engines to industrial cogeneration appears promising.

Cogeneration systems are in use to some extent in each of the three major market sectors of the U.S. economy: the industrial sector, the residential-commercial-institutional sector, and the utility sector. Several approaches to system design and use can be adopted by each of the three sectors, depending on the internal user's requirements and external variables. The most important internal user requirements influencing system choice are electrical demand, thermal demand, and operational cycle. External factors that should be considered in evaluating design options include the cost and availability of various fuels and hardware, relationships with the local utility, regulations affecting operation, and the proximity of the plant to users.

Three cogeneration approaches appear to be attractive to industrial users.

The first approach is to design a system that is capable of meeting plant peak-load requirements and is connected to the utility grid and sells excess electricity. The excess power is either purchased by the utility system for resale to customers or transmitted via the grid to another user. The use of the utility grid for selling excess electricity directly to another private user is known as "wheeling." The utility may be able to use the excess power it purchases from an industrial cogenerator to help meet its own demands, which it might otherwise have to meet with less efficient equipment. This approach provides the greatest flexibility and offers the greatest potential cost and fuel savings of the three industrial options if the utility and the industrial cogenerator work together. An example of this type of cogeneration scheme is the Eugene (Oregon) Water and Electric Board (EWEB)/Weyerhaeuser Company facility. The plant is located on a Weyerhaeuser paperboard plant site but owned by EWEB. The \$7.2 million "energy center" utilizes steam produced by Weyerhaeuser boilers that are fueled by liquid wastes containing lignin. The facility, which includes a steam-turbine generator capable of producing 50,000 kW, was expected to supply electricity to Burbank, Glendale, and Pasadena, California, until at least 1983, around which time EWEB expected to need the power. Other examples are found in California. The Pacific Electric and Gas Company purchases excess electricity from a lumber company that produces electricity using a steam turbine fueled with waste wood and from two chemical plants. one of which uses a gas turbine whereas the other uses a steam bottoming system. The Southern California Edison Company is involved in a similar arrangement.

A second approach pursued by industrial facilities is to build a cogeneration system connected to the utility grid to allow for the purchase of supplemental electricity from the utility when needed. In this approach, equipment is sized to meet the user's normal base-load electrical requirements and electricity is purchased from the grid to meet peak-load requirements. Supplemental thermal energy and some redundancy in standby equipment may be required, so capital costs may be somewhat higher than with the first option. A system of this type was recently installed in the Atlantic Gelatin Division plant of General Foods Corporation in Woburn, Massachusetts. Because of the operational and financial success of this plant, General Foods is considering cogeneration for other plants. Another example is found in Pittsburgh, where PPG Industries has a tie-in with the local utility company that enables it to purchase from 0 to 100 percent of its power needs. PPG's topping-system plant uses a wide variety of energy sources (coal, natural gas, and petroleum) to power steam turbogenerators, which provide on the average 75 percent of the electricity needed; the company purchases the rest of its power from the local utility.

A third industrial approach is to design a system that is independent of the utility grid. Although this approach eliminates the problems involved in reaching satisfactory agreements with a utility company, it requires overcapacity or redundant equipment to ensure reliability. Independent systems have traditionally been oversized to meet peak electrical requirements, with supplemental equipment included to meet thermal demand. This tends to be the most expensive of the three options, but it can be the most effective for a particular plant. An example of this system is the installation at Southland Paper Mills in Houston that ran from 1968 until 1974. It was a 52,000-kW combined-cycle cogeneration plant that was independent of the utility grid. Two 13,500-kW gas turbines and a 25,000-kW steam turbine supplied total power and process steam to the plant's newsprint and paper mills. In 1974, the mill expanded its operations and increased its power capacity to 100,000 kW. The company added a new 30,000-kW steam turbine and signed a contract with Houston Lighting and Power Company to supply the remainder of its electrical needs. During the summer peak season, when electricity prices rose, Southland purchased 27 percent of its power from the utility; during the winter it purchased 36 percent.

In the residential-commercial-institutional sector, grid-independent and grid-connected systems appear to be the most attractive cogeneration approaches. In either case, the system design is based on the same technical and sizing considerations that hold in the industrial sector.

At present, the cogeneration approach pursued most widely in this sector is the grid-independent system. Kings Plaza, a million-square-foot enclosed shopping mall in Brooklyn, New York, has an 11,000-kilowatt-electricity (kWe) rooftop cogeneration installation that employs five 2,200-kWe dual-fuel diesel-engine generators. Starett City, an urban development also in New York, meets all electrical power needs of 6,000 apartment units, a shopping center, a community center, and multilevel garages for 5,200 cars by use of two steam turbogenerators, which provide 57 million kWh of electrical energy annually.

The second and less common approach pursued in this sector is to design a system which is connected to the utility grid. An example of this type of system is Franklin Heating Station, a central cogeneration plant in Rochester, Minnesota. Franklin serves the base-load needs of a hotel and a medical clinic. It has an arrangement with the city to purchase up to 8,000 kW for emergency needs. Texas A&M University also has a system connected to the local utility grid.

To the author's knowledge, no cogeneration facilities in the residential-commercial-institutional sector sell excess electricity to other users.

The major approach available for designing utility cogeneration systems is for a utility to sell both thermal energy and electricity to facilities in the industrial or the residential-commercial-institutional sector. The cost and efficiency of the thermal energy distribution system is probably the most important technical consideration for a utility. Normally, to be economical, a utility's cogeneration operation must be within 10 miles of a substantial demand for process heat. In sizing and selecting the cogeneration components, the utility must often make tradeoffs between its main business of producing electricity and that of producing heat. A utility can design a cogeneration system to maximize the production of electrical energy, to maximize the production of heat, or to balance the production of electricity and heat to match demand, depending on what it foresees as the demand mix. An example of this type of system is the Atlantic City Electric Company's Deepwater Station in Wilmington, Delaware, which provides steam to the E.I. DuPont de Nemours Company for use in industrial processes. In another case, Applied Energy, Inc., a subsidiary of San Diego Gas and Electric Company, operates a cogeneration facility that supplies steam to two large military bases and delivers power to the utility's electricity-generation system. A third example is the Linden Station of the Public Service Electric and Gas Company of New Jersey, which provides 1-2 million pounds of steam per hour to an Exxon refinery.

Cogeneration has been practiced in the United States and in Europe since the late 1880s. As a matter of fact, the percentage of electricity cogenerated was much higher in the 1950s than it is today.

Over the years, the larger user of cogeneration in the United States has been the industrial sector. During the early part of this century most industrial plants generated their own electricity using coal-fired boilers and steam-turbine generators. Many of these plants used the exhaust steam for industrial processes. It has been estimated that as much as 58 percent of the total power produced by on-site industrial power plants in the early 1900s was cogenerated. Companies such as Dow Chemical Company and Republic Steel have used cogeneration since the early to mid 1900s.

When central electric power plants were constructed and the cost of electricity went down, many industrial plants began purchasing electricity and stopped producing their own. As a result, on-site industrial cogeneration accounted for only 15 percent of total U.S. electrical generation capacity by 1950. That percentage had dropped to about 5 percent by 1974. This change is shown more dramatically in Figure 9-1.

Other factors that contributed to the decline of industrial cogeneration were the increasing regulation of electric generation, low

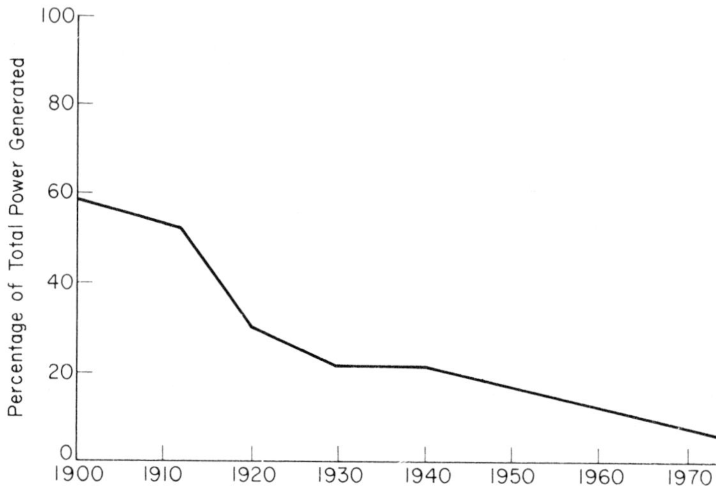

Fig. 9-1 Since 1900, industrial on-site power generation in the United States has declined relative to total generation.

energy costs which represented a small percentage of industrial costs, changing corporate income structures, advances in technology such as the packaged boiler, and tightening environmental restrictions. In addition, many companies decided that they should not be in the power production business. As cogeneration began to wane, industries began to forget the advantages it afforded, since for all practical purposes their energy needs could be met cheaply by purchased electricity.

Cogeneration is much more widely used in Europe. It supplies about 10 percent of the electricity in the Netherlands, 16 percent in France and West Germany, 18 percent in Italy, and less than 5 percent in the United States. Both Europe and the United States began to cogenerate in the late 1800s. Technological changes were implemented as they came along. Utility reliability was the first change to affect cogeneration practices. The United States had a reliable electric distribution grid by about 1920, and not until about 1945 was this accomplished in Europe. For that reason cogeneration tended to grow for a much longer period of time in Europe than it did in the United States. When Europe finally had reliable grids, the use of cogeneration began to decline somewhat. A second important reason for the difference between the success of cogeneration in Europe and the United States is that most European countries have fewer regulations restricting the sale of electricity. In fact, many of Europe's regulations encourage cogeneration. For instance, France's laws require that its state-owned utilities purchase surplus power from on-site generators. The industrial on-site generator can also use the grid to transmit its power between a number of its plants. Industry may also build its own transmission lines between plants if it can acquire a right of way. Many of the regulations that encourage cogeneration in Europe are absent or reversed in the United States.

The residential-commercial-institutional cogeneration market has never been very large in the United States, and currently accounts for less than 1 percent of the electricity consumed by that sector. The gas industry began to take an active interest in this area in the 1950s. Through the Group to Advance Total Energy (GATE) the industry played a major role in promoting cogeneration in this sector; there were 90 plants in operation by 1964 and almost 300 by 1967. These units were gas-fired diesels, combustion turbines, and oil-fired diesel engines. There are plants in Rochdale Village and Coney Island in New York, Regency Square in Jacksonville, Florida, and other locations throughout the United States. Many of these systems were unsuccessful. For the most part, maintenance problems, operating costs, and performance were the culprits. In many cases the cogeneration system was isolated from the grid by high standby charges, and reliability became a problem. Another problem was the continued reliable availability of fuel.

The European experience in the residential-commercial-institutional market has been somewhat limited. This is due mainly to the lack of natural gas or other easily usable fuels in this market. There are, however, plants at Orly Airport and at the Porte-Maillot convention center and shopping complex in Paris. These plants use primarily natural gas, but can be switched to diesel fuel if the need arises.

The U.S. utility market is also rather limited in its use of cogeneration. A number of utilities did supply cogenerated steam to large industrial users and urban areas in the early 1900s.

THERMODYNAMIC ANALYSIS

Regardless of the type of installation to be made, the requirements for both thermal and electrical use must be assessed. The total demand as well as the cyclic demand of each must be addressed; that is, a load profile for both thermal and electrical uses within the specific application must be developed. These load profiles will define the peak and average loads required for each type of energy and will enable the plant to be sized. The type of system to be employed depends on equipment availability, reliability, and maintainability, and on a rigorous analysis to determine from energy and cost savings which system is best.

Three systems are available for cogeneration: extraction or backpressure steam turbines, gas turbines, and diesels. Figure 9-2 compares the steam-turbine and gas-turbine cogeneration systems with the separate generation of steam and thermal energy. The diesel system is similar in arrangement to the gas-turbine system.

In the backpressure steam turbine, the fuel is burned in a boiler to produce steam at a pressure of 850-1,450 psi. The steam is used to drive a turbine which in turn drives a generator that makes electricity. The steam exhausted from the turbine is usually at a pressure of 50-300 psi and a temperature of several hundred degrees. Although the system does not produce as much electricity per unit of steam generated as a conventional Rankine steam cycle used in central power stations, the incremental heat rate for electricity production is about half that of the conventional system—that is, approximately 4,550 as compared with about 10,000 Btu per kWh of electricity.[2]

In the gas-turbine system, the fuel is burned at high pressure and is used to drive the turbine. This in turn drives the generator, which produces electricity. The exhaust gases in this system are usually rejected at temperatures of 600-1,100°F. The exhaust gas is then fed to a waste-heat boiler to produce steam. Although this process has an incremental heat rate for electrical production of 5,000-6,000 Btu/kWh, it produces 4-6 times as much electricity per unit of process steam as the back pressure turbine. Roughly 25-35 percent of the fuel used in this system is converted to electricity.

A variation of the gas turbine is the indirectly fired turbine. In this case, a heat exchanger placed between the combustor and the turbine inlet isolates the turbine from the fuel and therefore permits use of

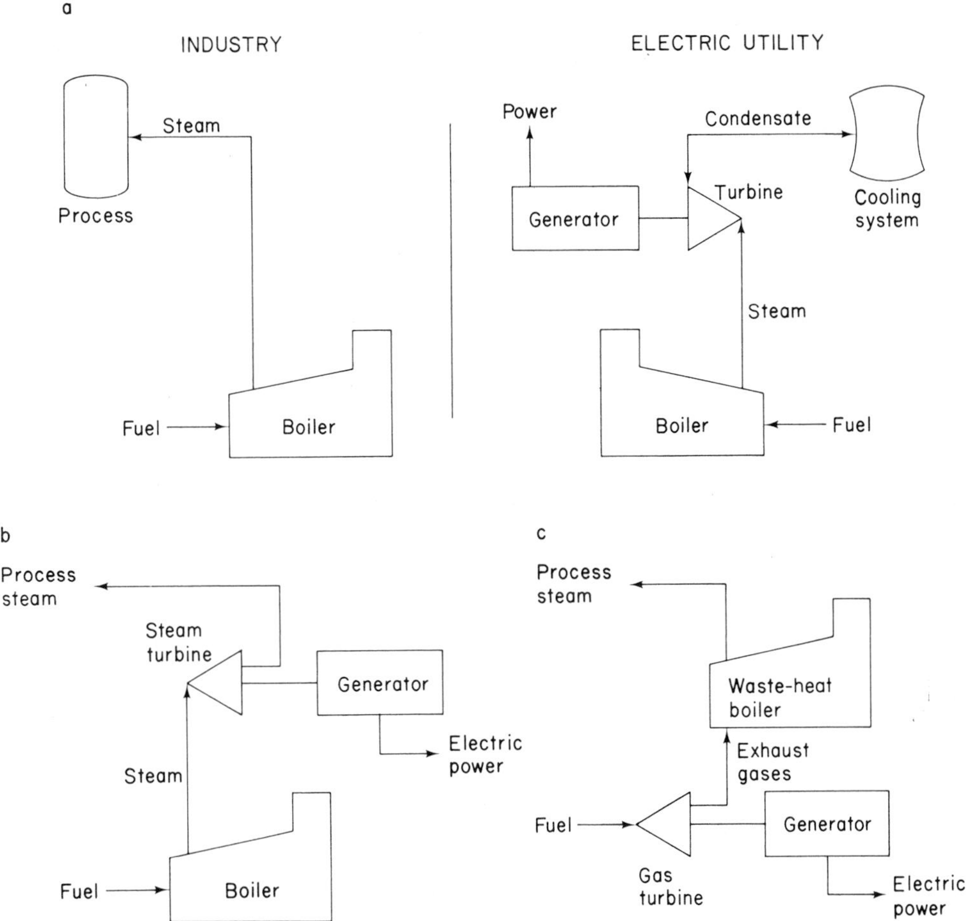

Fig. 9-2 Schematic diagrams of various power-generation schemes. (a) Separate steam and electric-power plants; (b) back-pressure steam turbine; (c) gas turbine and waste-heat boiler. Source: ref. 2.

a wide variety of fuels. For example, coal fired in a fluidized bed could be used as the fuel or heat source and would allow tremendous fuel flexibility. The heat rate of this system is somewhat higher than that of the gas turbine, but the other characteristics are similar.

Diesel engines have exhaust temperatures of 600-800°F. As with gas turbines, the exhaust gas can be used in a waste-heat boiler. This system converts approximately 40 percent of the fuel to electricity, but has an incremental heat rate of roughly 6,500-7,000 Btu/kWh. The system can, however, produce 5-8 times as much electricity per unit of process steam load as the steam turbine.

Process efficiency can be expressed in terms of enthalpies or in terms of thermodynamic availabilities (the so-called second-law analysis). The second-law efficiency is defined as

$$E = \frac{W_{min}}{W_{act}},$$

where W_{min} is the minimum amount of work required to accomplish a task and W_{act} is the amount of work actually expended in accomplishing the task. The values of W_{min} and W_{act} can be calculated for any process by evaluating the change in thermodynamic availability undergone by all materials in the task and the maximum work that could have been obtained from the fuel, respectively. (Manual 1 addresses the details of these calculations.)

The results of applying these calculations to representative cogeneration systems are shown in Table 9-1. The exact value of the efficiency depends on the condition of the heat generated (which differs from system to system) as well as on the system configuration. However, the values given in the table illustrate some important general differences among systems.

The last column in Table 9-1 shows the fuel energy savings ratio (FESR), defined as

$$\text{FESR} = \frac{\text{NF} - \text{CF}}{\text{NF}},$$

Fuel Fraction Converted to Electricity	Heat Rate[a] (Btu/kWh)	Second-Law Efficiency[b]	FESR[c]
0.16	4,550	0.40 (0.32)	0.21
0.30	5,400	0.48 (0.34)	0.29
0.36	5,900	0.50 (0.35)	0.30
0.35	5,750	0.47 (0.35)	0.25
0.13	4,550	0.42 (0.35)	0.17
0.30	5,700	0.50 (0.36)	0.27
0.35	6,950	0.48 (0.37)	0.24

Table 9-1 Fuel-utilization characteristics of cogeneration systems.

	Process Steam Pressure (psig)	Electricity Production Rate (kWh per million Btu of steam)
Steam turbine[d]	50	70
Gas turbine[d]	50	150
Gas/steam turbine[e]	50	320
Diesel[d]	50	380
Steam turbine[d]	150	50
Gas turbine[d]	150	200
Diesel[d]	150	405

Source: adapted from reference 2.

a. Fuel required to produce electricity, in excess of that required for process-steam generation alone, with 88 percent boiler efficiency assumed for process-steam production alone.

b. Value for separate generation of process steam and central-station electricity is shown in parentheses.

c. Fuel saved by cogeneration divided by fuel required to produce the same mix of steam and electricity separately, with heat rate for central-station power plant assumed to be 10,000 Btu/kWh and boiler efficiency for steam production assumed to be 88 percent.

d. See figures 3.1 and 3.3 in reference 8.

e. For the 73.2 MWe gas turbine (burning no. 2 distillate oil) described in table II of reference 11.

where CF is the cogeneration fuel energy (fuel required to cogenerate the mix of steam and electricity) and NF is the non-cogeneration fuel energy (fuel required to generate the same mix separately). An FESR value of 0.20 means that the cogeneration process uses 20 percent less energy than separate generation of the same power-steam mix.

The important characteristics of cogeneration systems as summarized in Table 9-1 are the following:

- Diesel and gas-turbine systems have higher thermodynamic efficiencies than steam turbine systems or the separate generation of electricity and steam.
- Diesel and gas-turbine systems require more energy per kWh of electricity generated than do steam turbine systems, but

significantly less than the 10,000 Btu/kWh common to central-station power generation.

- Diesel systems generate more electricity per unit of process heat than do gas turbines, while gas turbines are superior in this regard to steam turbines.

Ultimately, the choice of a system must be made on overall economic grounds from among systems capable of producing electricity and process steam in a ratio that would meet the plant's needs.

NONTECHNOLOGICAL AND ENVIRONMENTAL ISSUES

Typically, cogeneration is thought of as a technique to conserve energy and save money. Its use may also, however, yield a reduction in polluting emissions. Since one of cogeneration's attributes is to consume less fuel by performing two functions with the same equipment rather than with two different pieces of equipment, it seems logical that there would be a reduction in overall emissions. Unfortunately, cogeneration concentrates the total emissions at one location rather than at two locations, as is generally the case. This colocation of both sources can be a problem and has deterred some industries from installing cogeneration units.[3] Fuel supply may also be a problem for industries. Fuel quality may begin to deteriorate to heavy residual oils or coal. This deterioration, which is consistent with the policy stated in the National Energy Act, tends to aggravate the environmental problem.

In order to alleviate the environmental problem, the user must either clean up the fuel before it is burned, clean up the fuel while it is burning, clean up the exhaust gases, or increase the work and thermal output at the same emissions level. Cogeneration, of course, performs the last function.

If an open-cycle gas turbine is employed in a cogeneration scheme to make electricity and low-temperature process steam, it eliminates the need for a process-steam boiler through its use of a waste-heat boiler. As Figure 9-3 shows, the reduction in emissions actually comes from the elimination of the boiler. According to this simplistic view, the cogeneration application creates no more pollution than the less efficient systems already employed.[4] These results can be

a

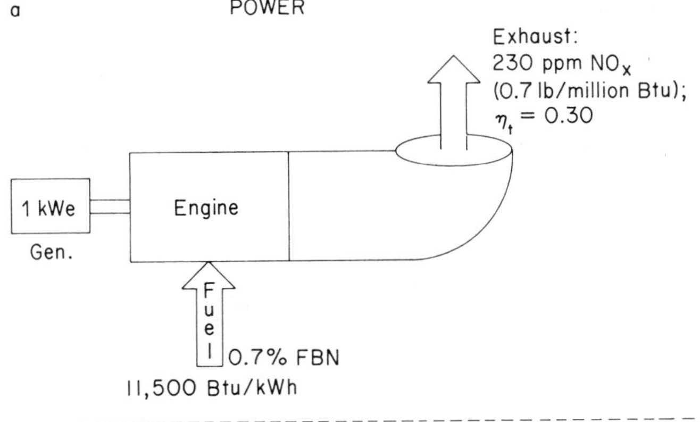

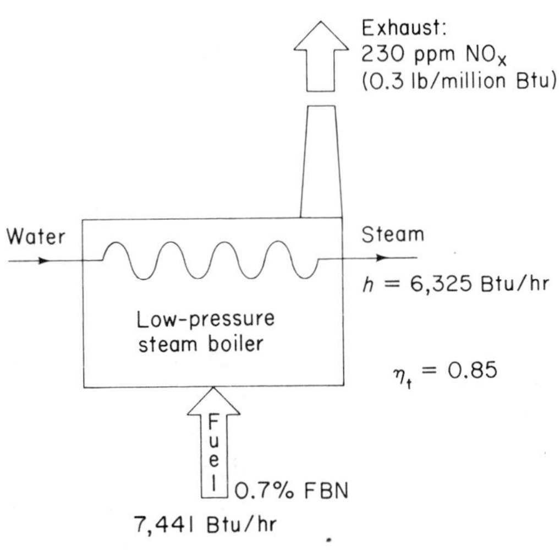

$$\text{lb NO}_x/\text{hr} = 0.007527_E + 0.002232_H = 0.00976$$

Fig. 9-3 Illustration of effects of cogeneration on emissions. (a) Isolated power and process heat production (no cogeneration); (b) integrated power and process heat production (cogeneration). Source: ref. 4. ***(Continued)***

Fig. 9-3 Continued

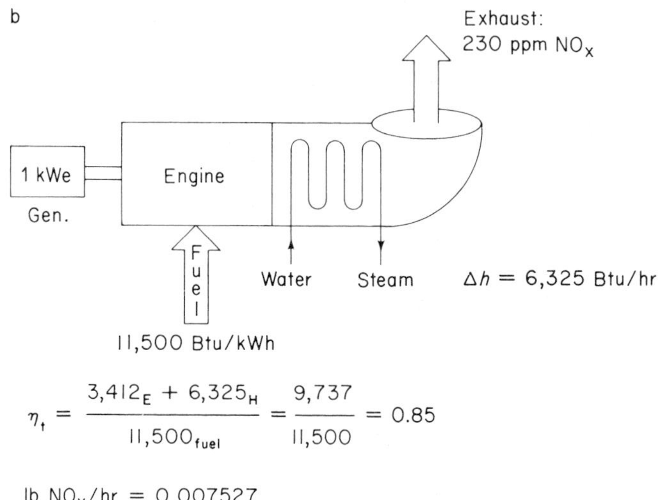

$$\eta_t = \frac{3{,}412_E + 6{,}325_H}{11{,}500_{fuel}} = \frac{9{,}737}{11{,}500} = 0.85$$

lb NO_x/hr = 0.007527

generalized as shown in Figure 9-4, which plots emissions per useful output. Superimposed on the curve of Figure 9-4 are ranges for engines making only electricity and those making electricity and process steam (cogeneration). This graph shows how the efficient use of fuel is tied directly to emissions.

Many of the cogeneration systems now on the market have good reliability and maintenance records. However, when a cogeneration system is installed, considerations as to its reliability inevitably come up. The reason for the concern is that an industrial firm is generally in the business of turning out a certain product, and any plant "down time," for whatever reason, affects the profit picture for at least that period. When the plant depends on a single machine, the reliability of that machine dictates the overall reliability. If the requirement is large enough, multiple machines can greatly improve the system's reliability. Concern for reliability has driven cogeneration users to attempt to interconnect with the utility grid in order to increase overall reliability. This practice may not work in the future if utility companies try to curtail use to large customers during high demand periods, as some of the large gas utilities have had to do.

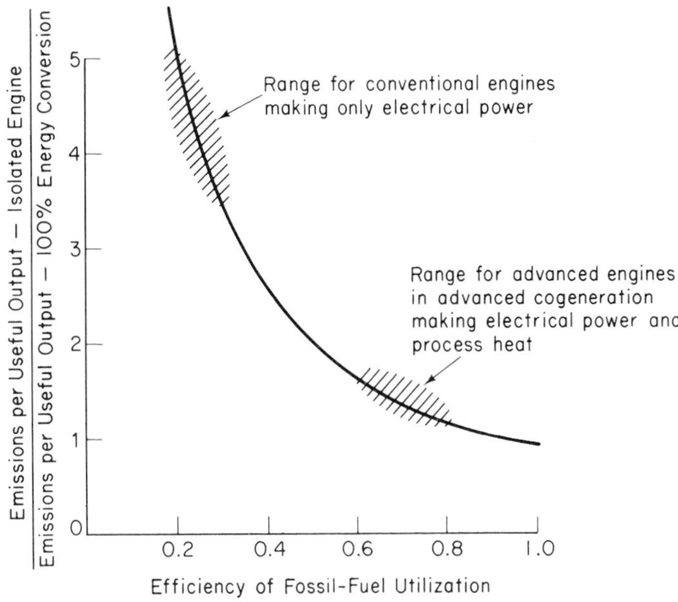

Fig. 9-4 Plot of impact of efficiency on emissions. Source: ref. 4.

Even though it may be advantageous to use a utility as a backup system in case of an emergency, charges for this service are normally very high. The charges are even higher if the industry requires a backup during the utility company's peak demand periods. This requirement forces the utility to own and maintain extra generation equipment that it will probably use for only a small portion of the time.

In many cases it may appear to be attractive to sell cogenerated electricity back to the utility grid. The prices paid for this power are generally fairly low, because it is considered "dumped power" which the utility cannot really count on. If the industrial firm is willing to sell what is referred to as "firm power," it must make contractual arrangements many years in advance in order to allow the utility to integrate this power into its generation and distribution planning. Many industrial concerns feel that predicting how much power thay can contract to sell over the next 10 years is impossible because of uncertainties in their own needs, and therefore do not want to write firm contracts.

Because of long-standing policies held by utilities, generation of electricity by industrial firms has been discouraged. Rate charges to industrial users have encouraged steady use of large blocks of electricity and discouraged standby use. Utilities have historically been unwilling to buy electricity from industrial firms producing excess electricity. One of the reasons for these policies is the fact that utilities base their profits on a rate of return on capital investment. This rate is dictated by the various regulatory agencies the utilities report to. Thus, utilities appear to have nothing to gain by encouraging industrial cogeneration, since it will not increase their rate base. For this and other reasons, such as capital availability, there is some question as to whether the industrial firm or the utility should own the cogeneration unit. The following points have been made in favor of utility-company ownership:

- With cogeneration units in the rate base, utilities should have a more positive attitude toward cogeneration.
- Utilities better understand how to operate and maintain power-generating systems, so that industrial "learning problems" with regard to power-plant operations would be avoided.
- Owning the plants, the utilities could more easily control their output and thereby better balance supply and demand for the utility system.
- Because regulated utilities are assured a fair rate of return on their investments, they are willing to accept a lower rate of return than private industry.
- An industrial firm may be deterred from owning a cogeneration unit capable of exporting power by the prospect of being encumbered by utility regulations.

In favor of industrial ownership, the following points have been offered:

- Problems of coordinating utility and industrial activities would probably be less.
- Scrutiny of rate negotiations between a utility and many industrial producers should give the Public Utility Commission a better data base for establishing utility rates to all customers.

(These points are not meant to be all-inclusive, but should give an idea of the types of issues being raised.)

At present, regulatory jurisdiction over privately held cogeneration facilities is unclear. Both the Federal Energy Regulatory Commission (FERC) and the state Public Utility Commissions (PUCs) have varying degrees of authority in utility regulation. If the cogeneration system is not connected to the grid, neither the FERC nor the PUC has any jurisdiction. However, there is concern about potential regulation of cogeneration facilities that (in the words of reference 3) "parallel but do not transmit power to the utility grid." These types of issues will continue to be a problem for cogeneration until all of the new questions can be laid to rest by the regulatory bodies involved. The propsects appear to be favorable for cogeneration, and many project that new regulations will encourage rather than hinder its prospects.

CHOOSING A COGENERATION SYSTEM

In the choice of a cogeneration system for a particular plant, a number of factors must be considered. It is first necessary to define the parameters of the plant, such as the electricity and heat loads, the process temperature, and the load factor (hours of operation per year). Figures 9-5 through 9-7 show representative values for a number of industries.

When considering the power-heat ratio, one must select the optimal energy-conversion system (ECS) for the particular plant. In sizing the ECS one must typically select either a power or a heat match. Ideally, the heat and power requirements would be at the same match point for a particular ECS, but this is almost never the case. For this reason, it must be decided whether to use auxiliary boiler heat or sell electricity (Figure 9-8), or to sell excess heat or buy electricity (Figure 9-9). (In actual cases selling excess heat is rarely considered because of the difficulties associated with heat losses in piping and with matching the other plant's load requirements.)

Figure 9-10 plots the FESR values for various cogeneration systems versus the power-heat ratio attainable. Also shown is the maximum FESR that could be achieved by a cogeneration system with a heat rate of 3,414 Btu/kWh. This figure is useful in choosing postulated cogeneration systems to meet specific plant needs. For example, suppose that the power-heat ratio for a plant is 0.2 (about 60 kWh per million Btu of process heat). Figure 9-10 shows that a steam-turbine system can produce this ratio at an FESR of about 17 percent. Alternatively, a closed-cycle turbine could be selected and excess electricity sold. Since a closed-cycle turbine operates with a power-heat ratio of 0.4-1.4, the amount of electricity that must be sold would be at least twice that consumed on-site if the system were sized to meet process-heat needs. As another example, suppose that the power-heat ratio was 1.5 (about 440 kWh per million Btu of process steam). Figure 9-10 shows that this ratio is too high for gas turbines but within the range for diesels. A turbine could be used if additional electricity were purchased.

Figure 9-10 is for general estimating only. Systems that perform outside the limits shown may be available. Use Figure 9-10 to get some idea of what kind of system is of interest for your application, then contact equipment vendors for more details on system capability.

The procedure for selection of a cogeneration system can be summarized as follows:

1. Determine the required plant heat load.
2. Determine the required plant electrical load.
3. Select a generator set that meets the required heat load, by referring to Figure 9-10.
4. Ensure that the kilowatt rating of the generator is close to the plant's required electrical demand.
5. Continue the selection process until a set or sets closely match the heat and electrical need.
6. If a total match is impossible, determine whether the plant should buy or sell power or buy or sell heat.
7. Select the final cogenator set.

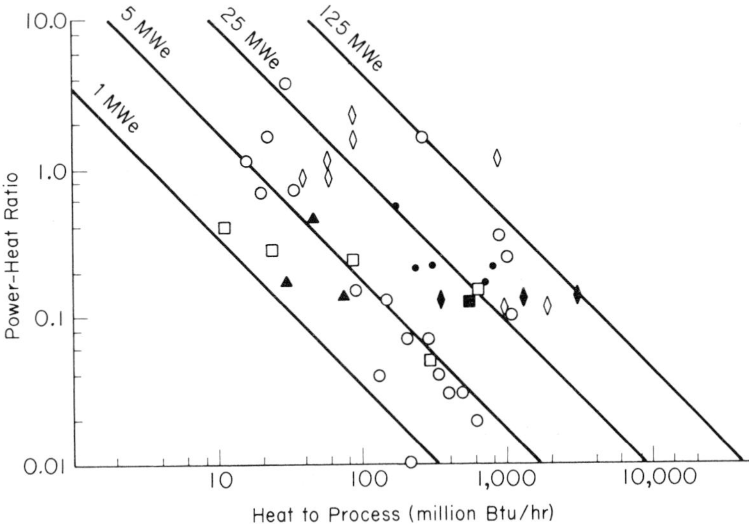

Fig. 9-5. Characteristics of industrial processes. (▲) Lumber; (■) textiles; (●) paper; (◆) food; (◇) primary metals; (○) chemicals; (△) cement and glass. Source: ref. 12.

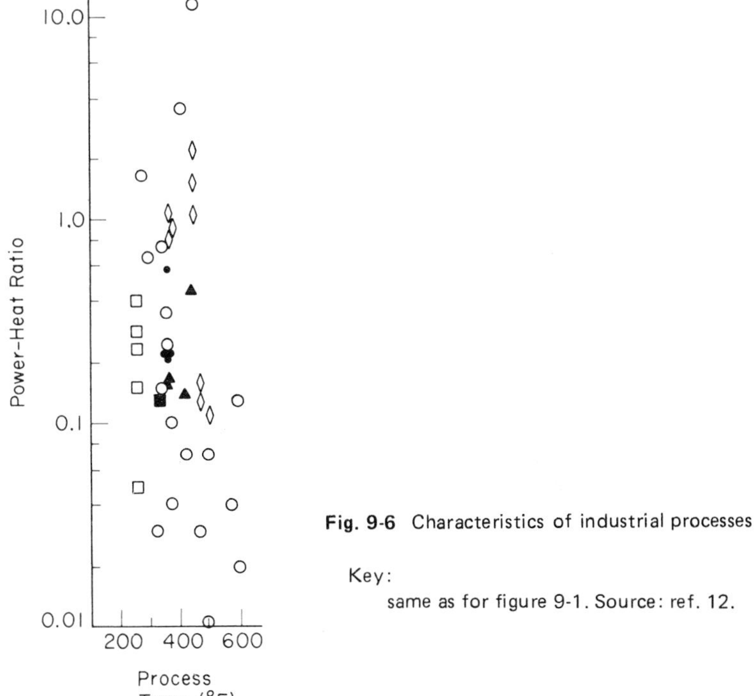

Fig. 9-6 Characteristics of industrial processes.

Key:

same as for figure 9-1. Source: ref. 12.

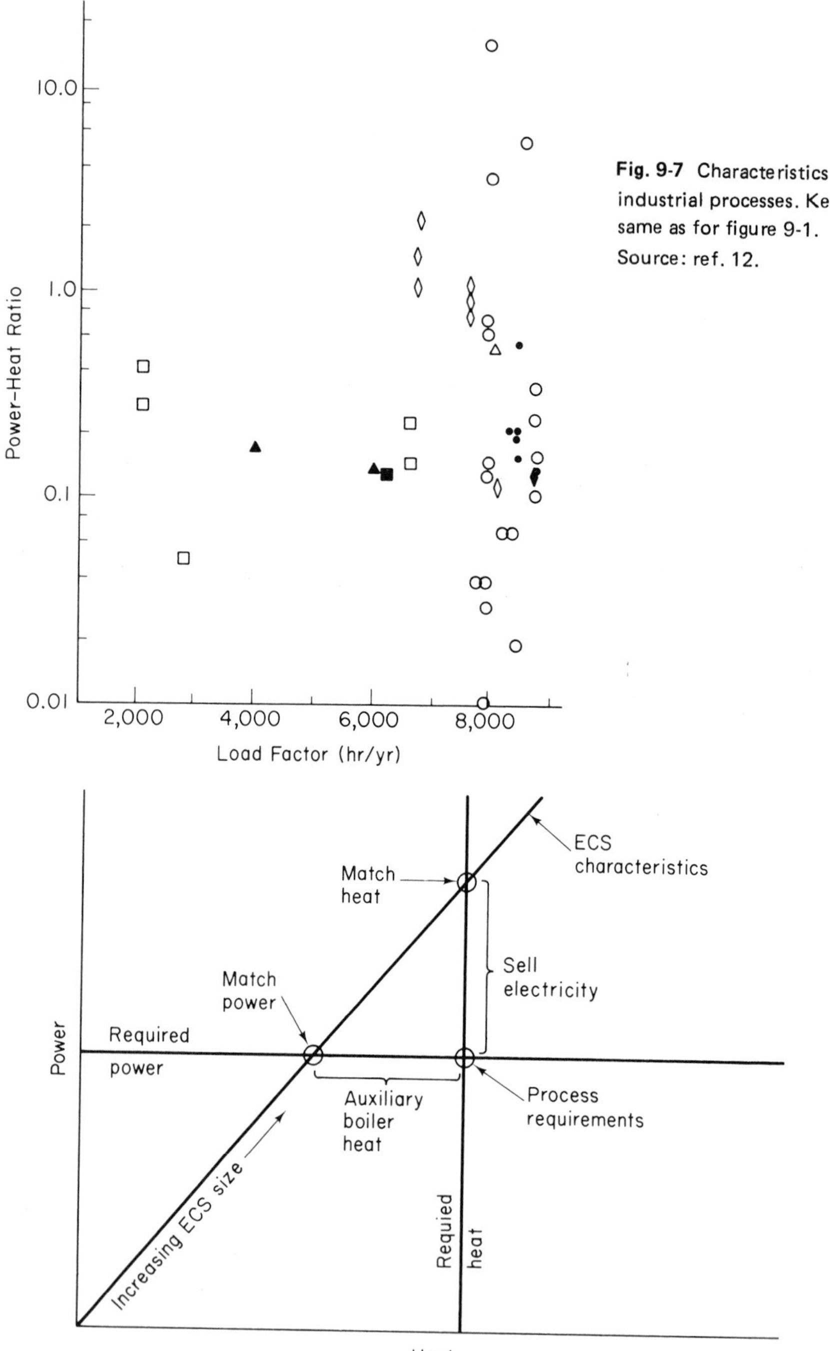

Fig. 9-7 Characteristics of industrial processes. Key: same as for figure 9-1. Source: ref. 12.

Fig. 9-8 Match between and energy-conversion system (ECS) when power-heat ratio of ECS is greater than required. Source: ref. 12.

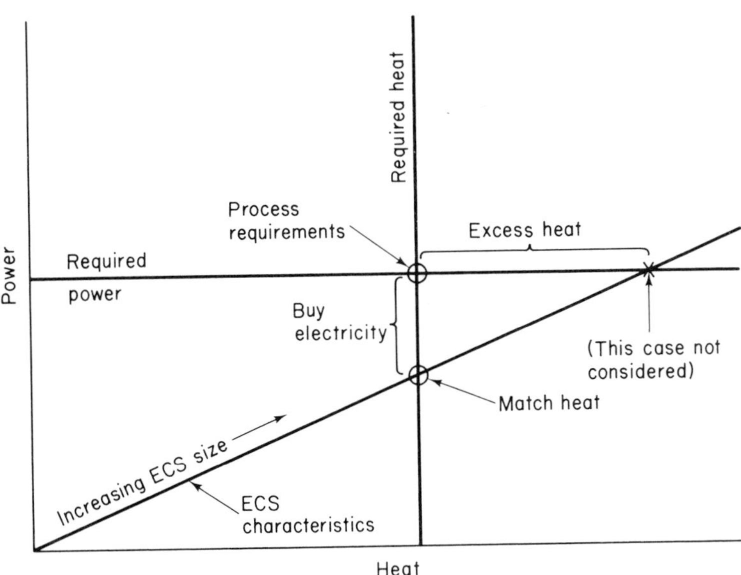

Fig. 9-9 Match between process and ECS when power-heat ratio of ECS is less than required. Source: ref. 12.

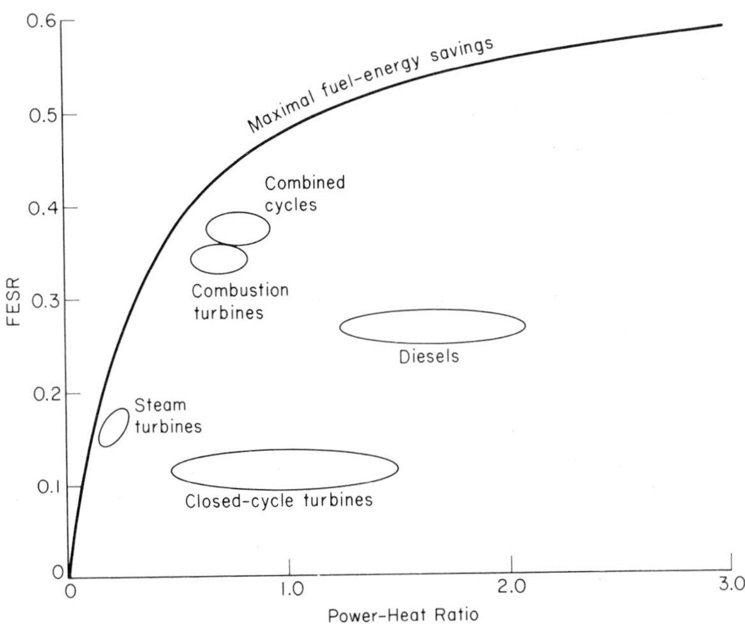

Fig. 9-10 Relationship between fuel energy-saving ratio (FESR) and power-heat ratio for a temperatures of 350°F.

10

Establishing a Maintenance Program for Plant Efficiency and Energy Savings

GOOD MAINTENANCE SAVES $

Energy losses due to leaks, uninsulated lines, dirt buildup, inoperable furnace controls, and other poor maintenance practices is directly translated into additional energy costs. Good maintenance saves in the plant's yearly operating costs.

In this chapter, you will see the results of a survey on maintenance effectiveness, look at ways to turn around the maintenance program, and learn to apply a preventative maintenance program to energy conservation.

WHAT IS THE EFFECTIVENESS OF MOST MAINTENANCE PROGRAMS?

Poor maintenance can be readily translated into inefficient operation of systems. Inefficient operation of systems spells energy waste. The problem is magnified by a survey[47] of 1000 plant engineers and senior maintenance supervisors that indicates the sad state of maintenance management. Findings of the survey indicate:

- Maintenance is not receiving management's attention
- Planning procedures are performed by:

  Superintendents–35%

Maintenance foremen–35%
Planners, schedulers–30%

- Few maintenance departments issue regular reports to management
- Few use work order (maintenance request) system
- Few maintenance departments analyze the cause of equipment backlog
- Most companies surveyed had started preventive maintenance programs, but most were dissatisfied with their progress
- Most companies had poor control over material and spare parts account

The aspect of specific interest to energy conservation is that preventative maintenance is not occurring satisfactorily. Only fifty-five percent of the maintenance departments have a tickler file or some other filing system for automatic work order generation for filter changes and other periodic preventive maintenance practices. Another factor which affects the energy conservation program is that most managers have little or no control over maintenance work backlog. This means that maintenance work tends to happen rather than being planned.

HOW TO TURN AROUND THE MAINTENANCE PROGRAM

Management Attention

The results of this survey indicate that the sad state of maintenance has perpetuated because management has paid little attention to it. If management wants to conserve energy, it must give maintenance top priority. Management has historically devoted much of its efforts to improving manufacturing productivity, and in the reporting, measuring and analyzing of manufacturing costs. Similar efforts to improve maintenance productivity are needed. Plant engineers need to evaluate their present practices to determine the status of their maintenance program. Periodic reports, quantitively stating key maintenance performance indicators, need to be issued to management.

Work Order Request

Effective maintenance insures that the activity is being controlled. A work order request system needs to be established and used. Work orders serve as an historical record of repairs and alterations made to key equipment. Since the work order is used to authorize maintenance work, it is the prime requisite for any maintenance control system.

Maintenance Planning

Planning of maintenance activity needs to come under the jurisdiction of a maintenance manager. The maintenance manager must have the same status within the organization as manager of production. In planning maintenance work, the following needs to be identified:

Scope of the job
Location of the job
Priority
Cause (why the work must be done)
Methods to be used
Material requirements
Manpower requirements (number and crafts).

In addition, the following should be implemented:

- Labor time standards should be established to cover recurring or highly repetitive tasks
- Reliable estimates should be established which cover nonrecurring (less than once a month) maintenance work
- Procedures for issuing maintenance reports should be established
- Reports should include productivity of labor force.

Applying a Preventative Maintenance Program to Energy Conservation

In the beginning of this chapter, we discussed first the sad state of maintenance and indicated some ways to turn this situation around. *Only* then can maintenance for energy conservation be considered.

TABLE 10-1 Energy Saving Survey Checklist

Department ______________________

Date ______________________

Supervised By ______________________

	Location #1	Location #2
Fuel, Gas or Oil Leaks		
Steam Leaks		
Compressed Air Leaks		
Condensate Leaks		
Water Leaks		
Damaged or Lacking Insulation		
Leaks of, or Excess HVAC		
Burners Out of Adjustment		
Steam Trap Operation Note: Each trap is to be tagged with date of inspection		
Cleanliness of Heating Surfaces Such as Coolers, Exchangers, etc.		
Hot Spots on Furnaces to Determine Deterioration of Lining		
Bad Bearings, Gear Drives, Pumps, Motors, etc.		
Dirty Motors		
Worn Belts		
Proper Viscosity of the Lubricative Oils for Large Electric Drives, Hydraulic Pumps (Proper Viscosity Minimizes Pump Drive Slippage)		
Cleanliness of Lamps		

The first step is to identify energy wastes that can be corrected by maintenance operations. Table 10-1 indicates an Energy Savings Survey which should be made. Let's look at some specific areas of maintenance as it applies to energy conservation.

STOP LEAKS AND SAVE

Steam Leaks

Leaks not only waste energy; they cause excessive noise as well. A case in point was a noise citation a plant received for steam discharges. The plant solved its noise problem by using steam traps and, in addition, it saved energy which paid for the installation.

Figure 10-1 illustrates the annual heat loss from steam leaks.

SIM 10-1

A maintenance survey of the plant indicates that steam leaks exist from 100 lines operating at 100 psig. The average size of the leak is estimated to be $\frac{1}{8}$ in. diameter. The value of steam in the plant is $4/10^6$ Btu. What is the annual savings from repairing the leaks?

ANSWER

From Fig. 10-1, the steam loss per line is 540 X $6/10^6$ Btu. The annual savings is then

$$540 \times 100 \times 6 = \$324{,}000 \text{ per year.}$$

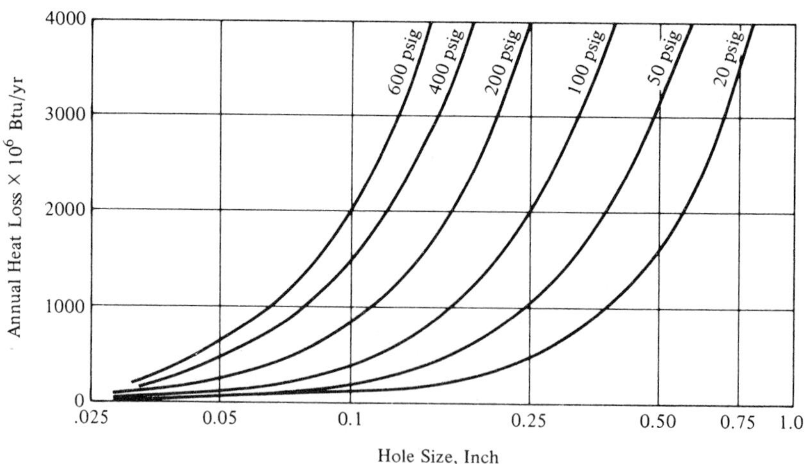

Fig. 10-1 Annual heat loss from steam leaks.

Steam leaks can be detected visually or by using acoustic or temperature probes. Leaks should also be checked at valves, heaters, and other equipment. Leaking or stuck traps, stuck by-pass valves discharging to the sewer and condensate systems should also be checked. A program should be established to regularly inspect for leaks, to shut off steam to equipment taken out of service, to repair steam leaks promptly, and to route piping such that leaks that develop are visible.

Leaks In Combustible Gas Lines

As with steam, leaks in combustible gases such as natural gas, methane, butane, propane, hydrogen, etc., is a direct waste of valuable energy, and causes the hazard of fire or explosions. Tables 10-2 and 10-3 summarize losses for various combustible gas piping systems.

TABLE 10-2 Natural Gas Loss per Year from Leaks in *Underground* Pipelines at Various Pressures. (KSCF)

Corrosion Hole Diameter, in	Line Pressure (psig) 0.25	5	25	60	100	300	500
1/64	1	4	10	20	30	80	140
1/32	2	6	20	35	60	160	250
1/16	10	40	100	200	320	900	1,500
1/8	50	200	600	1,200	1,800	5,000	8,200
1/4	250	1,200	3,300	6,500	10,300	28,300	46,500
1/2	1,400	6,600	18,800	37,300	58,000	156,500	263,000

Adapted from *NBS Handbook 115.*

TABLE 10-3 Natural Gas Loss per Year from Leaks in *Above Ground* Pipelines at Various Pressures. (KSCF)

Corrosion Hole Diameter, in	Line Pressure (psig) 0.25	5	25	60	100	300	500
1/64	5	26	69	136	212	581	953
1/32	21	102	277	544	846	2,330	3,810
1/16	85	409	1,110	2,180	3,390	9,300	15,300
1/8	341	1,640	4.430	8,700	13,500	37,200	61,000
1/4	1,360	6,540	17,700	34,800	54,200	149,000	244,000
1/2	5,450	26,200	70,900	139,000	217,000	595,000	977,000

Adapted from *NBS Handbook 115.*

SIM 10-2

A natural gas leak is detected in above ground piping. The natural gas pressure is 100 psig and the leak is from a $\frac{1}{8}$ in. diameter hole. What is the annual loss, considering the value of natural gas at $6.00 per million Btu?

ANSWER

From Table 10-3, leakage is 13,500 $\times$ 10^3 SCF. Since the heating value of natural gas is 1000 Btu/CF, then the heating value of the leak is:

$$13{,}500 \times 10^6 \text{ Btu}$$

The corresponding yearly energy waste is $81,000.

Leaks in Compressed Air Lines

Leaks in compressed air lines waste energy and cost yearly operating dollars. An air compressor supplies the plant air; thus, any air leak can be translated into wasting air compressor horsepower. Table 10-4 illustrates losses for various hole diameter openings.

SIM 10-3

A maintenance inspection of the plant compressed air system revealed the following leaks in the 70 psig distribution piping system.

Number of Leaks	Estimated Diameter In
5	$\frac{1}{4}$
10	$\frac{1}{8}$
10	$\frac{1}{16}$

What is the yearly energy loss, assuming an energy cost of $.060 per Kwh?

ANSWER

From Table 10-4,

Number of Leaks	Estimated Diameter	Cost of Power Wasted $/yr
5	$\frac{1}{4}$	17,550
10	$\frac{1}{8}$	8882
10	$\frac{1}{16}$	2199
	Total Energy Loss	$28,631

Air leaks occur at fittings, valves, air hoses, etc. A common way of detecting leaks is by swabbing soapy water around the joints. Blowing bubbles will indicate air leaks.

TABLE 10-4 Annual Heat Loss from Compressed Air Leaks.

Hole Diameter, in	Free Air Wasted (a), cu ft per year, by a Leak of Air at:	Fuel Wasted (b), MBtu/yr	Cost of Power Wasted (c), $/yr at Unit Power Cost of		
	100 psig		$0.020/KWH	$0.040/KWH	$0.060/KWH
$\frac{3}{8}$	79,900,000	2190	4370.00	8740	13110
$\frac{1}{4}$	35,500,000	972	1940.00	3880	5820
$\frac{1}{8}$	8,880,000	243	486.00	972	1458
$\frac{1}{16}$	2,220,000	60.6	121.00	242	363
$\frac{1}{32}$	553,000	15.1	30.30	60.6	90.90
	70 psig		$0.020/KWH	$0.040/KHW	$0.060/KWH
$\frac{3}{8}$	59,100,000	1320	2650.00	5300	7950
$\frac{1}{4}$	26,200,000	587	1170.00	2340	3510
$\frac{1}{8}$	6,560,000	147	294.00	588	882
$\frac{1}{16}$	1,640,000	36.6	73.30	146.60	219.90
$\frac{1}{32}$	410,000	9.2	18.40	36.80	55.20

(a) Based on nozzle coefficient of 0.65
(b) Based on 10,000 Btu fuel/KWH
(c) Based on 22 brake horsepower per 100 cu ft free air per min for 100 psig air and 18 brake horsepower per 100 cu ft free air per min for 70 psig air

From SIMs 10-1, 10-2, 10-3, the total energy waste is $433,631.00. *Only* when maintenance personnel *understand* that stopping leaks saves a significant amount on utility costs, will the program become effective.

PROPERLY OPERATING STEAM TRAPS SAVE ENERGY

Many existing plants remove condensate from steam lines by using tubing connected to the steam piping and discharged directly to the atmosphere. Figure 10-1 shows the loss of steam through a hole. Thus, all steam leaks, whether purposeful or not, should be avoided. A steam trap permits the passage of condensate, air, and non-condensible gas from the steam piping and equipment while preventing (trapping) the loss of steam. Air and non-condensible gases are undesirable in steam systems since they act as insulating blankets and

Check All Steam Traps

Why? To save steam . . . and to prepare your tracing and heating systems for winter operation.

By Whom? Plant Operators, Mechanics, and Fitters from the Field Service Units.

Frequency of Testing: Once/shift by plant operators not unreasonable, daily, weekly, etc.

Safety: Check all steam and condensate lines for insulation. When testing steam traps protect yourself from thermal burns.

Testing: Bucket Traps – Strong No. 141 or Armstrong 811

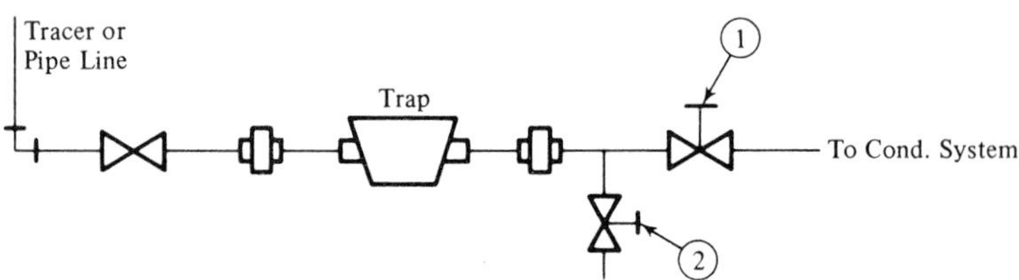

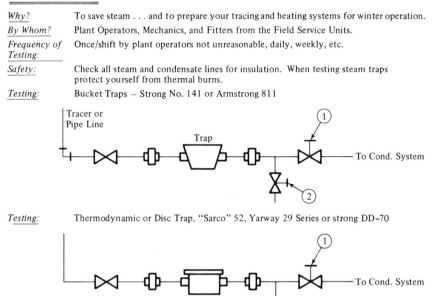

Testing: Thermodynamic or Disc Trap. "Sarco" 52, Yarway 29 Series or strong DD-70

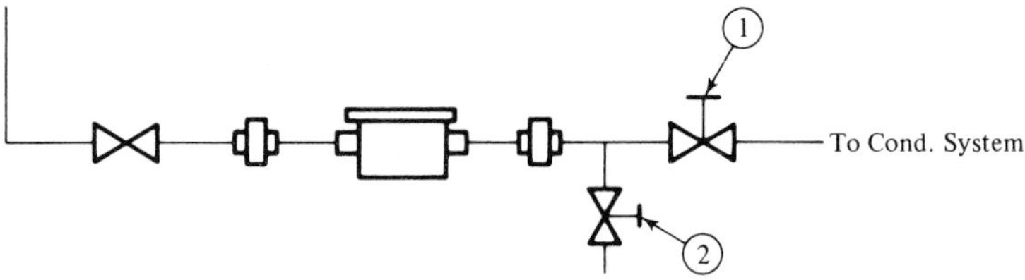

Close valve No. 1 – Open valve No. 2 – Trap should cycle and close. Should trap fail to operate, close valve No. 2 – open valve No. 1. Red Tag for further check. Should trap blow continuously, it could be worn out, undersized, on a cold system, or process start-up.

Blowing Traps Waste Steam!!

Fig. 10-2 Check all steam traps.
Source: Federal Power Commission, position paper No. 17.

reduce the equipment efficiency. Therefore, the use of steam traps is a must. Steam traps should be properly sized and maintained. An oversized or poorly functioning steam trap causes steam to be wasted.

Traps should be inspected once a week to determine the following:

1. Is the trap removing all of the condensate?
2. Is tight shut-off occurring after operation?
3. Is the frequency discharge in an acceptable range? (Too frequent discharge indicates possible under-capacity, while too infrequent discharge indicates possible over-capacity and inefficiency.)

SIM 10-4

A plant maintenance inspection indicated 10-100 psig steam traps were stuck open. The orifice in the trap is $\frac{1}{4}$ inch. Compute the yearly energy savings by repairing the trap, assuming the value of steam at \$6.00 per 10^6 Btu.

ANSWER

From Fig. 10-1, the steam loss is 2100 X 10^6 Btu/yr per trap.
Thus: energy savings = 2100 X 10 X 6.00 = \$126,000 per year.

Reduce Energy Wastes By Insulating Bare Steam Lines

Inspecting an existing plant will reveal that lack of insulation and damaged insulation is common on steam piping. In some instances, the line was originally insulated, but due to piping repairs or rerouting, the insulation was discarded. From Fig. 10-3, the energy loss due to bare steam lines is apparent.

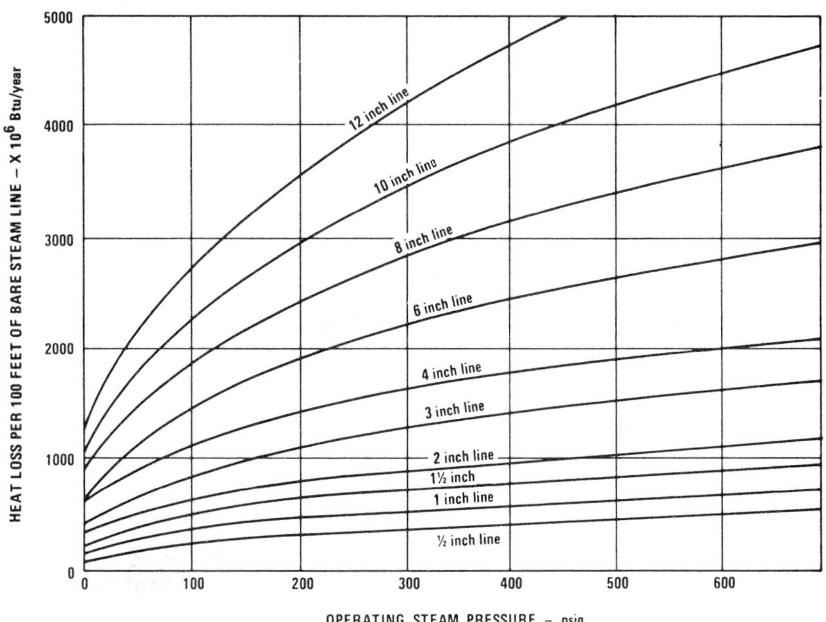

Fig. 10-3 Heat loss for bare steam lines. (Adapted from *NBS Handbook 115.*)

SIM 10-5

A maintenance inspection indicates uninsulated steam lines as follows:

Length	Size	Pressure
300	8″	100 psig
500	6″	50 psig

What is the yearly energy savings resulting from insulating the bare steam line? Assume a steam cost of \$6.00 per 16^6 Btu.

ANSWER

From Fig. 10-3

Length	Size	Loss	Energy Savings By Insulating Pipes
300	8″	5700×10^6 Btu	34,200
500	6″	5500×10^6 Btu	33,000
		Total yearly savings	\$67,200

Don't Overlook Steam Tracing

Steam tracing must be maintained as well as turned off when not required, such as in summer periods. Either automatic controls or improved operator attention is required in order to minimize steam waste.

EXCESS AIR CONSIDERATIONS

What is common about the following?

1. Improper temperature for fuel oil atomization.
2. Worn and obsolete combustion controls.
3. The burner is operated at a setting different than the design setting.
4. Shortage of natural gas supplies force a burner to be operated manually on the back-up oil system.

All of the above can lead to excess air and cause fuel to be wasted. Basically, worn and obsolete combustion controls or use of manual controls prevents operating the burner at the minimum excess air levels.

An improper temperature for fuel oil will cause the same result;

excess air. The solution of these items involves an analysis of the present operations, changing the temperature and design settings and a possible new combustion control system.

SIM 10-6

An existing natural gas burner is operating at a stack temperature of 800°F with a 5.5% oxygen content. A maintenance check indicates that the controls are worn and that excess air was difficult to control. What annual (8000 hr/yr) savings can be achieved by installing a $35,000 oxygen analyzer and control system to reduce the oxygen content to 2 percent? The flue gas flow is 180,000 scfh (standard cubic feet per hour). Assume a 10 year life and a rate of return before taxes of 15%. The price of natural gas is $6.00 per million Btu.

ANSWER

Using Fig. 5-8, the operating conditions are drawn. Fig. 5-8 illustrates a 4% fuel savings.

From Table 1-4, the heating valve of natural gas is 1000 Btu/cf.

q = 8000 hr/yr × 180,000 cf/hr × 1000 Btu/cf × 0.04

57,600 × 10^6 Btu per year

Thus, the annual savings is:

57,600 × $6.00 = $345,600

Annual owning cost = $35,000 × CR = $35,000 × 0.2 = $7000.00

$i = 15\%$

$n = 10$

Annual savings before taxes = $338,600

DIRT AND LAMP LUMEN DEPRECIATION CAN REDUCE LIGHTING LEVELS BY 50%

When lighting systems are designed, a light loss factor is used to compensate for lamp lumen depreciation based on lamp burning hours and dirt build up on lamps and reflecting surfaces. This means that lighting levels will initially be higher than needed and decrease to levels below requirements based on the maintenance program.

Figure 10-4 illustrates the effect of dirt on a reflector. Light must first pass through the layer of dirt to reach the reflector and then pass through again to escape from the luminaire to the working plane. The object of the lighting maintenance program is to use the light available to get the maximum output for the system. This means

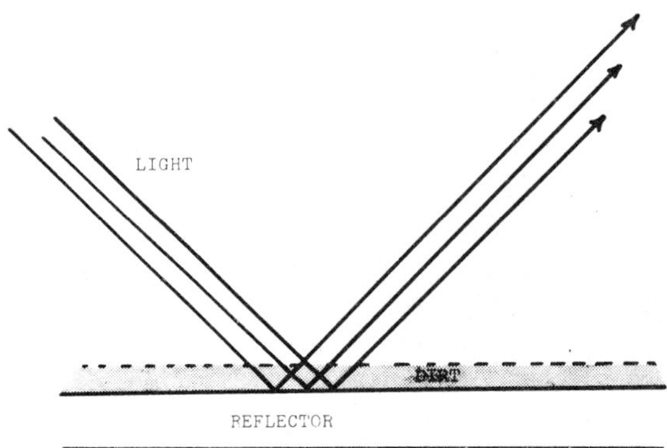

Fig. 10-4 The effect of dirt on a reflector.

that lamps should be replaced *prior* to burning out. A group relamp program at 70–80% of lamp life and a lighting cleaning program should be initiated. The results will be that luminaires can be removed from service without lowering the overall lighting level.

SUMMARY

How Much Is Saved By A Good Maintenance Program

From the discussion in this chapter, it is evident that *good maintenance* spells *energy savings*. Savings in excess of one million dollars per year is not uncommon if the maintenance program outlined in this chapter is followed. A good maintenance program provides for a quiet, safer, energy efficient plant where production downtime is minimized. Thus, the cost for the program will more than pay for itself.

11

Managing an Effective Energy Conservation Program

ORGANIZING FOR ENERGY CONSERVATION

By now it should be clear that energy management affects almost every major activity of a plant. It is involved in:

Electrical Engineering
Control Systems Engineering
Utility Engineering
Piping Design
Mechanical Engineering
Chemical Engineering
Heating, Ventilation, and Air Conditioning Engineering
Building Design
Environmental Engineering
Operations
Maintenance
Accounting and Financial Management

Each plant has assigned individuals who are responsible for one or more of the above functions. The problem facing management is how to organize the energy conservation activity so that all functions are moving in a common direction. The situation becomes more complex when several plants or outside consultants are involved.

Direction and coordination for the program needs to be provided. In most organizations, the function of the energy conservation coordinator/committee emerges.

In this chapter, you will see how to establish energy conservation goals, learn how to set priorities and how to improve communications and coordination, see how to use the critical path method to schedule energy conservation activities, and discover how to encourage the creative process.

TOP MANAGEMENT COMMITMENT

Energy conservation requires top management commitment. Table 11-1 indicates a checklist for top management. Formulating a committee and assigning a coordinator does not solve the energy conservation problem, but it does make individuals responsible for energy

TABLE 11-1 Checklist for Top Management.

A. Inform line supervisors of:
   1. The economic reasons for the need to conserve energy
   2. Their responsibility for implementing energy saving actions in the areas of their accountability
B. Establish a committee having the responsibility for formulating and conducting an energy conservation program and consisting of:
   1. Representatives from each department in the plant
   2. A coordinator appointed by and reporting to management
      Note: In smaller organizations, the manager and his staff may conduct energy conservation activities as part of their management duties.
C. Provide the committee with guidelines as to what is expected of them:
   1. Plan and participate in energy saving surveys
   2. Develop uniform record keeping, reporting, and energy accounting
   3. Research and develop ideas on ways to save energy
   4. Communicate these ideas and suggestions
   5. Suggest tough, but achievable, goals for energy saving
   6. Develop ideas and plans for enlisting employee support and participation
   7. Plan and conduct a continuing program of activities to stimulate interest in energy conservation efforts
D. Set goals in energy saving:
   1. A preliminary goal at the start of the program
   2. Later, a revised goal based on savings potential estimated from results of surveys
E. Employ external assistance in surveying the plant and making recommendations, if necessary
F. Communicate periodically to employees regarding management's emphasis on energy conservation action and report on progress

Adapted from *NBS Handbook 115*.

conservation. An individual does not become committed to a goal unless he believes in the cause.

In the face of frozen fuel allotments, shortages of raw materials and supplies, and the increasing fuel costs, commitment develops quickly out of the necessity of "survival."

WHAT TO CONSIDER WHEN ESTABLISHING ENERGY CONSERVATION OBJECTIVES

Energy conservation requirements must be translated into clear goals. Since funds are limited, the financial goals as well as the energy goals need to be defined. Typical energy conservation goals are illustrated in Table 11-2.

How To Set Priorities of Energy Conservation Projects

When setting priorities, it is necessary to make sure that competing energy conservation projects are evaluated on the same basis. This means that all life cycle cost analysis should be based on the same fuel costs and the same assumptions of escalation. In addition, the financial factors, such as depreciation method and economic life, need to be evaluated the same way. Table 11-3 summarizes the information to be furnished *prior* to the start of the life cycle cost analysis.

Competing projects can be ranked in order of the best rate of returns on investment, best payout period, or best ratio of Btu/year savings to capital cost.

TABLE 11-2 Typical Energy Conservation Goals.

1. Overall energy reduction goals
   (a) Reduce yearly electrical bills by __%.
   (b) Reduce steam usage by __%.
   (c) Reduce natural gas usage by __%.
   (d) Reduce fuel oil usage by __%.
   (e) Reduce compressed air usage by __%.
2. Return of investment goals for individual projects.
   (a) Minimum rate of return on investment before taxes is __________.
   (b) Minimum payout period is __________.
   (c) Minimum ratio of $\frac{\text{Btu/year savings}}{\text{capital cost}}$ is __________.
   (d) Minimum rate of return on investment after taxes is __________.

TABLE 11-3 Information Required To Set Energy Projects On the Same Base.

Fuel	Cost At Present	Estimated Cost Escalation Per Year	Energy Equivalent
1. *Energy equivalents and costs for plant utilities.*			
Natural gas	$______/1000 ft^3	$______/1000 ft^3	______Btu/ft^3
Fuel oil	$______/gal	$______/gal	______Btu/gal
Coal	$______/ton	$______/ton	______Btu/lb
Electric power	$______/Kwh	$______/Kwh	______Btu/Kwh
Steam			
______psig	$______/1000 lb	$______/1000 lb	______Btu/1000 lb
______psig	$______/1000 lb	$______/1000 lb	______Btu/1000 lb
______psig	$______/1000 lb	$______/1000 lb	______Btu/1000 lb
Compressed air	$______/1000 ft^3	$______/1000 ft^3	______Btu/1000 ft^3
Water	$______/1000 lb	$______/1000 lb	______Btu/1000 lb
Boiler make-up water	$______/1000 lb	$______/1000 lb	______Btu/1000 lb
2. *Life Cycle Costing Equivalents*			
A) After tax computations required		______	
Depreciation method		______	
Income tax bracket		______	
Minimum rate of return		______	
Economic life		______	
Tax credit		______	
Method of life cycle costing (annual cost method, payout period, etc.)		______	

Other factors enter into the decision making process. For example, if natural gas supplies are curtailed, projects reducing natural gas supplies will be given priority.

The Vital Element—Communication and Coordination

As obvious as it may seem, a company which fosters communications and coordination of energy conservation goals and information has a good chance of significantly reducing energy consumption.

A central energy conservation committee or plant conservation coordinator can only do so much. It is up to the individuals who are familiar with the detailed operations to make the greatest contributions. They need to be informed and to understand the energy con-

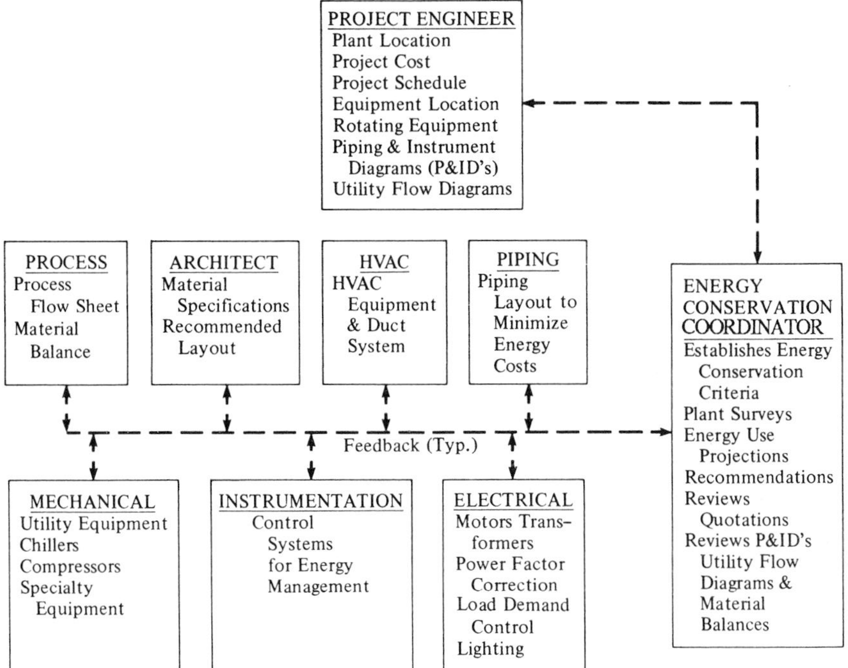

Fig. 11-1 Coordination of energy conservation information.

servation goals and the potential for saving yearly operating expenses. They need to know that energy conservation directly affects the profitability of the company.

Another aspect involved in coordination of energy conservation information is the flow of information generated by the project team assigned to a specific plant expansion. The project team is comprised of a project engineer and supporting design disciplines. An energy conservation coordinator assigned to the project insures that energy conservation goals are being implemented for the project. Figure 11-1 illustrated the coordination of energy conservation data.

USING THE CRITICAL PATH SCHEDULE OF ENERGY CONSERVATION ACTIVITIES

The critical path schedule for energy conservation takes into account that certain activities should take place prior to others. For example, before economic alternatives (life cycle cost analysis) is made, the objectives, financial data, utility rates, and plant survey need to have been completed. This schedule contains activities which affect plant expansions. For example, prior to purchasing equipment, energy consumption per item needs to be evaluated. Traditionally, bids received by competitive equipment manufacturers were compared on a first cost basis only. Today bids should be evaluated on a life cycle costing basis.

Key items to evaluate are:

Base Price
Delivery
Technical Specifications
Cost for Pollution compliance
Yearly Energy and Maintenance Cost

This means that during the quotation phase, equipment such as fans, air conditioners, pumps, compressors, etc., need to be evaluated with respect to energy consumption. This simple step will point up inefficient equipment: the penalty of operation prior to purchase.

ELECTRICAL SCHEDULING OF PLANT ACTIVITIES

As indicated in Chapter 4, the load demand controller is used to reduce peak electrical demand.

Scheduling of plant operations requires more than an electrical black box. As an example, the start up of a plant expansion can involve several start-up engineers working on several systems. A peak load demand can occur by testing several large motors at the same time. A little coordination can save substantially on the energy demand charge.

After the plant is in operation, the electrical energy curves should be analyzed. Careful analysis can indicate ways in which peak demands can be lowered by re-scheduling plant operations.

AN EFFECTIVE MAINTENANCE PROGRAM

Plant maintenance cannot be over-emphasized in any energy conservation program. As indicated in Chapter 10, management needs to take an active role.

CONTINUOUS CONSERVATION MONITORING

As discussed in Chapter 1, an energy conservation program must be continuously monitored in order to be controlled. This means that each operator needs to manage his own fuel usage. Adequate instrumentation is essential to the monitoring process.

Operator accountability can exist only if all utilities are metered. In planning what point in the process will be metered, consideration should be given to how the process will be operated and by whom. As an example, the fuel oil tank would come under the jurisdiction of the boiler plant operator. Any fuel take-offs to the process operator should be separately metered. This will enable each operator to control his usage, independent of other operations.

ARE OUTSIDE CONSULTANTS AND CONTRACTORS ENCOURAGED TO SAVE ENERGY BY DESIGN?

The type of contract may well influence the type of energy savings which will be realized. As an example, a competitive "lump sum" contract encourages the consultant to achieve the lowest first cost and ignore additional expenditures which will minimize future operating costs. This type of contract will not achieve the energy conservation goals unless the owner specifies the energy conservation requirements during the bid phase. Thus, the owner must be careful not to penalize creative energy conservation design during the conceptual phase.

ENCOURAGING THE CREATIVE PROCESS

Energy conservation engineering relies on creatively applying existing technology in new ways. Many examples were illustrated in this text, but this in no way can be considered the entire gambit. "Why" questioning must be promoted. A team of engineers representing several disciplines should meet periodically to "brainstorm" energy conservation concepts. "Do it the same way as the last project, use

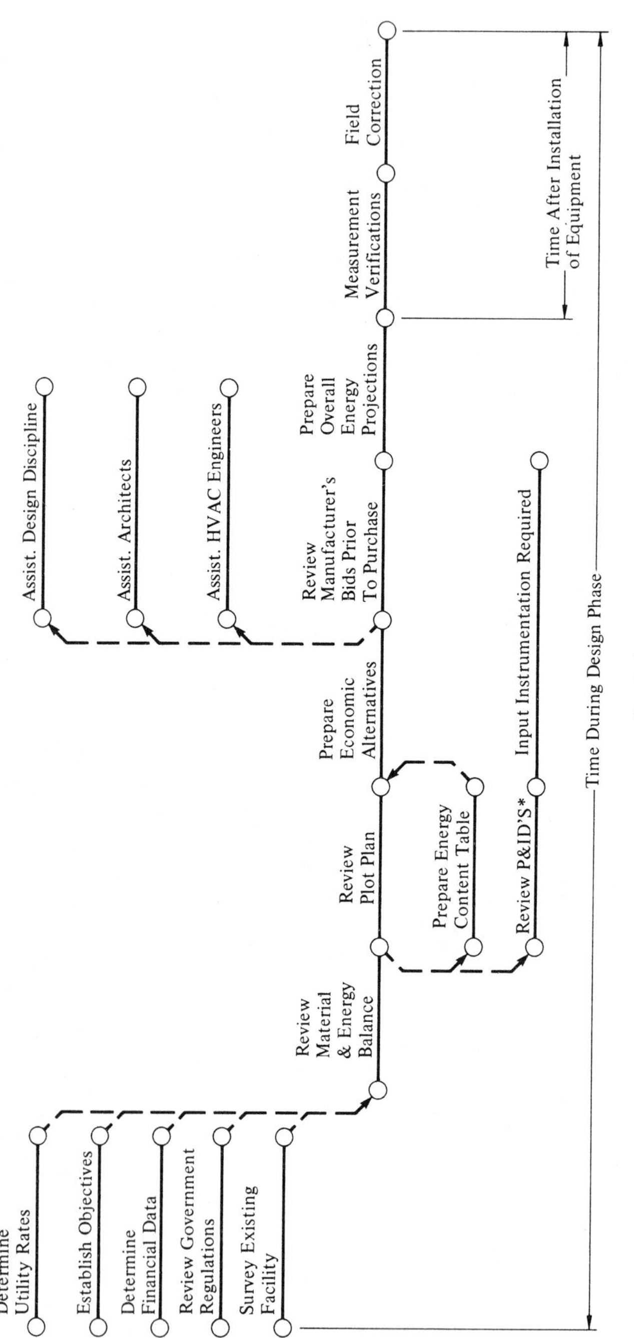

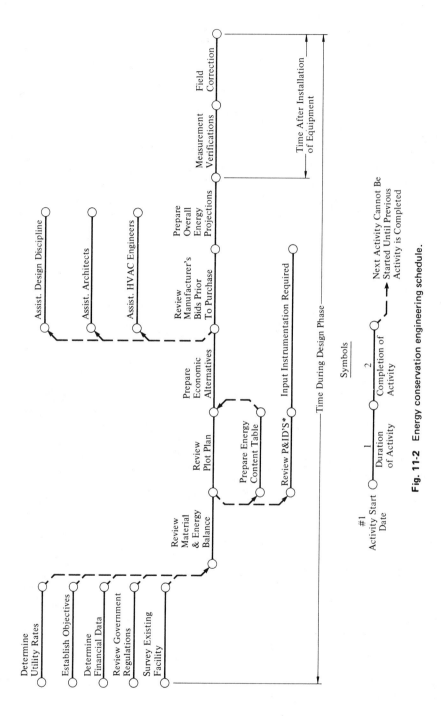

Fig. 11-2 Energy conservation engineering schedule.

rules of thumb, or Experience tells us . . ." should be replaced with "*Why* questioning."

Examples are:

1. Why can't vacuum pumps be used in this process instead of steam vacuum jets?
2. Can conventional warehouse heaters be replaced with infra-red heaters?
3. Can air coolers be used instead of water coolers?
4. Can no-vent equipment, such as vertical motors with special packing, be used to permit process steam to be returned to the boiler without steam traps?
5. Can loads be re-adjusted so that equipment runs at optimal levels of efficiency?
6. Can short runs on thermal process equipment be avoided?
7. Can spot cooling for personal cooling be used instead of area cooling?[14]
8. Can waste heat be recovered[6] by means of systems employing organic working fluids?
9. Can smooth coated pipe be used to reduce friction loss, thereby reducing horsepower?
10. Can pumps be eliminated by locating equipment to take advantage of gravity?

ENERGY EMERGENCY AND CONTINGENCY PLANNING

Energy managers need to deal with many unknowns. Experience of the last ten years indicates coal strikes, and fuel allocations do occur. How to plan for emergencies such as these is an important aspect of the energy manager's role.

In November of 1979, President Carter signed Public Law 96-102, Emergency Energy Conservation Act, giving the President the authority to establish energy demand targets for each state in case of a national emergency.

Risk Assessment

In order to establish the contingency plan, the energy manager needs to establish the risk associated with various scenarios.

There is a cost to business for doing nothing in the event an emergency does occur and no plan was put into effect. This cost is computed according to Equation 11-1.

$$C = P \times C_1 \tag{11-1}$$

where

C = cost as result of emergency where no contingency plans are in effect
P = the probability or likelihood the emergency will occur
C_1 = the loss in dollars as a result of an emergency where no emergency plans are in effect.

Thus risk assessment can help the energy manager determine the effect, money, and resources, required to meet various emergency scenarios.

An example of a simple contingency plan would be in the justification of a back-up fuel system. The risk assessment would indicate the probable cost of losing the primary system and the expected duration of the loss. The key is to develop the plan prior to the emergency occurring.

CONCLUSION

The energy conservation program, when properly managed, will pay for itself. Engineering "rules of thumb" are no longer sufficient because of increases in the cost of energy and shortages of raw materials. The decision making process is based on economics and local, state, and federal regulations. The guidelines and practices presented in this text will prove to save energy without decreasing the function of the system. The energy manager's role in the 1980s must be understood by the public and the government.

The answer to energy management is *not* to decrease the quality of life by living in darkness, but rather to apply the technologies available to increase the efficiency of the systems. The challenge facing the energy engineer/manager is great and the stakes are high.

12

Future Directions for Efficiency Improvement

Since its inception, the Office of Industrial Programs, U.S. Department of Energy, has supported over two hundred research and development projects covering a wide breadth of subject matter. Projects have been addressed to generic, wide-application technologies and to energy-intensive processes unique to single industries. Research and development activities have been initiated at all points along the path of technology development from basic and applied research, through proof-of-concept and process development testing and ending with commercialization.

This chapter describes some of the technical achievements of the Industrial Energy Conservation Program arising from several completed and ongoing R&D projects. These projects have successfully resolved existing technical needs faced by industry or addressed previously unknown or unaddressed opportunities for industrial energy conservation. Furthermore, most of the technologies have begun to be adopted by industry where they are now saving energy—a critical measure of the success of Program R&D. In the paragraph below, an overview of the larger technical areas addressed by the Program are described followed by brief descriptions of selected R&D projects and their accomplishments within each larger area.

WASTE ENERGY REDUCTION

Technologies that recover energy contained in industrial waste streams and waste materials, or that use the available energy in fuels more efficiently serve to extend or displace conventional energy supplies. Such technologies are applicable throughout industry and can have a marked impact on industrial energy consumption. Program activities in this area fall into three groups: waste heat recovery, combustion processes, and waste industrial materials utilization.

R&D aimed at waste heat recovery focuses on high-temperature (e.g., recuperators) and low-temperature (e.g., heat pumps) heat recovery. Recuperator R&D attempts to develop equipment that can operate in the environmental extremes of high-temperature processes and thus recover some of the 1.5 to 3.0 quads of thermal energy wasted in such processes. Similarly, R&D of heat pumps attempts to develop higher efficiency heat pumps that incorporate improved components or innovative approaches to upgrade an estimated 1.0 to 3.0 quads of energy contained in 180°-300°F process waste streams. R&D of combustion processes focuses on enhanced combustion efficiency, improved controls and expanded fuel flexibility. Projects addressed to waste materials utilization center on developing reliable, cost-effective processes to recycle or convert to fuels or feedstocks solid, fluid, and gaseous waste streams that might otherwise create costly environmental control problems.

High-Temperature Recuperator

Flue gases from high-temperature industrial furnaces are a large and still not fully utilized waste energy source. Recuperators (gas-to-gas heat exchangers in which hot flue gases preheat combustion air), are one common means of extracting this waste energy. Many conventional recuperators, however, are either limited to applications having relatively low-temperature and noncorrosive flue gases or must employ provisions for lowering the flue gas temperature. In either situation, substantial potential energy and cost savings are sacrificed. A new cross-flow, ceramic recuperator for use in high-temperature (1600°-2600°F) relatively clean exhaust environments was developed and its implementation accelerated in cost-shared projects. The "heart" of the recuperator is a ceramic (magnesium aluminum silicate) core composed of alternate layers of ceramic passages oriented at right angles to each other. Development and acceleration efforts have shown the recuperator to be applicable to a wide variety of industrial furnaces in several industries with resultant energy savings of 26 to 50 percent versus unrecuperated furnaces. As a result of the technology acceleration and the developer's marketing efforts the recuperator is achieving widespread industrial acceptance and implementation. Some 535 recuperators installed at about 180 sites are now saving over one trillion Btus and \$3.9 million per year.

High-Effectiveness Recuperator

Waste energy recovery from the hot flue gases of industrial furnaces is often limited by lack of space, inefficient heat exchangers, or high costs. A high-efficiency metallic plate-fin recuperator to overcome these limitations and thus promote waste energy recovery was developed and tested with IP support. The compact stainless steel plate-fin recuperator is designed to maximize the driving forces for heat exchange and to limit thermal stress. The plate-fin matrix consists of layers of corrugated sheet stock (fins) separated by plates. Tests showed the recuperator able to perform satisfactorily with clean exhausts at temperatures up to 1500°F and to sustain a recuperator effectiveness of 90 percent for over one year while requiring no maintenance. Success of the recuperator promoted an acceleration program to expand knowledge of the operational characteristics of the device. This effort, too, was successful; about 70 of the high efficiency recuperators were installed at industrial sites. By 1985 the high effectiveness recuperator is expected to be saving 20 trillion Btus per year.

Membrane Separation Oxygen Enrichment System

Moderate-to-large scale, inexpensive oxygen enrichment systems have the potential to improve the combustion efficiency of virtually all fuel burning equipment and thus achieve energy savings throughout industry through more efficient utilization of fuel resources. A promising oxygen enrichment technique being developed employs membrane separation technology to selectively separate an air stream into an oxygen-enriched stream to be used in combustion (achieving improved combustion efficiency and higher heat transfer rates than if ambient air were used) and a nitrogen enriched stream that could be used in processes requiring an inert atmosphere. A specially constructed polymer membrane element is packaged in a spiral-wound design (see Figure 12-1) to allow it to withstand the high flow rates and pressures required. A two-ton per day separation unit was suc-

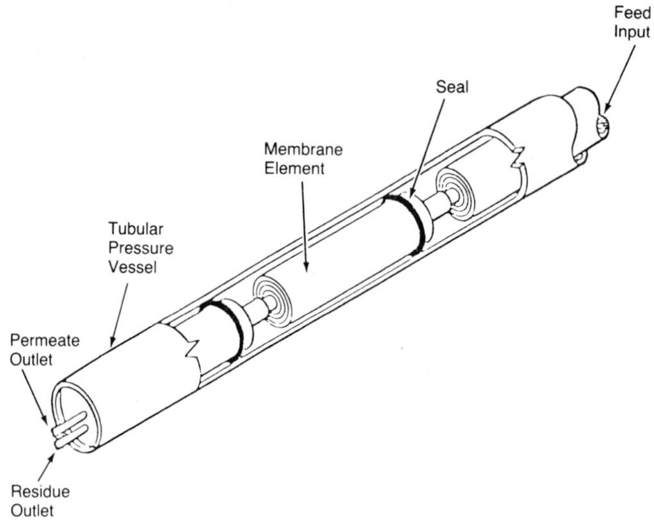

Fig. 12-1 Schematic Drawing of Spiral Wound Membrane Element.

cessfully tested at a copper tubing plant where use of combustion air enriched to about 30 percent oxygen resulted in natural gas savings of 15 to 25 percent. Work to improve the performance and reduce the cost of the membrane element is continuing.

High-Temperature Heat Pumps

Heat pumps currently used by industry are limited in the temperatures of the waste heat sources from which heat is transferred (e.g., cooling water or process effluents at 90°-175°F) and in the output temperature of the upgraded stream, restricting their usefulness (e.g., to heating water or space conditioning uses at 150°-220°F). As a result, the contained energy in many waste heat streams from industrial operations is not effectively utilized. Two prototype high-temperature heat pump systems have been successfully developed and tested. The first, which uses a two-stage compression process and methanol as a working fluid, was designed to use waste steam (10 psig, 240°F) from thermochemical pulping units to produce 50 psig steam with 20°F superheat. It can recover an estimated 190 billion Btus/year from 180-250°F waste streams. The second prototype unit was designed for energy recovery in the synthetic fiber industry. The unit uses a compressor unit integrated with an organic Rankine bottoming cycle that powers and drives the compressor to produce 25 psig, 270°F process steam from 10 psig, 240°F waste steam. Using this device some 170 billion Btus/year can be recovered from 200-240°F waste streams.

Plastic Wastes to Fuel

In tandem with the rapid growth rate of plastics production is the growing generation of plastic wastes from primary plastics producers, plastic fabricators, and discarded consumer products. A process to reclaim portions of this large waste stream by converting it into fuel oils and hydrocarbon gases has been successfully demonstrated. In the process (see Figure 12-2) plastic is pyrolyzed in a reactor coil heated by a fluidized bed furnace to heavy and light fuel oils and hydrocarbon gases which are then separated. The feasibility of the process was first established with atactic polypropylene (APP), a viscous waste by-product of polypropylene production, in a semi-works plant; a major polypropylene producer installed a commercial-scale plant that is now producing 2 million gallons of fuel oil per year from 17 million pounds of APP. Tests have shown the technology capable of reclaiming large-volume polyethylene and polypropylene plastics, and industrial interest in the process to reclaim mixed and specialty plastics has been generated. An estimated 0.2 quads/year of energy could be conserved by recovering the energy in industrial waste plastics.

Fuels from Cellulosic Waste

Another prevalent waste stream that has a large potential energy value is cellulosic waste such as sawdust, wood chips, paper, and agricultural and forest residues. A promising process for conversion of these wastes to high-grade, transportable liquid hydrocarbon fuels is under development. The process consists of two phases: high-temperature flash pyrolysis followed by indirect catalytic liquefaction of the off-gases. About 50 industrial waste feedstocks have been tested in a laboratory unit (10 lb/hr) including several integrated continuous runs from solid feedings to liquid product collection. Yields of 40 gallons/ton of feedstock have been reached in the

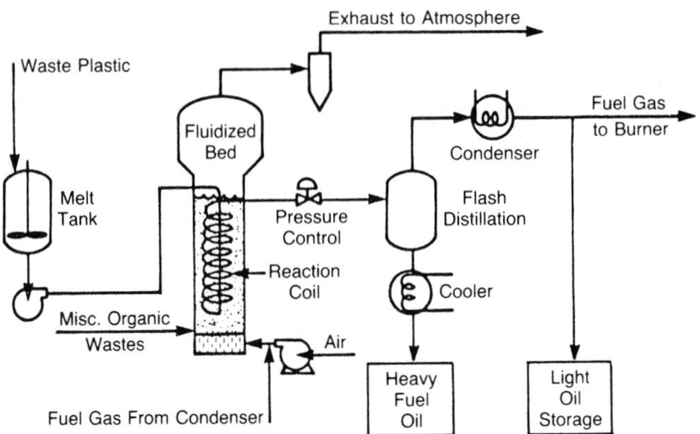

Fig. 12-2 Plastic Wastes to Fuel Process.

laboratory unit. Yields of about 60 gallons/ton are expected of the process in its current state of development.

Design of a 10-ton/day test facility is underway. The energy saving potential of cellulosic waste is high, amounting to 0.3 quads by the year 2000.

COGENERATION

Industrial cogeneration (the simultaneous generation of steam and electricity) offers vast potential for energy conservation. At present this potential is underutilized. Program activities to promote more and more appropriate use of cogeneration are aimed at advanced cogeneration systems that incorporate load flexibility, high power-to-thermal ratio prime movers, fuel flexibility including the use of coal and coal-derived duels, more efficient and lower cost waste heat engines, and direct thermal to electrical energy conversion technologies. Separate focus is given to cogeneration systems based on topping cycles and those based on bottoming cycles.

Low-Speed Diesel Cogeneration System

Low-speed, two-stroke diesel engines, used extensively in marine applications, are not being used by industry although they have potential advantages over alternative cogeneration system components such as a higher electrical to thermal output ratio than steam turbine systems. In addition, when fired with residual oil this engine offers a relatively compact size, greater load flexibility, lower fuel costs, and lower maintenance costs when compared to presently used cogeneration options. To prove the advantages of such a system IP has co-funded the design, installation and evaluation of the first industrial cogeneration system in the world with a low-speed, large-bore diesel engine as the prime mover.

The system (shown in Figure 12-3) is sized to generate a net power output of 23.3 MW, 160,000 lb/hr of 225 psi saturated process steam, and 262,000 lb/hr of 170°F boiler feedwater using:

- a low-speed, two-stroke diesel engine fired with low-sulfur, low-ash No. 6 fuel oil coupled to an electrical generator;
- a supplementary oil-fired waste heat boiler that reburns the diesel's exhaust for steam production; and
- heat exchangers for recovering waste heat from the engine cylinders, pistons and turbochargers to provide hot water.

The system has been installed and is operating successfully at a chemical plant where evaluations are continuing. This single installation is achieving net annual energy savings of 1.5 trillion Btus. At $5 per million Btus, this equates to cost savings of $7.5 million per year.

Biphase Concentration

Pressurized two-phase (gas and liquid) flow streams are found in many industries. Conventional turbines used to recover the energy of single-phase streams cannot withstand the forces released during the flashing of two-phase streams and thus the contained energy of these streams is largely lost. A simple and reliable biphase turbine concentrator has been developed that can efficiently generate either electrical or shaft power from two-phase streams while simultaneously separating the two streams. Power is generated by expanding a single- or multiple-component two-phase stream through a nozzle where the pressure and thermal energy is converted to kinetic energy. High centrifugal forces created as the stream exits the nozzle and impinges on the interior of a rotating cylinder serve to separate the

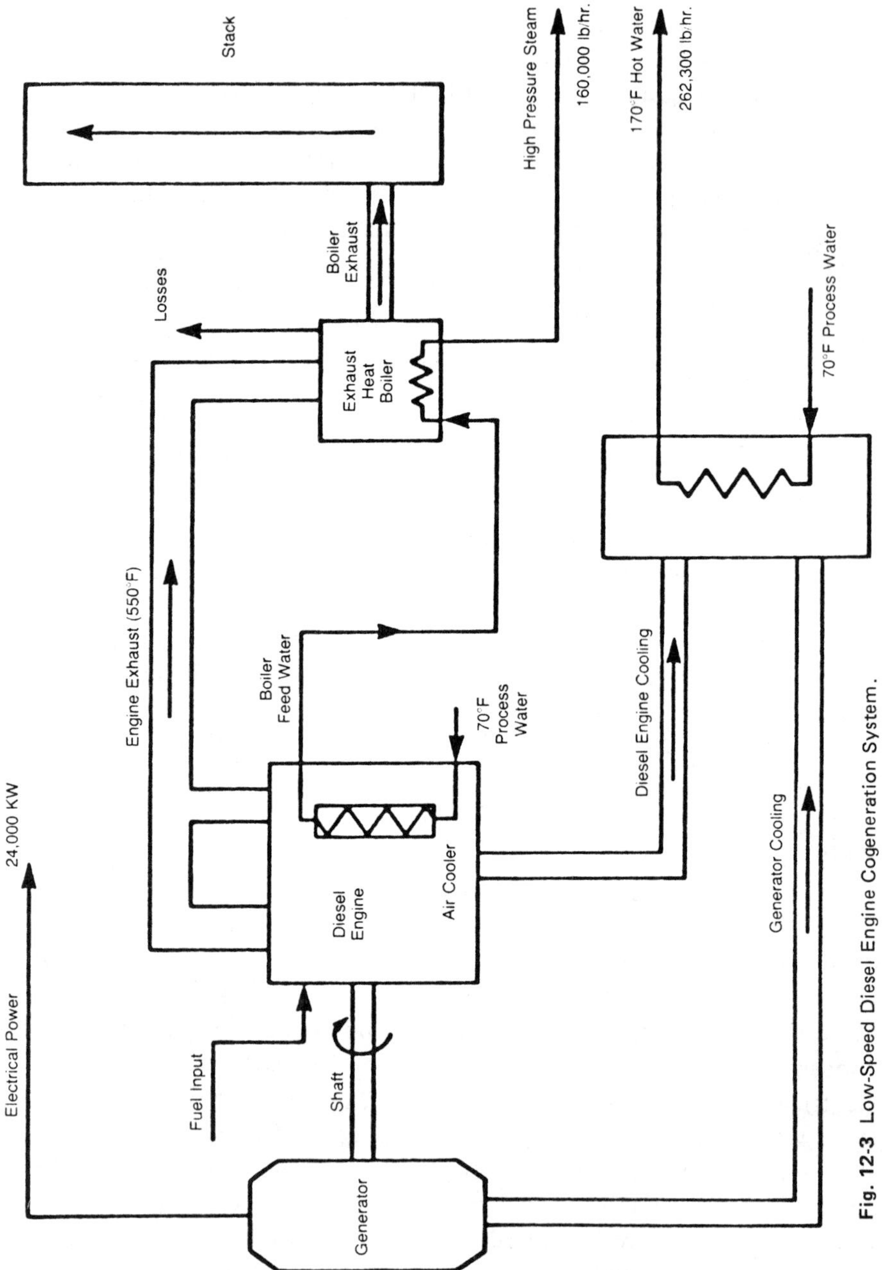

Fig. 12-3 Low-Speed Diesel Engine Cogeneration System.

gas and liquid phases. Well-head oil production with gas/oil separation is viewed as the most promising near term industrial application of the concentrator; design, fabrication and testing of a unit for this purpose are underway. If applied to all applicable industrial two-phase streams, the biphase turbine could yield energy savings as high as 0.35 quad/year.

Coal-Fired Cogeneration System

Use of coal or coal-derived fuels in gas turbine and other cogeneration systems in place of the premium natural gas or distillate oils currently used can provide large cost benefits to industry and can be an important means of more fully tapping the cogeneration potential. Direct firing of coal in gas turbines is constrained by the erosive and corrosive nature of coal combustion products as they pass through the turbine. IP is investigating an alternative approach that uses a coal-fired atmospheric fluidized bed air heat (AFBAH) system to heat compressed air contained in tubes to 1400°F and higher prior to delivery to the turbine. The system can be designed to provide various ratios of thermal energy output to electric power; steam can be produced by circulation through the walls of the fluidized bed; turbine exhaust can be used to preheat combustion air for the fluidized bed or to generate steam in a waste heat boiler. In an initial R&D project a 'reference cycle' configuration design was developed and in a subsequent effort alternative AFBAH configurations were identified and screened with respect to cost, risk and potential applications.

INDUSTRIAL PROCESS EFFICIENCY

The objective of this Program area is to improve the energy efficiency and productivity of processes used by the most energy-intensive industries. Such improvements not only conserve large amounts of energy, but promote a stronger, more flexible industrial base and increase the competitiveness of U.S. industry in world markets. R&D activities, while addressed to specific industries, generally fall into generic clusters including process electrolysis, carbothermic processes, comminution, materials processing, sensors and control systems and separation systems. These processes, many of which were developed when energy was relatively inexpensive, are major consumers of energy and are characterized by low energy efficiencies.

Advanced Slot Forge Furnace

Slot forge furnaces, which heat metal stock to about 2300°F prior to forging, can have thermal losses amounting to about 90 percent of the energy input from the fuel as a result of hot combustion gases leaving the furnace, conduction through the furnace walls and roof, and heat radiation through the slot. A high-performance slot forge furnace that minimizes these and other heat losses has been developed, tested, and demonstrated. The furnace design (see Figure 12-4) incorporates features that offer energy savings of 50 percent or more, while retaining the simplicity of a conventional slot furnace. The key feature is a ceramic recuperator that heats incoming combustion air with furnace exit gases. Other energy saving modifications include recirculation burners, improved temperature and air/fuel ratio controls, slot closure doors that open only for metal stock insertion or removal, and lightweight furnace wall insulation. Some 12,000 furnaces used in the forging industry could benefit from the improved technology. Potential energy savings are on the order of 0.02 quad/year. In part due to an acceleration effort, 30 units at 12 forging plants are now in operation.

Methanol-Based Carburization

Carburization is a method of heat-treating steel parts in a carbon-containing atmosphere to impart improved strength, hardness and wear resistance to the parts. With IP funding, a more energy efficient methanol-based carburization process was developed and tested in batch and continuous furnaces. The process improves energy-efficiency by changing the chemical composition of the atmosphere. Pure methanol, enriched with natural gas, is broken down to produce an atmosphere containing one-third carbon monoxide (CO) and two-thirds hydrogen (H_2) that compares with a conventional atmosphere containing 20 percent CO, 40 percent H_2 and 40 percent nitrogen. The higher concentrations of CO and H_2 accelerate the

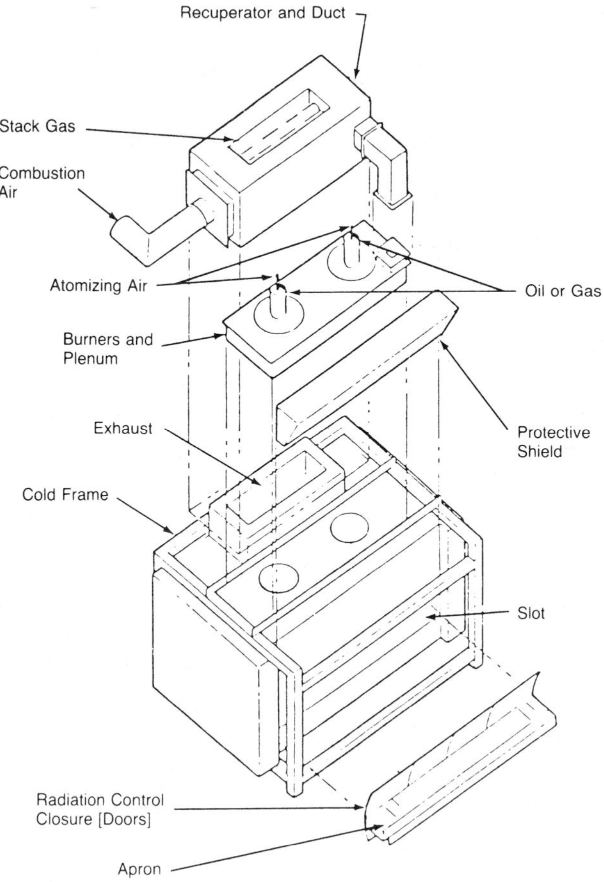

Fig. 12-4 Exploded View of High Performance Slot Forge Furnace.

rate of carburization and allow a reduction in the cycle time which saves energy. The advanced process is now in operation in 200 furnaces and is saving over one trillion Btus annually. Total potential savings are estimated at 10 trillion Btus annually.

Cupola Furnace Modification

The combustion process in cupola furnaces, used to melt scrap or iron ingots to produce grey iron, is somewhat incomplete producing excessive (8-15 percent) carbon monoxide (CO) concentrations in the exhaust gas. To meet environmental and safety requirements, energy-intensive natural gas-fired afterburners are used to ignite and combust this CO. An alternative to after burners that takes advantage of the fuel value of carbon monoxide and its ability to ignite at low temperatures in an oxygen-rich environment has been developed and successfully tested. The technique, which reduces final exhaust gas CO concentration to less than one percent, works by injecting air into the exhaust gases below the charging door of the furnace where the CO can ignite at temperatures existing in the stack. Two cupola furnaces have been successfully retrofitted to date and more installations are expected. If applied throughout the metallurgical industry about 9 billion cubic feet of natural gas could be saved each year in addition to saving a roughly equivalent amount of energy in the form of coke.

High-Temperature Remote Inspection

IP is funding the development and testing of a promising hot inspection system for steel slabs that uses the principles of computer-image processing and pattern recognition techniques developed by the aerospace industry. The system would eliminate the energy-intensive reheating step between the production of intermediate steel shapes (slabs, blooms and billets) and the subsequent processing of those shapes in various mills that require between 2 and 4 million Btus/ton of steel. The system (see Figure 12-5) uses two linear scanning array cameras linked to a computer which compares the image of the slab to the characteristics of known steel-slab surface defects and generates appropriate displays. Potential energy savings of such a system are considerable, amounting to about 0.25 quad per year not including potential savings from the technology in the non-ferrous metal industries.

Inert Anodes

One aspect of a major IP program to improve the efficiency of electrolytic processing is the development of permanent, chemically inert anodes for Hall-Heroult electrolytic reduction cells. Conventional carbon anodes have two disadvantages. First, they are energy intensive to produce (they are made from molded petroleum pitch

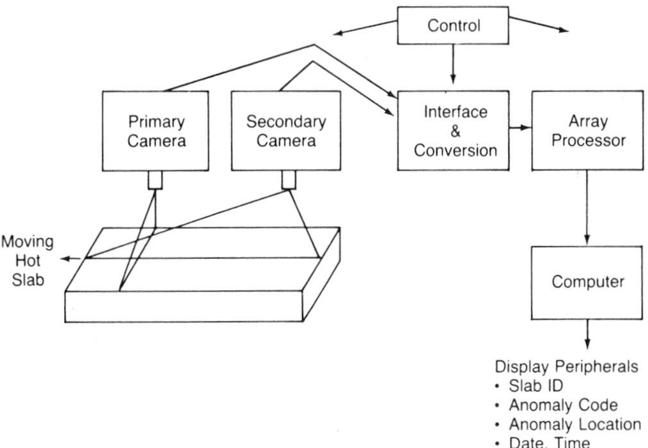

Fig. 12-5 Hot Steel Inspection System.

and petroleum coke baked at high temperatures) and are consumed during cell operation. Second, their replacement causes turbulence in the cell requiring that a relatively large gap be maintained between the cell anode and cathode that in turn contributes to low cell energy efficiency (only 35-45 percent). A permanent anode would save petroleum resources used in production of carbon anodes as well as electricity by narrowing the anode-cathode gap; combined savings of about 23 percent when compared to conventional cells are expected. Thus far in the project, great progress has been made in identifying adequate materials for inert anodes; materials research is continuing and anode fabrication and the design and testing of small-scale cells is underway. Since Hall-Heroult cells consume 5 percent of all of the electric power consumed in the U.S., this and other efforts to improve the efficiency of such cells can have large and important benefits to the aluminum industry.

Foam Processing of Textiles

Conventional wet processing of textiles (e.g., desizing, bleaching, dyeing, printing and finishing) use water to distribute pigments or other chemicals throughout the fabric. Wet processing is energy intensive since the fabric is successively wetted and dried as different treatments are applied. Working with an industrial partner IP has advanced a more energy-efficient process from laboratory investigations through pilot plant trials and commercial production, and has accelerated the adoption of the process by industry. The process replaces some of the water with foam in the dyeing, printing and finishing of a variety of fabrics reducing the energy required for drying as well as reducing water consumption and pollution control requirements and increasing plant productivity. At least 68 foam processing units are now in operation and over one billion yards of fabric have been processed. If applied throughout the textile industry, foam processing could save fully half of the energy used in the wet processing of textiles or about 52 trillion Btus/year.

Innovative Coil Coating Ovens

In coil coating ovens, solvent-based paints on metal coils are continuously cured (evaporation of solvent and setting of resins) in the presence of large volumes of dilution air required to keep solvent concentrations within safe limits. Exhaust from the ovens is heated in thermal incinerators to burn the contained solvents before release to the atmosphere. An innovative coal coating system has been tested and shown to be capable of reducing natural gas requirements by 74 percent. The innovative system (see Figure 12-6) works by recycling half of the oven exhaust (at 600°F) to zone incinerators where the heating value of the solvent is recovered and by ducting the exhaust from the zone incinerators (at 900°F) back to the curing oven as a source of thermal energy to vaporize more solvent. More energy is saved in pollution control, first by reducing the volume of gases in half and second by replacing the thermal incinerators with afterburners and waste heat boilers to produce process steam. Since being retrofitted on an operating coil coating line, the technology is rapidly being adopted by the coil coating industry with some 23 units now in operation. The system is applicable as well to other curing processes and if retrofitted to 10 percent of all curing ovens could save 63 trillion Btus/year.

Computer Control of Curing Ovens

Curing ovens (operating at temperatures of 150°-900°F) are used in many industries to evaporate and set solvent-based coatings. To meet safety requirements large volumes of dilution air are added to

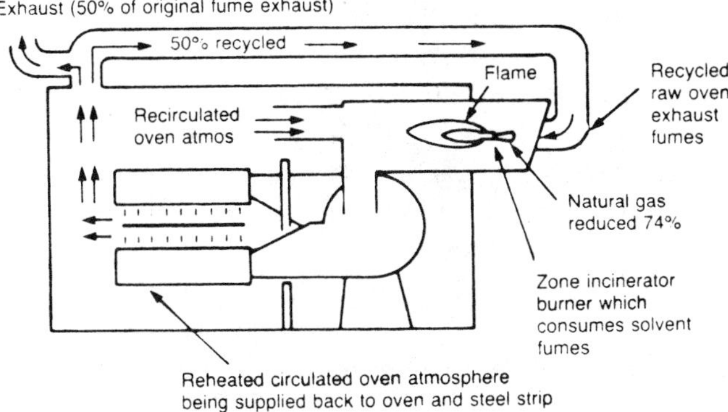

Fig. 12-6 Innovative Coil Coating Oven.

the oven to maintain solvent concentrations within their lower explosive limits (LEL); to meet environmental regulations, thermal incinerators are used to burn the solvents in the oven exhaust. As dilution air flow increases, energy consumption in both the ovens and thermal incinerators increases dramatically. In an IP-sponsored project, computer control of curing ovens was shown to be highly effective in reducing oven energy consumption. Computer control permits the ovens to operate at a higher percentage of the LEL than previously, reducing the amount of dilution air required and thus the energy needed to heat oven air and achieve pollution control. Energy savings are possible in curing ovens used in the production of cans, metal coils, paper, fabrics, automobiles and other products. If used on 1000 of the 14,000 industrial curing ovens in the U.S. annual savings of 20 trillion Btus could result.

Freeze Crystallization of Black Liquor

Separation of concentration processes (e.g., thermal evaporation) are among the most widely used and energy intensive processes in industry. An example is the concentration of black liquor (a by-product of the Kraft pulping process which is concentrated to recover pulping chemicals and its energy content). A new component separation process, freeze crystallization, which offers potential energy savings over conventional processes, is being investigated for black liquor concentration. Freeze crystallization removes heat from solution until one component crystallizes, at which point the crystals are removed, washed, and remelted. Advantages of the process are that latent heats of crystallization are 10-15 percent of those of vaporization and that the separation can be accomplished in a single stage. Several process approaches have been successfully tested in the laboratory, and a pilot plant is to be built and tested leading to the adoption of the technology by private industry on a commercial scale.

Energy Efficient Fertilizer Production

Fertilizers, vital to agricultural productivity, are energy intensive to produce and their production consumes large amounts of energy. In this very successful project a prototype reactor, developed by the Tennessee Valley Authority, that substantially reduces energy used when producing many fertilizers was tested and further developed. The reactor is a pipe cross reactor (see Figure 12-7) within which raw materials are reacted to produce fertilizer. With this design the reactor heat in conjunction with increased air flow causes the final product to have less than one percent moisture, reducing or eliminating the drying function required of conventional fertilizers that contain 5-7 percent moisture before drying. Some 28 of the innovative reactors are now in operation in the fertilizer industry, saving about 2.2 trillion Btus per year.

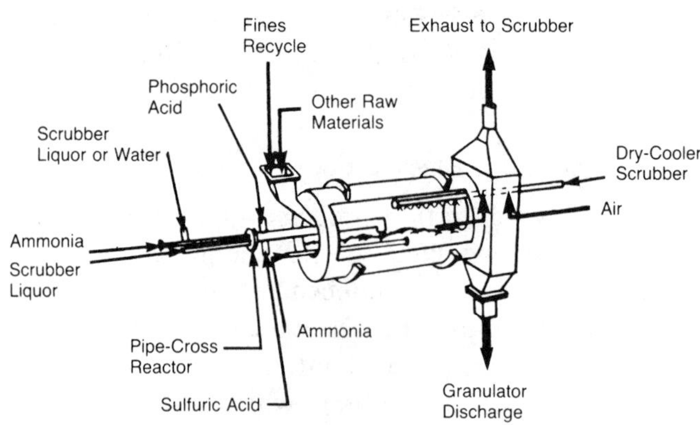

Fig. 12-7 Pipe Cross Reactor for Granulation Fertilizer Plants.

Potential Energy Savings

Energy conservation activities in the industrial sector encompass three phases. The first phase, the housekeeping phase, involves minimal cost and risk, and results in energy savings of 3-7 percent per unit of product. The second phase, the re-equipping or retrofit phase, works with existing processes and technologies and can accomplish an additional 5-10 percent savings. Here the risk is still small, but capital investment requirements can be large. The largest improvement in energy use can only be achieved through major process change entailing very high risks and major capital investments. When these investments are made, however, energy improvements of 25 to 65 percent are possible depending on the particular industry and the alternatives available. The primary barriers to the introduction of these new technologies are the availability of proven, profitable technologies and investment capital to the industries.

Industry has improved the efficiency of its energy use over the last decade largely as a result of easy-to-implement housekeeping measures and inexpensive retrofits made in response to rising energy costs. Projections of future industrial energy use assume that increasing costs will continue to promote efficiency improvements but at a slower rate than in the past. The inherently low energy efficiency of many processes still in place in industry is such that there is considerable room for further and large energy efficiency improvements in specific processes and for industry as a whole. Achieving these improvements will be more technically difficult and costly, however, as they require expensive equipment retrofits and major process changes.

Figure 12-8 shows estimates of additional energy savings (over those implicit in the projections of energy use) that are technically possible in the year 2000 for various industrial groups using the most advanced technologies which can be developed and assuming that the technologies are adopted instantaneously once available. The "technical savings potential" could not realistically be achieved in the year 2000, however, since industrial equipment is not normally replaced before its useful lifetime is over.

Figure 12-9 shows the "technical savings potential" in terms of the industrial fuels saved by taking into account the fuel mixes in each industry. It also shows the "achievable savings" in the year

SIC	Industry Group	Projected Energy Use in 2000 (Quads)	Potential Decrease in Specific Energy Use in 2000	Technical Savings Potential in 2000 (Quads)
28	Chemicals	9.7	36%	3.5
29	Petroleum	5.1	20%	1.0
33	Primary Metals	4.0	50%	2.0
26	Paper	2.8	45%	1.3
32	Stone, Clay & Glass	2.8	40%	1.1
20	Food	2.0	45%	0.9
22	Textile Mills	0.5	50%	0.3
34	Fabricated Metals	0.6	41%	0.2
35	Machinery	1.2	41%	0.5
37	Transportation Equipment	0.9	41%	0.4
	Other Mfg. Industries	3.7	35%	1.3
	Non-mfg. Industries	6.2	30%	1.9
	TOTAL	39.6	36%	14.4

Source: Office of Industrial Programs

Fig. 12-8 Technical Energy Savings Potential by Industry Segment.

Fuel Type	2000 Projected Usage (Quads)	2000 Technical Savings Potential (Quads)	2000 Achievable Savings (Quads)
Coal	4.6	2.00	0.84
Natural Gas	8.6	4.00	1.48
Oil	8.2	3.04	1.07
Electricity	4.7	1.92	0.90
Electrical Losses	11.1	4.54	2.12
Other	2.4	−1.14[(1)]	−0.39[(1)]
Total	39.6	14.35	6.02

[(1)] Negative energy savings means that waste-derived energy is being substituted for coal, oil, gas and electricity.

Source: Office of Industrial Programs

Fig. 12-9 Summary of Achievable Fuel Savings.

2000 that could realistically occur given agressive and successful technology R&D and taking into account the time required for technical innovations to come into use after becoming available.

Major Technology Needs

The Office of Industrial Programs (IP) has identified twelve major technology areas for which R&D can contribute to energy savings. These technology areas were identified from discussions of conservation needs with industry representatives, workshop meetings and trade association conferences, studies by consultants and DOE laboratories, and Program evaluations by DOE's Energy Research Advisory Board and the National Materials Advisory Board of the National Research Council. Each of the twelve areas is briefly described below along with current IP-sponsored R&D projects addressing those needs. Figure 12-10 identifies the industry segments impacted by advances in these technology areas.

- *Process Electrolysis.* Electrolytic reduction processes are used in the production of aluminum, magnesium, chromium, manganese, chlor-alkalis and other industrial materials. Such processes typically operate at energy efficiencies of less than 50 percent and are major consumers of energy, accounting for more than 2.5 quads of energy annually in aluminum, magnesium and chlor-alkali production alone. To reduce the energy used for electrolysis, better understanding of the process mechanisms and improved materials for electrolytic cell construction are needed.

  Several IP projects are underway to improve the energy efficicncy of electrolytic cells. One fundamental project entails the use of Raman spectroscopy to diagnose sources of current inefficiency in molten salt electrolytic systems so that techniques to suppress reactions causing inefficiencies can be developed. A second fundamental effort, aimed at development of improved electrodes, will examine dissolution behavior of electrode materials during electrolysis and develop and test improved candidate electrode materials. A project that complements inert anode work (described in the previous section)

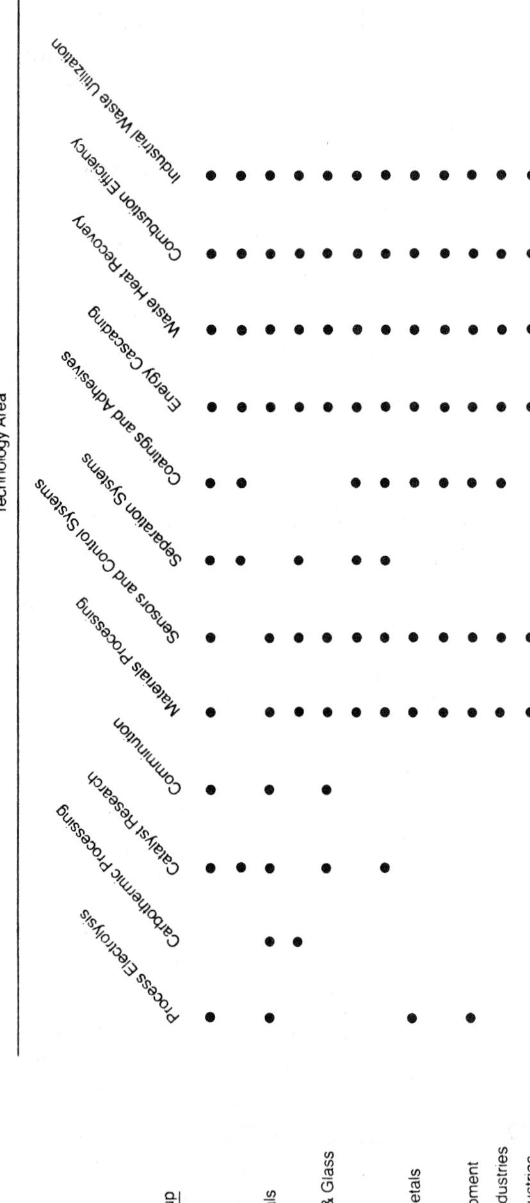

SIC	Industry Group	Process Electrolysis	Carbothermic Processing	Catalyst Research	Comminution	Materials Processing	Sensors and Control Systems	Separation Systems	Coatings and Adhesives	Energy Cascading	Waste Heat Recovery	Combustion Efficiency	Industrial Waste Utilization
28	Chemicals	•		•	•	•	•	•	•	•	•	•	•
29	Petroleum			•				•	•	•	•	•	•
33	Primary Metals	•	•	•	•	•	•			•	•	•	•
26	Paper		•			•	•	•		•	•	•	•
32	Stone, Clay, & Glass			•	•	•	•			•	•	•	•
20	Food					•	•	•	•	•	•	•	•
22	Textile Mills			•		•	•	•	•	•	•	•	•
34	Fabricated Metals	•				•	•		•	•	•	•	•
35	Machinery					•	•		•	•	•	•	•
37	Transp. Equipment	•				•	•		•	•	•	•	•
	Other Mfg. Industries					•	•		•	•	•	•	•
	Non-mfg. Industries					•	•			•	•	•	•

Technology Area

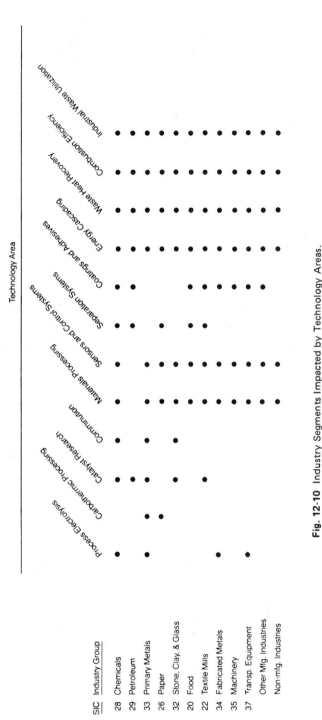

Fig. 12-10 Industry Segments Impacted by Technology Areas.

is the development of a fixed titanium diboride cathode insert for aluminum reduction cells. Such an insert, by providing a wettable surface to aluminum, would avoid aluminum surging and permit a reduction in the anode-cathode gap, increasing cell energy efficiency. The successful combination of an inert anode and wettable cathode could achieve the long-range goal of the aluminum industry for the development of a bipolar cell.

- *Carbothermic Processing.* Carbon, typically in the form of coke, is a second widely used agent for reducing metals contained in ores to elemental form. Carbothermic processes are at the heart of the iron and steel industry; of the 3.5 quads of energy used each year by the steel industry, 2.3 quads are in the form of coke. R&D on cokemaking, blast furnace operation, and direct steelmaking processes could yield large energy savings in the iron and steel industry. Carbothermic processes could also be used to advantage in place of electrolytic reduction processes; a notable example is the direct reduction of aluminum using coke (or coal) that could displace large amounts of electrical energy.

  A fundamental understanding of coke consumption in the blast furnace that could ultimately reduce fuel use, costs and imports of coke is the purpose of one IP-sponsored R&D project addressed to carbothermic processes. The project entails the use of advanced instrumentation (e.g., Raman spectroscopy) to analyze the behavior of coke in the blast furnace, specifically the effects of gas atmosphere, slag making materials and iron ore on the consumption rate of coke. A second ongoing project is the direct reduction of aluminum using coke or coal. An objective of this project is to research techniques to reduce the temperatures in the smelting furnace necessary for the aluminum reaction to proceed.

- *Catalysis.* Catalysis represents the prime mechanism for energy efficiency improvement in chemical reactions, such as chemical producers, petroleum refiners, and pharmaceuticals. Catalysis can lower operating temperatures and pressures, and increase reaction rates to achieve energy efficiency and productivity improvements. Energy reductions as high as 75 percent in major processes through the use of improved catalysts indicate the importance of catalysis technology in these highly energy-intensive industries, which consumed over 11 quads in 1983.

Since development of catalysts for specific reactions is a highly proprietary activity in the private sector, IP-sponsored research is currently directed toward fundamental, long-term studies of catalysis for generic use. For example, IP is sponsoring a project to develop a widely applicable monolithic porous catalyst support system that provides high heat transfer rates and thus effective temperature control necessary for efficient catalytic chemical reactions. The catalyst support system can replace the expensive, inefficient steps to add or remove heat in systems currently using solid state catalysts. A second promising basic research project is attempting to develop more efficient catalysts by depositing thin films of catalytically active amorphous metal on various types and shapes of substrate materials.

- *Comminution.* Fragmentation and grinding account for over 30 billion kWh of annual energy use primarily in ore body material removal and final finishing of materials such as cement and paint solids. These processes are highly inefficient–typically less than one percent. A fundamental understanding of fracture physics and the practical application of such understanding to comminution technologies can lead to large energy efficiency improvements.

  Several IP-sponsored R&D projects are underway to advance the state of comminution technology. Among them is a project to reduce wear and corrosion of grinding media in ball mills that is expected to extend ball life through use of new alloy materials and matching of ball materials to specific ores. A second project attempts to characterize a new energy-efficient autogenous grinding mill concept. In another project, computer simulation models of particle size classification technology will be developed and verified to enable users to design optimal configurations and operating conditions for two-stage classifiers.

- *Materials Processing.* New and innovative methods for reducing energy consumption, increasing productivity and controlling product quality in the materials processing industries can have immediate and wide ranging benefits to the industrial sector. For example, the processing of steel from ingot and slab to strip requires a series of energy-intensive process steps which could be avoided if thin strip could be cast to near final size with a minimum of reheating and rolling.

  Development of the technology necessary for high-speed direct casting of thin strip steel from molten metal is the goal of one applied research project. Improved steel-making methods combined with improved sensing and control devices provide the basis for this major improvement in steel production technology. The purpose of a second IP project directed at more efficient materials processing is to determine the feasibility of processes to recover materials contained in waste products of steelmaking such as mill dust and sludge.

- *Sensor and Control Systems.* Industrial process energy efficiency can be vastly improved by using automatic control systems. A limitation on the use of such systems is the lack of appropriate sensors and systems which will operate in the environmental extremes of industrial processes. Many energy-intensive industrial processes, for instance blast furnaces and Tomlinson boilers, are not controllable in real-time for lack of appropriate sensors and control systems.

  IP is pursuing several advanced sensor and control concepts. For instance, a project to develop an advanced electron-optic spectral analysis system for rapid and accurate quantitative analysis of liquid metal melts will be continued. This system avoids periodic sampling and laboratory analysis and permits more continuous throughput and easier alloying. Another effort focuses on R&D of sensors and control for the pulp and paper industry including sensors for lignin in paper pulp for on-line digester control, humidity measurement and control in drier hoods, and measurement of temperature profiles in recovery boilers. An automatic sensing/control device for accurate application of nitrogen (also pesticides and water) to soils is the goal of another project in this area.

- *Separation Systems.* An important and widely used step in industrial operations is the separation or concentration of components in a solution or mixture. This step (typically thermal evaporation or distillation) is usually the most energy consuming on any manufacturing site. Process research and engineering development is needed on less energy intensive processes such as membrane separations, freeze concentration, solvent extraction, critical fluid extraction, and advanced drying concepts. Some process applications in which these technologies could save energy include black liquor concentration in the paper industry, hot food processing wastewater concentration, dilute soluble food process stream concentration, and chemical and petrochemical stream separations.

  A number of R&D projects are attempting to make separation processes more energy efficient. A strong focus is given to membrane systems including: an electro-osmotic membrane cell for azeotrope separation, a ceramic membrane support substrate system, study of membrane fouling problems, non-destructive evaluation of membrane films, and membrane separation of hydrocarbon liquids and gases. Several projects are addressed to black liquor concentration/combustion including fundamental combustion studies, use of freeze concentration, and black liquor gasification. Some additional projects are addressed to critical fluid extraction, development of imbibitive polymers for absorption of chemicals from a solution/mixture, and development of a direct contact tubeless evaporator.

- *Coatings and Adhesives.* Development of improved material coatings that protect against extreme temperatures, corrosion, or excessive wear can have positive impacts on energy use throughout industry. They can, for example, displace more energy-intensive materials, allow more continuous process operations at environmental extremes, increase equipment lifetimes, and reduce weight requirements. Similarly, novel adhesives can be used to advantage in many applications. In addition, the coating processes themselves can be made less energy intensive. Examples are processes that coat surfaces with solvent based materials (e.g., paint), and then remove the solvent by thermal evaporation that could be modified or replaced with more energy efficient processes.

An example of IP coating work is use of magnetron sputtering, a form of vapor deposition, to apply amorphous metal coatings to complex steel surfaces to impart corrosion resistance. Another project in this area will develop a technique to apply chemicals to substrates (e.g., textile fabrics) without using a solvent, such as water, but using an electrostatic charge to apply the chemicals instead. Another solventless coating project will develop ultraviolet curable coatings that eliminate drying operations of conventional coating systems along with solvent recovery/disposal problems.

- *Energy Cascading.* Increased use of industrial cogeneration could save several quads of energy each year. To fully realize the potential of cogeneration, R&D is needed to improve load flexibility and load matching characteristics of cogeneration systems, particularly to obtain systems with higher electrical-to-thermal output ratios. R&D is also needed to promote fuel flexibility in cogeneration systems (particularly increased use of coal, synthetic fuels and residual oils) and to improve the performance, reliability and economics of low temperature waste heat driven cogeneration components. Development of economical direct thermal to electrical conversion devices to supply low voltage d.c. power for electrolytic and other processes is needed.

  IP is currently supporting a comprehensive mix of R&D projects directed towards these areas. Work will continue on a coal-fired Brayton cycle system employing an external atmospheric fluidized bed air heat system and on a large low-speed diesel engine, which will be modified and tested with coal/water slurry fuel. Other projects are directed toward the organic Rankine waste heat driven bottoming cycle, specifically the development and evaluation of tailored organic fluids and improved system components. Other projects address the use of an array of thermionic diodes installed in burner walls, and the use of molten alkali metal (sodium potassium) heat engines for the production of d.c. power.

- *Waste Heat Recovery.* Vast amounts of waste heat over a wide range of temperatures are generated as a result of industrial operations. Since the contained energy of a major portion of waste heat streams, both high- and low-temperature, is largely wasted, there are ample opportunities for increased waste heat recovery. Before this energy can be recovered, economical and reliable hardware must be developed, specifically hardware that incorporates improved materials and engineering designs. Further, prototype testing of heat pumps, recuperators and other heat exchangers that operate in the fouling, corrosive, and temperature environments of industrial waste streams is needed.

  Current IP efforts to promote waste heat recovery cover a mix of advanced materials and components development, system design and fabrication, and prototype testing. An example is the application of fluidized bed technology to non-fired industrial heat recovery purposes where severe fouling problems may occur and where high heat transfer rates are desirable. Similar efforts are underway to promote the use of advanced ceramic heat recovery technology for use in high-temperature, corrosive environments. Various new initiatives in heat pump R&D are underway including R&D on chemical heat pumps (absorption, heat of reaction concepts); high-temperature gas cycle heat pumps; and on heat pumps based on the heating effects of materials subjected to a magnetic field.

- *Combustion Efficiency.* Combustion processes are a form of energy conversion used by all industrial sectors. Energy consumed in combustion processes exceeds 20 quads annually. There is a need for technologies that enhance the efficiency and expand the fuel flexibility of industrial combustion systems. Such improvements require the development of improved combustion hardware and operating methods. Specific areas where technological innovations are needed include: improved burner configurations, coal-fired retrofit combustors, oxygen enriched combustion, burner efficiency controls, and combustion analysis and sensing devices.

A spectrum of R&D projects are underway that address technical needs relating to combustion efficiency. One group of projects attempts to develop novel solid fuel (coal) combustor retrofit systems to replace oil- and gas-fired burners on industrial boilers and process heat operations. Another group of projects is aimed at practical techniques to supply oxygen-enriched combustion air to burners that will increase combustion efficiency and product throughput. Other projects entail theoretical studies of combustion fundamentals and application of sensors and diagnostics to enhance combustion efficiency and control.

- *Industrial Waste Utilization.* A variety of solid, liquid, and gaseous waste streams that often represent costly environmental problems for industry could be economically recycled or converted into useful fuels or products, if suitable technology existed. Innovative mechanical, biochemical, and thermochemical processes for economically converting wastes into fuels and feedstocks, and for recovering and recycling organics and inorganics, are needed to foster the development of private activity in this area. If cost effective and reliable technologies are developed, over 8 quads/year could be saved by the utilization of industrial waste streams.

  In response to the diverse technical needs and opportunities in this area, IP supports R&D of generic waste utilization technologies and technologies to utilize specific solid, liquid or gaseous waste streams. Examples of current activities include continuation of research on a cellulosic waste-to-fuel process focusing on developing catalysts to increase indirect liquefaction yields; prototype tests on a compression drying process for wood residues for more efficient wood combustion; and a comprehensive program in industrial wastewater utilization encompassing fundamental studies in fixed film anaerobic reactor configurations, recovery of metals from metal finishing wastewater sludges, and studies on the anaerobic degradation of polynuclear aromatics. Other projects address techniques to use waste tires, asphalt roofing wastes, hydrocarbon residues, and waste industrial gases.

Future Program Activities

Each of the technology areas just described encompasses a variety of Program activities which can be undertaken in order to achieve potential energy savings. Activities can range from applied research, through development and testing, to technology transfer of the new knowledge resulting from the activities. Specific projects can address generic technologies or technologies specific to the most energy-intensive industries.

An important planning document used by IP in the conduct of the Program is the Energy Conservation Program Planning Document prepared in conjunction with other DOE Conservation Program Offices and covering a five-year planning horizon. This 'five-year plan' serves as a management-oriented, long range planning tool. It includes statements of objectives, planning methodologies, background data, assessments of R&D needs, mechanisms for private sector inputs, and estimates of resource and time requirements for pursuing candidate R&D projects.

Future program activities will be in line with current IP strategies. Working closely with private industry to promote and accelerate the implementation of energy-saving technologies, IP will continue to:

- Characterize energy usage in the industrial sector at large, individual energy-intensive industries, and specific industrial processes. Assess barriers to energy conservation by private industry.
- Identify existing and candidate R&D efforts that can be brought to bear on using energy more efficiently in producing industrial goods.
- Establish priorities for R&D activities and select the most appropriate activities emphasizing high risk, high potential R&D the private sector will not pursue alone.
- Manage and monitor R&D projects and transfer project responsibilities to the private sector when appropriate.
- Transfer results of R&D to the appropriate industrial audiences.

13

Evaluation of a High-Temperature Burner-Duct-Recuperator System

ACKNOWLEDGMENT

This technical case study was written for the U.S. Department of Energy by Energetics, Incorporated, Columbia, Maryland, under DOE Contract No. DE-AC01-87CE40762. Project-related information was supplied by the Babcock & Wilcox Company, Lynchburg Research Center, Lynchburg, Virginia.

This chapter was prepared as an account of work sponsored by an agency of the United States government. Neither the United States government nor any agency thereof, nor any of their employees, makes any warranty, express or implied, or assumes any legal liability or responsibility for the accuracy, completeness, or usefulness of any information, apparatus, product, or process disclosed, or represents that its use would not infringe privately owned rights. Reference herein to any specific commercial product, process, or service by trade name, trademark, manufacturer, or otherwise does not necessarily constitute or imply its endorsement, recommendation, or favoring by the United States government or any agency thereof. The views and opinions of authors expressed herein do not necessarily state or reflect those of the United States government or any agency thereof.

SUMMARY

The U.S. Department of Energy's (DOE) Office of Industrial Technologies (OIT) sponsors research and development (R&D) to improve the energy efficiency of American industry and to provide for fuel flexibility. Working closely with industry, OIT has successfully developed more than 25 new technologies that are saving industry approximately 45 trillion Btu of energy annually. Another 100 projects are in various stages of development from laboratory research to field tests.

OIT has funded a multiyear R&D project by the Babcock & Wilcox Company (B&W) to design, fabricate, field test, and evaluate a high-temperature burner-duct-recuperator (HTBDR) system. This ceramic-based recuperator system recovers waste heat from the corrosive, high-temperature (2170°F) flue gas stream of a steel soaking pit to preheat combustion air to as high at 1700°F. The preheated air is supplied to a high-temperature burner developed by the North American Manufacturing Company. The HTBDR system could save industry as much as 43 trillion Btu each year, worth $170 million at an average energy price of $4 per million Btu.

The B&W R&D program, which is now complete, involved several activities, including selecting and evaluating ceramic materials, designing the system, and developing and evaluating the prototype. In addition, a full-scale unit was tested at a B&W steel soaking pit in Koppel, Pennsylvania. The full-scale system consisted of a modular single-stage ceramic recuperator, a conventional two-pass metallic recuperator, a high-temperature burner, fans, insulated ducting, and associated controls and instrumentation. The metallic recuperator preheated combustion air to about 750°F before it passed to the ceramic module.

During 1400 h of field testing, the ceramic stage performed well with no material-related problems, no air-to-flue leakage, minimal pressure drop, and good thermal stability. Using flue gas at 2170°F, the system preheated the combustion air to as high as 1425°F. This temperature was lower than expected, however, because flue gas leaked from the soaking pit, resulting in lower than expected flows and temperatures through the ceramic recuperator. Despite these problems, fuel savings were measured at 17%-24% compared with the previous metallic recuperator system. The savings would be higher when compared with an unrecuperated furnace—as much as 41%.

B&W performed an economic analysis that indicated an acceptable payback period of 2-3 years for HTBDR installations in unrecuperated furnaces. The payback period would be longer at currently low natural gas prices for installation in systems with existing metallic recuperation. In both cases, return on investment would be high over the projected 10-year life of the system.

This joint project successfully developed and demonstrated a ceramic-based, high-temperature recuperator technology that is now ready for commercial use. The primary steel, copper, lead, zinc, aluminum, and glass industries that use high-temperature combustion processes could apply this technology. The high-temperature burner that North American Manufacturing Company developed also represents a significant achievement, and it is now being marketed to industry.

This technical case study describes the DOE/B&W recuperator project and highlights the field tests of the full-scale recuperator system. The document makes results of field tests and data analysis available to other researchers and private industry. It discusses project status, summarizes field tests, and reviews the potential effects the technology will have on energy use and system economics.

INTRODUCTION

When direct heating is provided by combustion, large amounts of energy are lost with the combustion products in the exhaust gas. In fact, losses range from 30% to 65% for processes ranging from 1000°F to 2400°F. These processes are common, particularly in the primary steel, copper, lead, zinc, aluminum, and glass industries.

Recuperators installed in the furnace flue can minimize energy loss by recovering heat from the exhaust. These gas-to-gas heat exchangers preheat the combustion air supplied to the furnace burner. Recuperation directly increases furnace efficiency by reducing fuel consumption. Conventional metallic recuperators can preheat combustion air to as high as 1300°F, saving 20%-30% of the fuel. Preheating combustion air to higher temperatures would save more, but conventional recuperators and burners cannot withstand the higher temperatures because of material and design constraints. In fact, high-temperature flue gas streams are normally diluted with ambient air to lower the temperature to acceptable levels (below 1600°F).

Recent advances in ceramic science now offer solutions to the temperature and durability limitations of recuperators and burners. Today's ceramics have high-temperature and corrosion-resistant properties that make them well suited to recuperator applications, including improved fracture resistance, increased strength, low gas porosity, and greater chemical stability. In addition, new fabrication techniques and innovative housing designs can help minimize gas leakage in ceramic devices. The high cost of ceramic materials, however, is a potential drawback.

In a multiyear R&D program funded by DOE, B&W developed and evaluated a ceramic-based, HTBDR system that can operate in corrosive, high-temperature flue gas streams. The recuperator is designed to recover waste heat from the flue gas stream of steel soaking pits (deep, rectangular or circular furnaces used to heat steel ingots to between 2150°F and 2450°F before further forming operations). Efficiency improvements to soaking pits could provide energy savings for the steel industry, but the corrosive components of pit flue gases and the cyclical variation in gas temperature (in the 1200°-2400°F range) have kept energy-efficiency measures from being implemented.

The B&W recuperator uses the 2170°F gas stream from the soaking pit flue to preheat combustion air to as high as 1700°F. The system then supplies the preheated air to an innovative, high-temperature burner that can use air preheated to 2050°F. The HTBDR increases the fuel and cost savings from recuperation by more fully extracting and using the available heat in waste exhaust gases.

The HTBDR system can be added to existing metallic recuperators. This preheating takes advantage of both ceramics, which perform best in high-temperature environments, and metals, which are suited to mid- and low-temperature environments, to obtain high preheat temperatures, reliability, and cost-effectiveness.

The B&W recuperator system uses a modular, ceramic, cross-counterflow heat exchanger to recover heat (Figure 13-1). The heat exchanger is based on a bayonet design (tube-in-tube) using 50 ceramic tube assemblies. Each tube assembly consists of two tubes—one tube open on both ends inserted into a second tube that is closed at the lower end, bayonet style. The tubes are suspended vertically from an air-cooled, insulated, metal plenum into a horizontal flue leading from the furnace. Heat is recovered from the hot flue gas by radiant and convective heat transfer to the suspended ceramic tube assemblies. After the combustion air is preheated in a metallic recuperator, it flows into the upper plenum of the ceramic stage, down the inner ceramic tube, up the annulus between the tubes to the lower plenum, and finally out the recuperator to the burner.

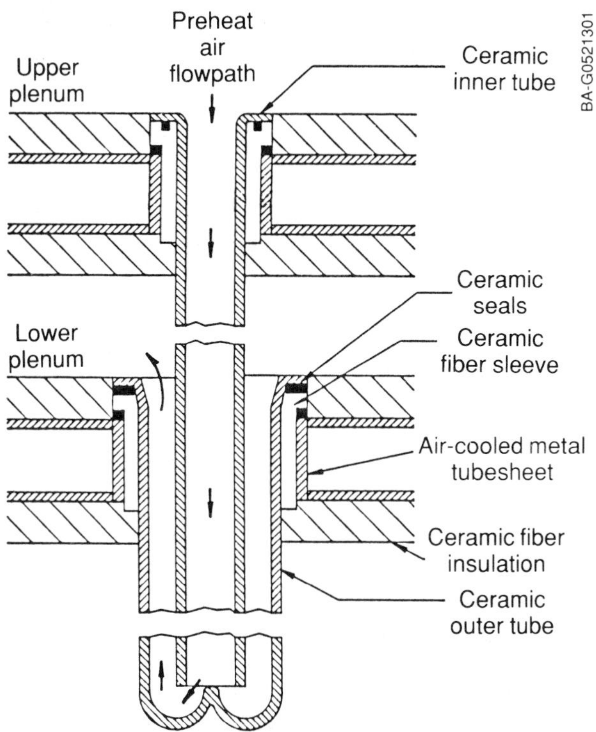

Figure 13-1. Ceramic recuperator design

Figure 13-2 shows the complete ceramic recuperator design, including how the tubes are supported by metallic tube-sheets. The tubes in this design rest on individual sleeves and seals of ceramic fiber and are not constrained from movement caused by thermal expansion. This sleeve-seal design reduces mechanical stresses on the ceramic tubes and provides an adequate seal to prevent air-side to flue-side leakage. The benefits of this ceramic recuperator design include high operating temperatures, low flue-side pressure drop, unconstrained structural ceramics, and simplified maintenance.

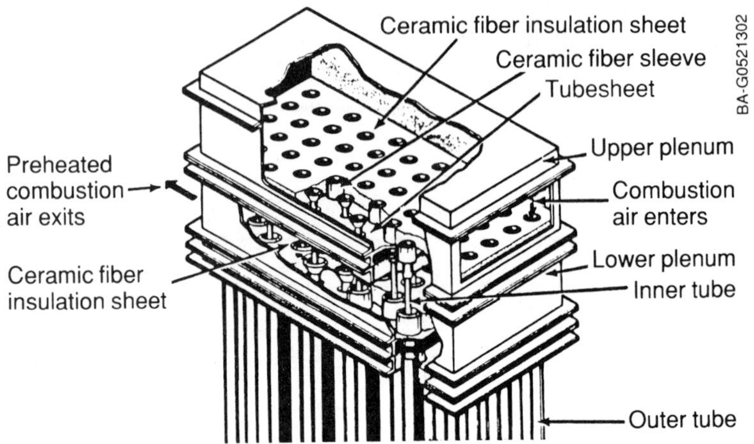

Figure 13-2. Ceramic stage of HTBDR system

PROJECT DEVELOPMENT AND STATUS

Since its beginning in October 1981, the HTBDR project has progressed from a preliminary design phase (1981-1982), through prototype development and testing (1982-1985), to fabrication and field testing of a full-scale system at a steel soaking pit in Koppel, Pennsylvania (1985-1988). The R&D program concluded in 1988, and a final report on the project was recently published. This section summarizes the major project activities leading to the field tests. The field tests and economic analysis are described in later sections.

Ceramic Materials Selection

Ceramic materials are used in three areas of the HTBDR system: the tubes, the sleeves and seals, and the burner. The tubes are a particularly critical component, requiring good thermal shock resistance, corrosion resistance, high strength, and thermal stability. Consequently, an initial step of the project was to select and evaluate suitable materials.

Eight grades of commercially available tubular silicon carbide (SiC) were evaluated. In these tests, samples of the candidate materials were exposed to flue gas in a soaking pit while it was operating. None of the materials tested showed catastrophic degradation or significant loss in strength, and only minor micro-structural changes occurred. Based on material properties, availability, and cost considerations, siliconized SiC (Norton NC-430) and recrystallized SiC (Norton CS-101) were selected for the outer and inner tube elements, respectively.

The design of the ceramic recuperator features a compliant two-component sleeve-seal arrangement for supporting the tubes, which allows some tube movement (to minimize mechanical stresses) and provides an adequate seal between the air side and the flue side. Construction materials for the seals were evaluated for strength, creep, compliance, and leakage at elevated temperatures. The sleeve material selected is 70-lb/ft^3 bonded ceramic fiberboard, which provides the required rigidity. The seals are of ceramic fiber paper, which provides pliancy. (See Figure 13-3.)

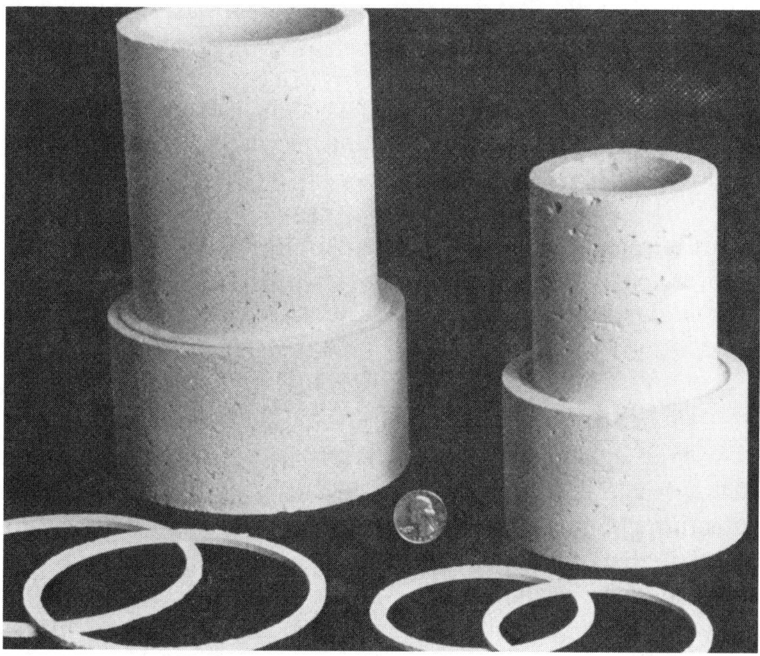

Figure 13-3. Refractory collars and sealing rings

Preliminary Design

B&W assembled a multidisciplinary team from several divisions to design the HTBDR system. During this effort, individual tube assemblies were characterized for performance along with the entire recuperator assembly. Design analyses covered the following areas:

- Thermal and fluid-flow models to characterize heat transfer and pressure-drop performance of the individual ceramic bayonet elements and to design the recuperator stages for given operating conditions and constraints

- Finite-element stress analysis to predict stresses and thermal expansion in the different components, particularly the ceramic tubes and sleeves and metallic tubesheets and plenums

- Flow-induced vibration analysis to determine whether the outer tubes suspended in the flue would vibrate excessively at certain flue gas velocities

- Design of the control interface between the existing furnace, burner, and recuperator and instrumentation and data collection systems to measure HTBDR effectiveness

- Mechanical design and arrangement of the system, integrating results from the other technical analyses.

See Figure 13-4.

High-Temperature Burner

B&W evaluated three high-temperature burner designs for use with the HTBDR. A variable-heat-pattern (VHP) design by North American Manufacturing Company was finally selected because its operation is simpler than that of other, more complex designs using internal moving parts to handle high-temperature air. The burner selected has dual fuel capacity (natural gas and no. 2 fuel oil) and meets the following requirements for the soaking pit application:

- Operation using 2000°F preheat air

Figure 13-4. External view of the HTBDR

- Long-flame/short-flame capability to maintain a pit temperature variation of less than ±60°F
- Long-flame/short-flame capability at high-fire and low-fire rates with at least a 6:1 turndown ratio
- A firing capacity of 18 million Btu/h with an over-fire capacity of 21 million Btu/h.

The burner design incorporates a pneumatically operated air deflector that adjusts the flame length for pit temperature control. The air deflector assembly is mounted on the burner inlet port and is made of SiC materials resistant to high temperatures. At low fire, a steam jet is used to lengthen the flame and improve pit temperature control. This feature provides the high 6:1 turndown ratio. The burner design did not specifically provide for NO_x reduction because of the uncertainty in predicted NO_x levels.

Prototype Evaluation

Following the preliminary design, B&W designed, constructed, and operated a prototype recuperator system to simulate the maximum temperature conditions anticipated. The objective was to verify the thermal and pressure-drop models; to assess system leakage, materials performance, and tube vibrational characteristics; and to estimate the performance of a full-scale system.

The prototype recuperator design (Figure 13-5) used a single ceramic stage with seven heat exchanger tube assemblies in a hexagonal array. The sizes of the plenums, tubesheets, and ceramic components were identical to similar elements in the full-size recuperator except that the ceramic tubes were 30 in. shorter, about 47 in. long. An electric air preheater and a metallic heat exchanger were coupled in series with the ceramic recuperator

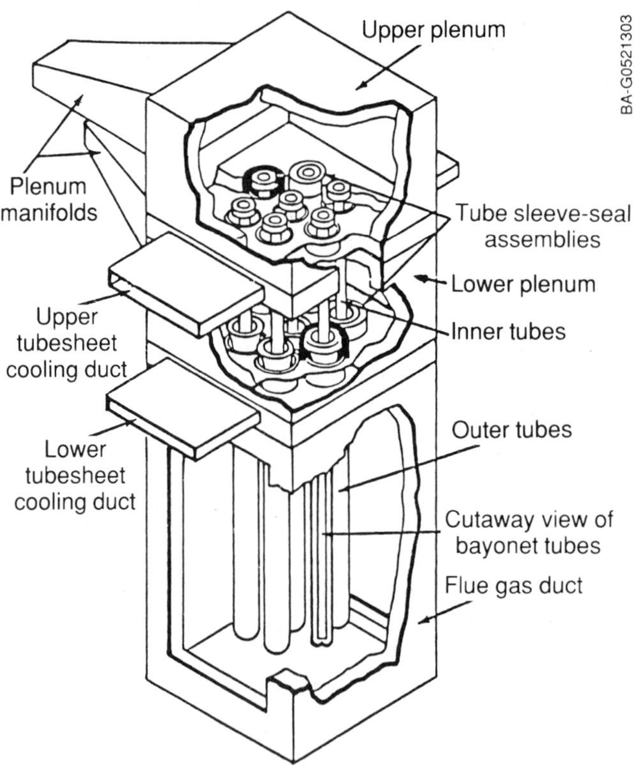

Figure 13-5. Cutaway view of prototype recuperator

to preheat air for the ceramic stage (Figure 13-6). The ceramic recuperator was inserted into a flue carrying hot combustion gases from a gas-fired kiln. The entire system was heavily instrumented. The prototype operated for 500 h with temperatures that went as high as 2300°F.

The prototype tests indicated that flow-induced vibration of the tubes was not a problem and that the materials of construction were adequate based on an inspection of the components after the test. Analysis of the thermal and fluid flow data indicated the following:

- Heat transfer in tube assemblies was as good or better than predicted. Pressure drops were as predicted.
- System performance was better when air flow was directed down the inner tube rather than up.
- Heat losses from the system and air leakage under high-fire conditions were excessive. Several changes were made to correct these problems in the final design.

FIELD TESTING

Following the prototype tests, B&W fabricated, installed, and tested a full-scale HTBDR system in a steel soaking pit at its steel-making facility in Koppel. The facility and the HTBDR system (as installed) are described briefly here, followed by the system performance results.

Host Facility

The field-test site was a 120-ton-capacity steel soaking pit with the following inside dimensions: 27 ft long, 9 ft wide, and 14 ft deep. Natural gas-fueled burners fire down the length of the pit from the top of the burner wall. The hot gases then circulate around the pit, heating the steel ingots, before exiting into a horizontal flue section. In the original setup (before the HTBDR was installed), the flue gases were immediately diluted with ambient air to lower their temperature to below 1650°F before they entered an American-Schack canal-type recuperator. The metallic recuperator preheated combustion air to about 900°F.

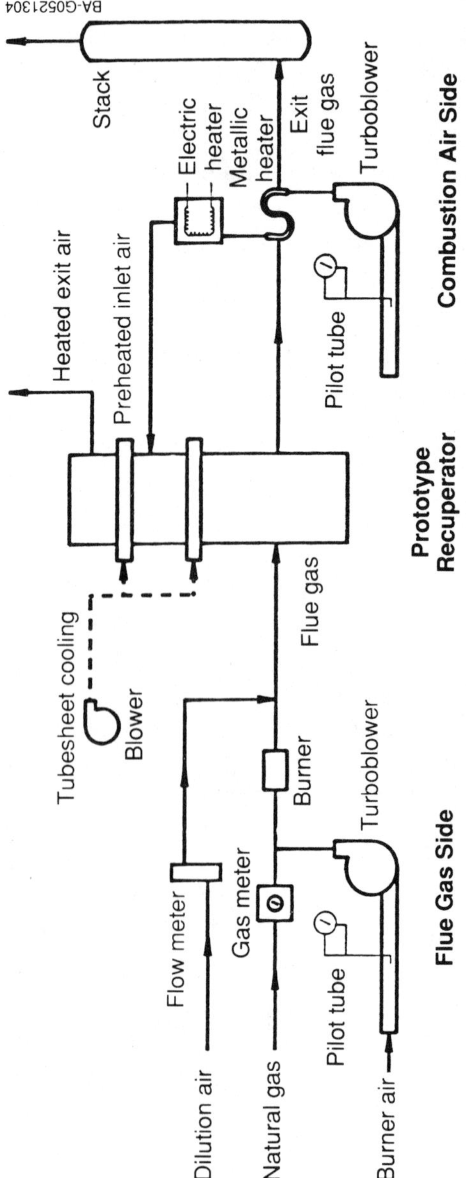

Figure 13-6. Schematic of prototype test setup

The combustion system for the soaking pit operates at low or high fire and short or long flame settings. The pit operation consists of loading the steel into the pit at 1600°F. High fire is used for about 4 to 5 h until the pit reaches soaking temperatures. Low fire is then used to hold an average soaking temperature of 2250°F for the remainder of the 20-h average cycle. The combustion system modulates the length of the flame to establish temperature uniformity between the ingots at opposite ends of the soaking pit.

HTBDR System Design

The full-scale HTBDR design (Figure 13-7) consists of a single ceramic stage added in series to the existing metallic recuperator and associated ducting, fans, and instrumentation. The existing three-burner combustion system of the soaking pit was replaced with one large North American Manufacturing Company burner (described in the preceding section).

The ceramic recuperator stage is a cross-counterflow heat exchanger composed of 50 tube-in-tube (bayonet) assemblies supported by insulated, air-cooled tubesheets and sealed with Kaowool™ fiber-based seals. Figure 13-8 illustrates assembly of HTBDR tubesheet. Detailed modeling and performance curves for the individual bayonet elements formed the basis for the final heat-exchanger design; engineers verified the design by prototype testing. The design takes maximum advantage of radiant heat transfer and minimizes air- and flue-side pressure drops, thermal stresses, and fouling. All the outer tubes and the five inner tubes nearest the soaking pit are of ceramic (silicon carbide class) material. The remaining inner tubes are of steel. The front tubes are coated with zircon. Table 13-1 gives the specifications for the ceramic stage.

In addition to the recuperators and burner, the HTBDR system includes

- a combustion air fan
- a tubesheet cooling fan to prevent the metallic tubesheets from overheating
- a control system designed to control pit temperature, pressure, and combustion air/fuel ratio and to protect the recuperators from over-temperature exposure
- an extensive instrumentation and automatic data acquisition system to collect detailed temperature, flow, pressure, and gas composition data.

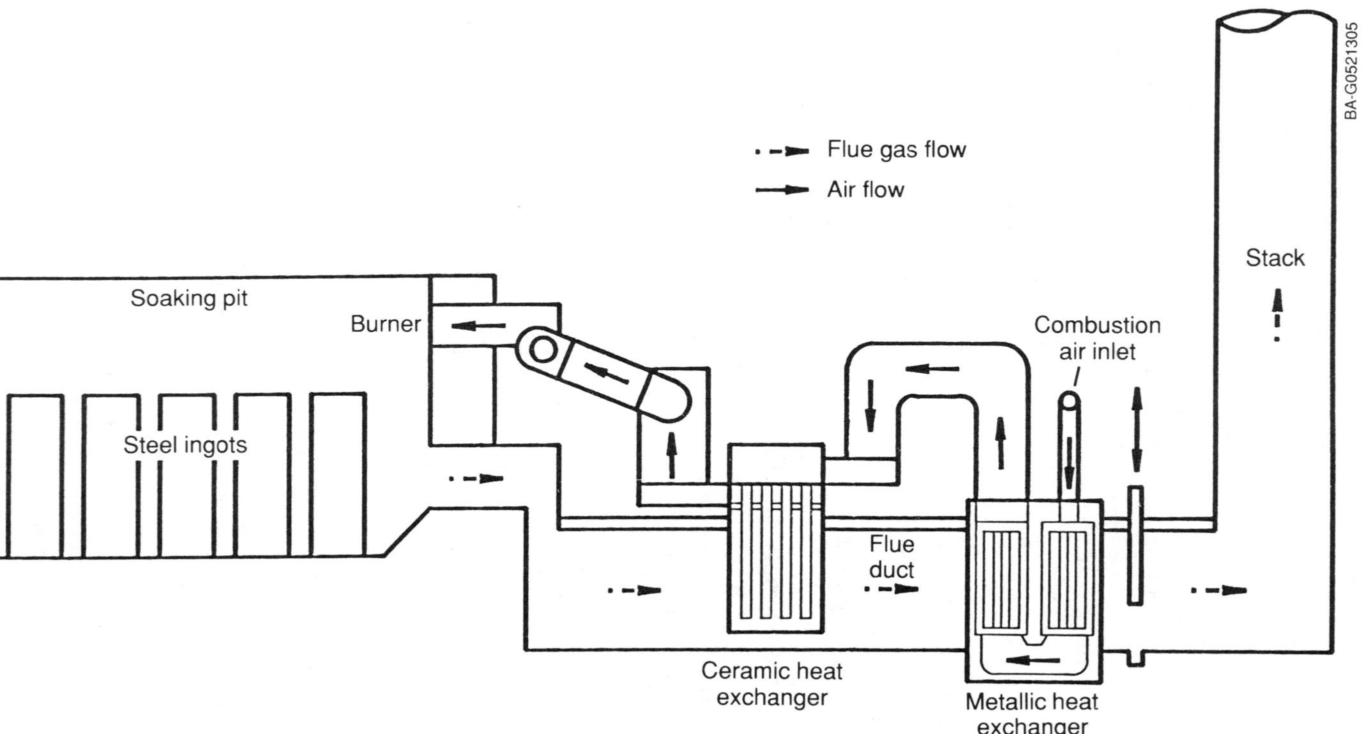

Figure 13-7. Schematic of full-scale HTBDR system

Figure 13-8. Assembly of HTBDR tubesheet

Table 13-1. Single Ceramic-Stage Specifications (50 tube assemblies/pass)

Tube arrangement	5-4-5 hexagonal array
Tube spacing	6 in. on center
Tube-tube gap	3/8 in.
Overall dimensions	36 x 76 x 40 in.
Cross-section of flue	32 x 69 in.
Outer Tubes	
Length	76-7/8 in.
Wall thickness	1/4 in.
Outside diameter	3-1/2 in.
Material	Norton NC-430 (siliconized silicon carbide)
Inner Tubes	
Length	96 in.
Wall thickness	1/4 in.
Outside diameter	2-1/4 in.
Material	Norton CS-101 (ceramic) AISI 446 (metallic)

System Installation

B&W fabricated and assembled the ceramic recuperator stage and tested it for leaks before installing it at the site. Leakage rates were projected to be less than 1% at operating temperatures. The soaking pit was modified somewhat before and during installation. These modifications included widening and relining the flue, installing the new burner, integrating new controls with existing pneumatic controls, installing monitoring equipment, removing dilution air and bleed air ducts (no longer required), and fabricating ducting at the site to connect major HTBDR components.

System Performance

The field tests were aimed at verifying the predicted performance and materials response to actual operating conditions. The initial startup runs yielded preheat air temperatures of 1100°-1200°F, much lower than the 1640°F predicted for the low-fire design case. Analysis of the system data indicated good performance of the ceramic stage but revealed several problems related to the age and condition of the existing soaking pit. These included massive air leakage into the flue duct, flue gas leakage out of the pit, excessive heat loss from the flue, poor pit pressure control, and preheat air temperatures from the metallic stage that were lower than predicted (813°F versus 1040°F). The flue gas temperatures into the ceramic stage were much lower than originally predicted (1850°F versus 2250°F).

To correct these problems, B&W improved the insulation of the flue duct between the pit and ceramic stage, reduced the gaps between ceramic tubes and the flue walls and floor, redesigned the damper system to reduce air infiltration and improve pit pressure control, and sealed leaks.

Steady-State Testing

B&W conducted a series of steady-state tests to characterize system performance. Steady-state data (summarized in Table 13-2) were taken after the pit had operated for 24 h under low-fire conditions, and 1375°F preheated air had been produced. The maximum preheated air temperature reached for any steady-state run was 1420°F.

The preheat air temperatures obtained were still lower than the 1640°F predicted by the HTBDR performance model, even after considering actual duct heat losses, pit energy loss, and poor metallic recuperator performance. Analysis of test data, including component and system energy balances, indicated convincingly that the lower-than-desired performance was caused primarily by flue gas loss (as high as 20%) out of the soaking pit. Velocity measurements made in the flue duct supported this conclusion. Computer performance model runs, assuming 20% flue gas leakage, yielded results close to measured test data.

Table 13-2
Summary of HTBDR Performance
(Steady-state tests, low fire)

Flow Rates (lb_m/h)	
Natural gas	187
Air	6340
Flue gas	5140
Temperature (°F)	
Ceramic Stage	
Air in	815
Air out	1375
Flue gas in	2170
Flue gas out	1605
Metallic Stage	
Air in	100
Air out	815
Flue gas in	1545
Flue gas out	830
Pressure Drop (in. of water)	
Air Side	
Metallic stage	1.4
Ceramic stage	6.3
Burner	0.7
Flue Side	
Metallic stage	0.05
Ceramic stage	0.30

Production Cycle Testing

To determine the production fuel savings, B&W monitored the energy used during a typical production cycle. The production run consisted of heating 15 steel ingots (82 tons) from room temperature to 2300°F and soaking for 10 h. The initial pit temperature was 1600°F. Temperature and fuel flow data from this run are shown in Figure 13-9.

The soaking pit with the HTBDR consumed 24% less fuel than those with only a metallic recuperator at the test site consumed (1.9 million

Btu/ton versus 2.5 million Btu/ton, respectively). Compared with the same pit in its original configuration (a 1982 energy audit showed that the soaking pit with only a metallic recuperator consumed 2.3 million Btu/ton of fuel), a 17% fuel savings was achieved. Projections show that if there were no flue gas leaks, the potential fuel savings could reach 41% for the HTBDR installed on an unrecuperated furnace and 24% for the ceramic recuperator as an addition.

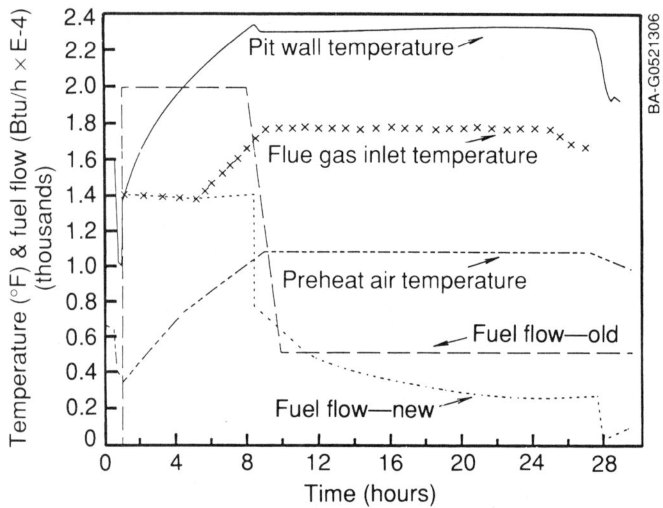

Figure 13-9. HTBDR production-cycle test fuel flow and temperatures

Materials Performance

B&W ran a special 10-h test to verify materials performance at design conditions. An electric heater was installed in the combustion air duct between the metallic and ceramic stages to increase the air temperature into the ceramic stage to 1250°F. The flue gas inlet temperature was 2150°F for the test. No difficulties were experienced in operating the unit, and the maximum preheat air obtained was 1606°F.

During the entire field test, the HTBDR operated approximately 1400 h with the flue gas temperature as high as 2250°F. In contrast to the old soaking pit, the ceramic recuperator operated as designed with good materials performance. There were no observable tube recession or significant corrosion, no detectable leakage through the seals, and no problem with the tubesheet design. The silicon carbide tubes had normal glazing, and there were no tube failures. B&W expects the ceramic stage to last at least 10 years.

APPLICATIONS

This section discusses the economics of the B&W HTBDR system, potential applications of the technology, energy savings, and commercial prospects.

System Economics

Following successful field tests at its steel soaking pit site, B&W evaluated the HTBDR system costs, fuel savings, and investment payback for installations on both recuperated and unrecuperated furnaces.

The installed system costs for a ceramic recuperator alone, a metallic recuperator alone, and a combined system (as tested) were estimated at $314,000, $94,500, and $344,000, respectively. These costs include the high-temperature burners, ducting, fan, controls, and installation. Retrofit costs are site-specific. The ceramic stage accounts for $194,000 of the cost, with the ceramic tubes contributing the most. For example, each ceramic outer tube costs approximately $1,400.

Based on the design and field test data, B&W estimated the annual fuel savings and simple payback on investment with the HTBDR for the steel soaking pit application with and without existing recuperation. Table 13-3 presents the savings for the system installed in an unrecuperated furnace as a function of three natural gas prices, three preheat air temperatures, and three annual operating hours. A gas price of $3.35/1000 ft^3 represents the existing price at the time of the field tests. The three preheat air temperatures represent the original predicted temperature for a single ceramic stage (1725°F), the revised predicted temperature for a single ceramic stage after accounting for actual system heat losses (1640°F), and the highest actual preheat air temperature achieved (1425°F) during tests.

With full operation (8300 h/yr), payback of investment is less than 2.5 years for all preheat air temperatures and gas prices. With field test site gas prices and full operation, the payback is between 2.1 and 2.5 years, depending on the preheat air temperature. Paybacks become less attractive as the operating hours decrease. Fuel savings and payback are strongly affected by fuel cost and operating hours and moderately affected by preheat air temperature.

Table 13-3
Annual Fuel Savings with HTBDR in Unrecuperated Furnace

Natural Gas Price ($/1000 ft^3)	Preheat Air Temperature (°F)	Savings, $K/yr (payback in years)		
		8300 h/yr	6240 h/yr	4160 h/yr
3.35	1725	158 (2.1)	119 (2.9)	79 (4.3)
3.35	1640	154 (2.2)	116 (3.0)	77 (4.5)
3.35	1420	140 (2.5)	105 (3.3)	70 (4.9)
6.0	1725	284 (1.2)	213 (1.6)	142 (2.4)
6.0	1640	276 (1.3)	207 (1.7)	138 (2.5)
6.0	1420	251 (1.4)	189 (1.8)	126 (2.7)
9.0	1725	425 (0.8)	320 (1.1)	213 (1.6)
9.0	1640	413 (0.8)	311 (1.1)	207 (1.7)
9.0	1420	377 (0.9)	284 (1.2)	189 (1.8)

Figure 13-10 shows the projected fuel cost savings for the soaking pit site over the expected 10-year life of the HTBDR. The cumulative savings are shown for three gas prices, based on the design preheat air temperature of 1725°F and 8300 operating hours per year. Payback for each case occurs when the savings line intersects the system base price. Figure 13-10 indicates that payback is acceptable with all three gas prices and that the projected savings over the 10-year life will exceed the unit cost.

Figure 13-11 shows the projected fuel cost savings for the case where a metallic recuperator is already in use. The savings are from adding the ceramic stage alone. At current gas prices, there is no reasonable payback. Adding a ceramic recuperator would be justified if gas prices rise to $6/1000 ft^3 or if ceramic tube prices decrease significantly, bringing down the ceramic unit cost. Additional savings might also be achieved through improving the quality of the product (reducing scaling).

Potential Applications

DOE estimates that the B&W HTBDR technology could save industry 43 trillion Btu each year by 2010. The technology can be applied to industrial furnaces that produce dirty flue gases at temperatures above 1500°F and have floor or trench exhausts. These conditions exist in the primary steel, copper, lead, zinc, aluminum, and glass industries. Although the HTBDR was developed for a steel soaking pit, the design is flexible and can be adapted for other applications. However, the technology is not suitable for applications such as those with chloride or fluoride contaminants in the flue gas or with flue temperatures that exceed 2250°F. New designs and materials will be required for these applications.

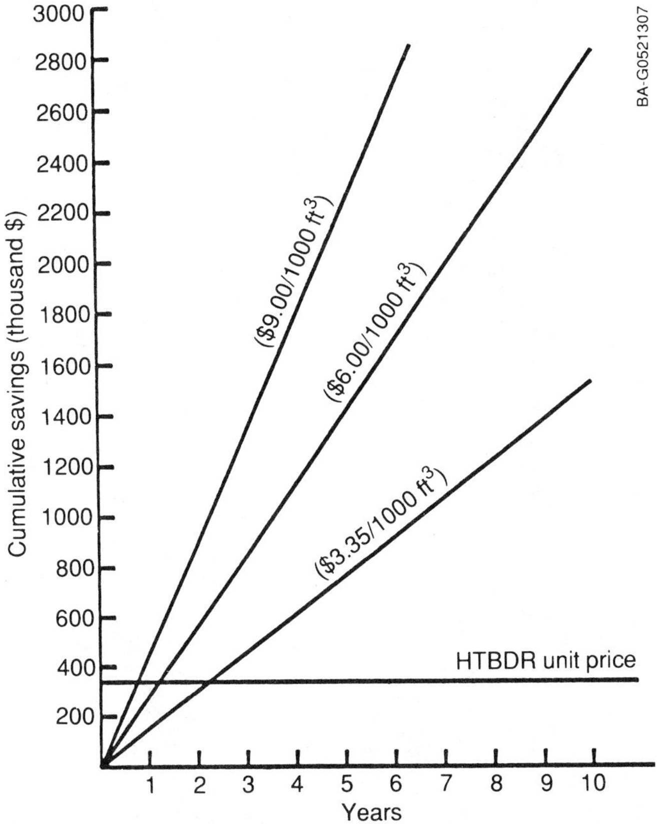

Figure 13-10. Projected savings for an HTBDR placed in an unrecuperated furnace

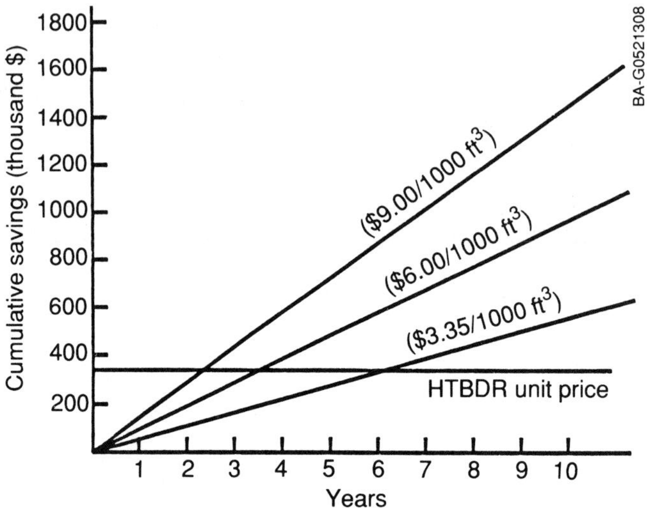

Figure 13-11. Projected savings for an HTBDR placed in a recuperated furnace

CONCLUSION

The HTBDR system developed through this DOE-funded program incorporates the benefits of a systematic research and design process complemented by a thorough prototype test and a field test of a full-scale system under typical industrial operating conditions. Through this program, considerable insight was gained into the challenges and technical problems associated with implementing structural ceramics into high-temperature industrial waste streams. Today, the HTBDR is a proven, technically sound technology and awaits industrial acceptance. B&W is currently looking for industrial applications. With low fuel prices and paybacks of less than 3 years being the norm, markets are restricted to sites where no recuperation is currently used. However, as fuel prices increase and costs of silicon carbide ceramic tubing decrease (as expected), the prospects for the HTBDR technology will improve.

References

1. Business Makes Room for the Energy Executive, J. R. Stock and D. M. Lambert, *Business*, November–December, 1980.
2. 1972 Statistical Review of the World Oil Industry, British Petroleum Company.
3. Patterns of Energy Consumption in the United States, Office of Science and Technology, Executive Office of the President, January, 1972.
4. The Technical Aspects of the Conservation of Energy for Industrial Processes, J. Burroughs, Federal Power Commission National Power Survey Technical Advisory Committee on Conservation of Energy, Position Paper No. 17.
5. Solar Energy: Its Time is Near, W. Morrow, Jr. *Technology Review*, December, 1973.
6. Rankine–Cycle Systems for Waste Heat Recovery. R. Barber, *Chemical Engineering*, 1974.
7. Heat Recovery Combined Cycle Boosts Electric Power with No Added Fuel, *Electrical Consultant*, November, 1974.
8. Energy Conservation in Existing Plants, J. C. Robertson, *Chemical Engineering*, January 21, 1974.
9. Solar Energy System Reduces Life Costs, Louis C. Sullivan, *Specifying Engineer*, August, 1974.
10. Energy Self-Sufficiency; An Economic Evaluation, *Technology Review*, May, 1974.
11. More Cooling at Less Cost, W. Farwell, *Plant Engineering*, August 22, 1974.
12. The Heat Pump: Performance Factor, Possible Improvements, E. Ambrose, *Heating/Pump/Air Conditioning*, May, 1974.
13. Energy, Economy and Industrial HVAC Systems, A. Hallstrom, *Specifying Engineer*, December, 1974.

14. A New Approach to Factory Air Conditioning, F. Dean, *Heating/Piping/Air Conditioning*, June, 1974.
15. Insulation Saves Energy, R. Hughes and V. Deumaga, *Chemical Engineering*, May 27, 1974.
16. Energy Conservation in New Plant Designs, J. Fleming, J. Lambrix and M. Smith, *Chemical Engineering*, January 21, 1974.
17. Fuel Savings Through No-Vent Systems, M. Stout, *Chemical Engineering*, September 30, 1974.
18. Heating Fuel Conservation, R. Meister, *Plant Engineering*, July 11, 1974.
19. Conserving Utilities–Energy is New Construction, C. Schumacher, B. Girgis, *Chemical Engineering*, February 18, 1974.
20. How About A Steam Turbine Drive?, J. Behl, *Plant Engineering*, July 26, 1974.
21. How to Conserve Energy in the Design of Plants, G. Moor, *Specifying Engineer*, September, 1973.
22. Power Demand Control Uses Energy Better, B. Murphy and R. Putman, *Specifying Engineer*, December, 1974.
23. Ways to Save Energy in Existing HVAC Systems, W. Landman, *Specifying Engineer*, January, 1975.
24. Application Guide Weathertron Heat Pump, General Electric Publication No. 23-3047-2.
25. ASHRAE 1972 Hand Book of Fundamentals
    Chapter 2–Heat Transfer
    Chapter 11–Applied Heat Pump Systems
    Chapter 20–Design Heat Transmission Coefficients
    Chapter 21–Heating Load
    Chapter 22–Air Conditioning Cooling Load
26. ASHRAE 1974 Application Handbook
    Chapter 59–Solar Energy Utilization for Heating and Cooling
27. ASHRAE 1973 Systems Handbook
    Chapter 7–Heat Recovery Systems
    Chapter 43–Energy Estimating Methods
28. Trane Air Conditioning Manual.
29. How to Save Refinery-Furnace Fuel, D. Cherrington, H. Michelson, *Oil and Gas Journal*, September 2, 1974.
30. Conserve Energy by Design, Applications Engineering Manual, Trane Company.
31. Organizing for Energy Conservation, R. Kulwiec, *Plant Engineering*, February 20, 1975.
32. Energy Conservation Program Guide for Industry and Commerce, NBS Handbook 115.
33. ASHRAE Standard 90P, Energy Conservation in New Building Design, 1975.
34. Thermodynamics of Engineering Science, S. L. Soo, Englewood Cliffs, N. J., Prentice Hall, Inc. 1959.
35. Economic Analysis for Engineering and Managerial Decision Making, N. Barish, New York, McGraw-Hill Book Company, Inc., 1962.

36. Thermodynamics, New York, F. Sears, Reading, Mass., Addison-Wesley Publishing Company, Inc., 1952.
37. Which Air Conditioning System *Really* Saves Energy?, I. Naman, *Specifying Engineer*, August, 1974.
38. Keeping That Electric Bill Under Control, A. Wright, *Plant Engineering*, June 19, 1974.
39. Steam–Its Generation and Use, The Babcock's-Wilcox Company, 1963.
40. Energy–Crisis and Conservation, A. Field, *Heating/Piping/Air Conditioning*, February, 1974.
41. Saving Energy in Department Store M/E Systems, H. Argintar, *Electrical Consultant*, October, 1974.
42. An Efficient Selection of Modern Energy–Saving Light Sources Can Mean Savings of 10% to 30% in Power Consumption, H. Anderson, *Electrical Consultant*, April, 1974.
43. Electric Power Distribution for Industrial Plants, The Institute of Electrical Electronic Engineers, 1964.
44. Flow of Fluids Through Valves, Fittings and Pipe, Crane Technical Paper No. 410, 1970.
45. Organizing for Energy Conservation, Ray Kulwiec, *Plant Engineering*, February 20, 1975.
46. Life Cycle Costing in an Energy Crisis Era, Joseph P. Keefe, *Professional Engineer*, July, 1974.
47. The Sad State of Maintenance Management, Albert B. Drui and Paul T. Suul, *Plant Engineering*, June 12, 1975.
48. Evaluating the Effects of Dirt on HID Luminaries, A. C. McNamara and Andy Willingham, *Plant Engineering*, April 17, 1975.
49. Electrical Consulting: Engineering and Design, A. Thumann, Atlanta, Ga., Fairmont Press, 1972.
50. Secrets of Noise Control, A. Thumann and R. K. Miller, Atlanta, Ga., Fairmont Press, 1974.
51. Lighting Handbook, 1971, Westinghouse Electric Company.
52. Design Concepts for Optimum Energy Use in HVAC Systems, Electric Energy Association.
53. Square D Bulletin SM-451-CPG-1-73, Demand Control.
54. The Engineering Basics of Power Factor Improvement, *Specifying Engineer*, February, 1975; May, 1975.
55. A New Look at Load Shedding, A. Thumann, *Electrical Consultant*, August, 1974.
56. Improving Plant Power Factor, A. Thumann, *Electrical Consultant*, July, 1974.
57. An Efficient Selection of Modern Energy–Saving Light Sources Can Mean Saving of 10% to 30% Power Consumption, H. A. Anderson, *Electrical Consultant*, April, 1974.
58. The International Solar Energy Society Conference of 1970, Paper No. 4135, R. V. Dunkle and E. T. Davey.

59. Solar Energy Utilization for Heating and Cooling, National Science Foundation, NSF 74–41.
60. Low Temperature Engineering Application of Solar Energy, ASHRAE 1967.
61. Climatic Atlas of the U. S., reprinted under the title: Weather Atlas of the United States, Gale Research Co. 1975.
62. The DOE Industrial Energy Conservation Program DOE/CS/40402-T4.
63. Cool Storage System Performance presented at 8th World Energy Engineering Congress by D. Eppelheimer, The Trane Company.
64. Verderber, R.R. and O. Morse. Lawrence Berkeley Laboratory, Applied Science Division. "Performance of Electronic Ballasts and Other New Lighting Equipment (Phase II: The 34-Watt F40 Rapid Start T-12 Fluorescent Lamp)." University of California, February, 1988.
65. Morse, O. and R. Verderber. Lawrence Berkeley Laboratory, Applied Science Division. "Cost Effective Lighting." University of California, July, 1987.
66. Hollister, D.D. Lawrence Berkeley Laboratory, Applied Science Division. "Overview of Advances in Light Sources." University of California, June, 1986.
67. Verderber, R. and O. Morse. Lawrence Berkeley Laboratory, Applied Science Division. "Performance of Electronic Ballasts and Other New Lighting Equipment: Final Report." University of California, October, 1985.
68. Lovins, Amory B. "The State of the Art: Space Cooling." Rocky Mountain Institute, Old Snowmass, CO, November, 1986.
69. U.S. Department of Energy, Assistant Secretary, Conservation and Renewable Energy, Office of Buildings and Community Systems. "Energy Conservation Goals for Buildings." Washington, DC, May, 1988.
70. Rubinstein, F., T. Clark, M. Siminovitch, and R. Verderber. Lawrence Berkeley Laboratory, Applied Science Division. "The Effect of Lighting System Components on Lighting Quality, Energy Use, and Life-Cycle Cost." University of California, July, 1986.
71. Manufacturing Energy Consumption Survey: Consumption of Energy 1985 Energy Information Administration, November 18, 1988.

Index